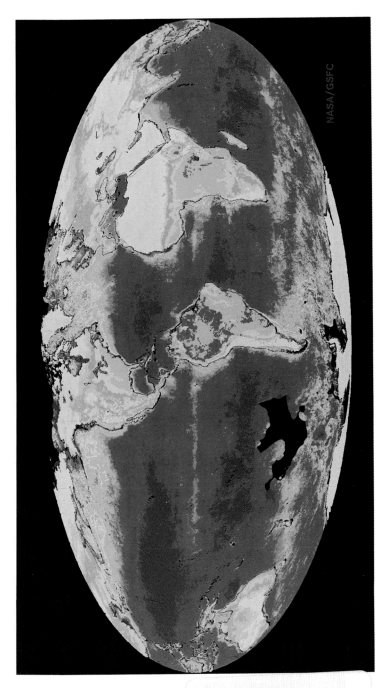

NASA/GSFC

Plate 1 Global distribution of chlorophyll estimated from the colour of the oceans as observed by satellites having colour scanners. Regions in the middle of the ocean have the lowest biomass of phytoplankton and presumably the lowest productivity. Higher productivity is found along the equator and in bands lying poleward of the mid-ocean gyres. The highest values are found in the coastal upwelling regions off the west coasts of the Americas and north and south Africa. Some of the high values in coastal and northern regions may not be due to chlorophyll but suspended organic and inorganic particles, or ice. Reproduced from the original colour image, courtesy of NASA and the University of Miami.

D0336139

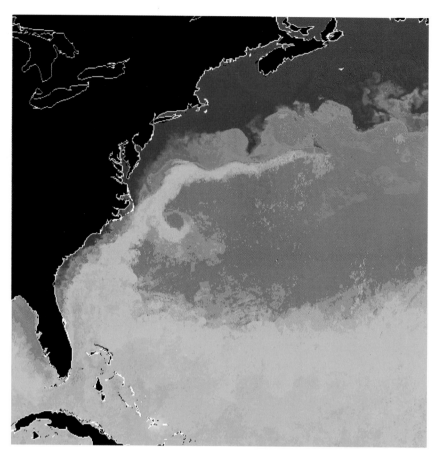

Plate 2 Distribution of surface temperature on the western North Atlantic Ocean on 19 January 1989. Highest temperatures are in the warm subtropical waters carried north in the Gulf Stream, where they mix with the colder northern waters. Note the evidence of ring formation on each side of the Gulf Stream. Derived from satellite data, courtesy of NASA and the University of Miami.

DYNAMICS OF MARINE ECOSYSTEMS
Biological–Physical Interactions in the Oceans

DYNAMICS OF MARINE ECOSYSTEMS

Biological–Physical Interactions in the Oceans

K. H. MANN & J. R. N. LAZIER

DEPARTMENT OF FISHERIES AND OCEANS
BEDFORD INSTITUTE OF OCEANOGRAPHY
DARTMOUTH, NOVA SCOTIA

BOSTON

BLACKWELL SCIENTIFIC PUBLICATIONS

OXFORD LONDON EDINBURGH

MELBOURNE PARIS BERLIN VIENNA

©1991 by
Blackwell Scientific Publications, Inc.

Editorial offices:

238 Main Street, Cambridge
 Massachusetts 02142, USA
Osney Mead, Oxford OX2 0EL, England
25 John Street, London WC1N 2BL
 England
23 Ainslie Place, Edinburgh EH3 6AJ
 Scotland
54 University Street, Carlton
 Victoria 3053, Australia

Other editorial offices:

Arnette SA
2, rue Casimir-Delavigne
75006 Paris
France

Blackwell Wissenschaft
Meinekestrasse 4
D-1000 Berlin 15
Germany

Blackwell MZV
Feldgasse 13
A–1238 Wien
Austria

First published 1991

Set by Times Graphics, Singapore
Printed and bound in the
United States of America
by BookCrafters, Ann Arbor, Michigan
92 93 94 5 4 3 2

DISTRIBUTORS

USA
 Blackwell Scientific Publications, Inc.
 238 Main Street
 Cambridge, MA 02142
 (*Orders:* Tel: 800 759-6102
  or: 617 876-7000)

Canada
 Oxford University Press
 70 Wynford Drive
 Don Mills
 Ontario M3C 1J9
 (*Orders:* Tel: 416 411-2941)

Australia
 Blackwell Scientific Publications
 (Australia) Pty Ltd
 54 University Street
 Carlton, Victoria 3053
 (*Orders:* Tel: 03 347-0300)

Outside North America and Australia
 Blackwell Scientific Publications Ltd
 c/o Marston Book Services Ltd
 PO Box 87
 Oxford OX2 0DT
 (*Orders:* Tel: 0865 791155
  Fax: 0865 791927)

Library of Congress
Cataloging-in-Publication Data

Mann, K.H. (Kenneth Henry), 1923–
 Dynamics of marine ecosystems:
 biological-physical interactions
 in the oceans/K.H. Mann,
 J.R.N. Lazier.
  p.   cm.
 Includes bibliographical references
 and index.
 ISBN 0-86542-082-3
 1. Marine ecology.
 2. Biotic communities.
 I. Lazier, J.R.N.
 II. Title.
 QH541.5.S3M25   1991
 574.5'2636–dc20

Dedicated to the memory of
GORDON A. RILEY
A good friend and colleague
who was thinking and writing about
these same topics 50 years ago

Contents

Preface

In an earlier book by the senior author entitled *Ecology of Coastal Waters: A Systems Approach*, marine ecosystems were described in terms of their characteristic primary production, whether by phytoplankton, seaweed, mangrove, marsh grass or seagrass. Estimates were presented of the annual mean values for primary production, and pathways of energy flow were traced through the food webs. One chapter was devoted to water movement and productivity. A reviewer commented that the book was too much about mean flows and not enough about variance.

Reflecting on this, it was clear that much of the variance in marine productivity is a function of water movement. Decomposition and liberation of nutrients tend to take place in deep water or on the sea floor and water movement is needed to bring those nutrients back up into the euphotic zone for use by the primary producers. Tides give water movement a diurnal and fortnightly periodicity, while seasonal changes in solar heating impose changes in the mixed layer on scales of months to a year. Long-term climatic cycles impose their own variations on water movement and hence on biological productivity. The theme for this book began to crystallize as an expansion of the earlier chapter on water movement and productivity considered at a range of temporal and spatial scales.

While the ideas were developing, remarkable changes were occurring in oceanography. More and more, biological oceanographers were teaming up with physical and chemical oceanographers to study marine ecosystems in their totality. Physical oceanographers were increasingly able to explain to their colleagues what was going on in gyres, at fronts, on banks or in estuaries, and the biologists were developing the instrumentation needed to obtain continuous records of biological variables to supplement the spot samples that had been characteristic of biological oceanography for decades. Satellite observations of ocean colour were giving large-scale perspectives on ocean productivity undreamed of by earlier generations. The feeling emerged that marine ecology was coming of age. It was developing a new maturity based on the integration of disciplines, and in the process yielding important new understanding about ecosystem function. We therefore decided to use the theme of physical processes and productivity as a starting point for an account of recent developments in marine ecology, in which physics, chemistry and biology are inter-related aspects of the dynamics of marine ecosystems.

Formal courses in oceanography tend to separate marine biology and marine physics. Our aim is to emphasize the links between the two subjects by presenting in each chapter the relevant physical processes along with the biology. Because the reader is expected to have a more complete background in biology than in physics the two subjects are written from slightly different viewpoints. The presentation of the physics is fairly elementary and emphasizes the important physical processes, while the presentation of the biology emphasizes the recent development of the field. To assist the physical presentation we have used some mathematical symbols and equations simply because they are part of the language of the subject and provide a useful shorthand for presenting ideas. It is for example much easier and more precise to write the symbol for the derivative of a variable than to write it out every time it is used. In a further attempt to present the physics in manageable portions some of the details required for a more advanced understanding have been separated into boxes which may be skipped on first reading. The references for both the physics and biology, though numerous, are by no means exhaustive and where good reviews exist we have drawn attention to them and left the reader to find the original sources.

In view of the current concern about the role of the oceans in climate change, we believe that there will be an increasing need to understand the integrated biological–physical functioning of marine ecosystems. We therefore hope that professional researchers in the various disciplines of oceanography will find this book of value in broadening their understanding of marine ecology, as an aid to defining those research programs that will be needed if we are to anticipate the consequences of global change.

In covering such a broad field we have relied heavily on the advice and assistance of many colleagues. For biological material Glen Harrison, Steve Kerr, Alan Longhurst, Eric Mills, Trevor Platt and Mike Sinclair have been particularly helpful, while on the physics and chemistry side we have enjoyed the advice of Allyn Clarke, Fred Dobson, David Greenberg, Ross Hendry, Edward Horne, Peter Jones, Hal Sandstrom, John Loder, Neil Oakey and Stuart Smith. We thank Mark Denny, Mike Keen and Jim McCarthy for helpful comments on various parts of the manuscript, and we particularly thank our editor, Simon Rallison, for his most helpful advice and guidance at all stages of this project. We wish to thank Betty Sutherland and her library staff in the Bedford Institute of Oceanography for expert assistance with the literature and for suffering more or less continuous occupation of part of the library over an extended period. We also wish to thank Steve McPhee, Jim Elliott and Mike Sinclair of the Science Branch of the Department of Fisheries and Oceans for supporting us in our endeavours and our wives Isabel Mann and Catherine Lazier for their encouragement, enthusiasm and patience.

K. H. M., J. R. N. L.

DYNAMICS OF
MARINE ECOSYSTEMS
Biological–Physical Interactions in the Oceans

1

Marine Ecology Comes of Age

Marine ecology, as traditionally understood, is the study of marine organisms and their relationship with other organisms and with the surrounding environment. The subject parallels similar studies of organisms on land but while terrestrial organisms are relatively easy to observe and manipulate, marine organisms are much more inaccessible and this has led to a slower growth of knowledge. The physical factors leading to fertile and infertile areas are very different on land than in the ocean. The nutrients required by land plants are generated nearby from the decaying remains of previous generations, but decaying matter in the ocean tends to sink and leave the sunlit euphotic layer where plants grow. The nutrients supplied by the decay are thus unavailable for plant growth unless some physical mechanisms bring the nutrients back up to the surface. This book is largely about those mechanisms and the resulting biological phenomena.

It is now possible to add an extra dimension to marine ecology. Instead of putting the organisms at the centre of the picture and considering them in relationship to other organisms and the environment, it is possible to work with marine ecosystems in which physical, chemical and biological components are equally important in defining total system properties. Those properties include production of living organisms such as fish, but flux of carbon dioxide as determined by both physical and biological processes may be more important in the context of climate change.

Interest and research activity in marine ecology are intensifying. There are many reasons, and four may be mentioned:

1 The physical processes underlying some of the large-scale biological phenomena are now better understood. For example, the mechanisms underlying the major coastal upwelling systems of the world and their variations in time, the connection between variations in the Peruvian upwelling system and the southern oscillation in the atmosphere, or the physical mechanisms giving rise to the biologically important tidal fronts.

1

2 There have been important advances in our ability to make continuous, fine-scale biological measurements by means of automated sensors feeding into computers. It is now possible to collect biological data with a coverage and resolution comparable with the best physical data. Examples are the measurement of chlorophyll by means of satellites or with towed fluorimeters, or the electronic counting of zooplankton. The satellite image in Plate. 1 shows the global distribution of chlorophyll in surface waters, and reveals a great deal about the incidence of upwelling and the exchange of gases with the atmosphere.

3 The need to understand marine ecological processes influencing the greenhouse effect and other aspects of world climate is becoming more urgent. The flux of carbon dioxide from the atmosphere into surface waters and on down into the deep ocean, as a result of biological processes, is believed to be an important part of the mechanism of climate change. In this connection, the development of the spring bloom in temperate waters or the formation of patches of high productivity in otherwise unproductive tropical oceans are fields of active research with promising futures.

4 The increased understanding of fundamental processes is beginning to influence the management of the ocean's living resources. The way in which year to year changes in the ocean's characteristics affect the growth and survival of fish larvae, and the effects of changing weather patterns on the distribution of fish are two topics that will receive a great deal of attention in the coming decades.

For all of these reasons, marine ecology has changed rapidly and may be said to have come of age. The dominant theme of this book is that physical processes create the conditions for many important biological processes; the biology cannot be understood in isolation. One good example of this close link is found in shelf-sea tidal fronts, which have been known for many years to be biologically important. As a result of elegant analysis the location of these fronts can now be predicted with simple calculations involving the speed of the tidal current and the depth of the water. Another example is the discovery that spring phytoplankton blooms are generated when organisms are confined to the euphotic layer by vertical stratification developed in the water by the sun's warming. In this volume, the connections between the physical and biological processes are emphasized and brought into focus more sharply than before.

The nature of the relationships between physical and biological processes is subtle and complex. Not only do the physical processes create a structure, such as a shallow mixed layer, or a front, within which biological processes may proceed, but they influence the rates of biological processes in many indirect ways. Discussion of this relationship has most often been in terms of energy flow. Biologists often model food web relationships in terms of the flow of solar energy, captured in photosynthesis by the phytoplankton and passed from organism to organism by means of feeding transfers. The physical phenomena such as currents, turbulence and stratification also rely on solar energy, transmitted to the water directly as heat or indirectly as momentum from the wind. These two fluxes of solar energy are in one sense quite distinct: organisms do not use the energy of water motion for their metabolic needs. In another sense, they are inter-related. Water movement alters the boundary layers around organisms, transports nutrients and waste products, assists migrations, and influences the rate of encounter between planktonic predators and their prey. Stratification causes the retention of planktonic organisms in the upper layer of the ocean, making light more available but limiting access to inorganic nutrients. Water temperature has a profound influence on the rates at which biological processes proceed, and differences in water motion, from place to place, largely determine the kinds of organisms colonizing those places. From a biological point of view, the physical energy is termed auxiliary energy, which means literally 'helping energy'.

LENGTH SCALES

In approaching the subject it is useful to have a feeling for the dimensions of the organisms and phenomena to be discussed (Fig. 1.01). Ocean basins are typically 10,000 km wide and confine the largest biological communities. The average depth of the ocean is 3800 m but the depths of the euphotic layer ($\simeq$100 m) and the mixed layer ($\simeq$100 m) are more often critical to open-ocean biological processes.

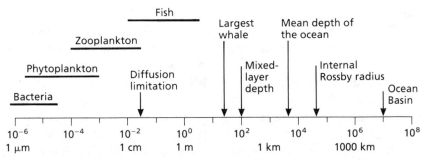

Fig. 1.01 The size scale from 1 μm to 100,000 km, showing some characteristic size ranges of organisms and physical length scales.

The Coriolis and gravitational forces give rise to the Rossby internal deformation scale or radius, a frequently encountered length scale in physical/biological oceanography (see Section 5.2.3). It arises in flows of stratified water when a balance between the two forces is established. This scale, which varies strongly with latitude, is the typical width of ocean currents such as the Gulf Stream, the width of the coastal upwelling regions or the radius of the eddies in the ocean.

The viscous or Kolmogoroff length is the scale where viscous drag begins to become important and where viscosity starts to smooth out turbulent fluctuations in the water (see Section 2.2.6). The scale represents the size of the turbulent eddies where the viscous forces are about equal to the inertial forces of the turbulent eddies. The scale also indicates an important change in the methods of locomotion and feeding. Organisms larger than $\simeq 1$ cm are not affected by viscous drag while for the smallest organisms swimming is akin to a human swimming in honey. Because of the change in the turbulent motions the smallest organisms must depend on molecular diffusion for the transfer of nutrients and waste products. For the larger animals nutrients and wastes are moved rapidly by turbulent diffusion which is not affected by viscosity. These topics are developed in Chapter 2.

TIME SCALES

As a first approximation, time scales change in direct proportion to length scales. On the global scale, the thermohaline circulation may take 1000 years to complete a circuit. On the ocean basin scale, the major gyres may require several years to complete a circuit. Eddies and gyres spun off from the major currents have lifetimes of weeks to months, and as energy cascades through smaller and smaller scales of turbulence, the characteristic time for rotation decreases to seconds at the smallest scale.

It seems that while physical features determine the spatial scales of ecological processes, it is the organisms that determine the time scales. While the life span of a large marine mammal may be close to 100 years, those of fish are more like 1–10 years, and zooplankton may complete a generation in a few days or weeks. Phytoplankton have doubling times of the order of days, and bacteria of hours. It follows that small organisms are likely to undergo more rapid fluctuations in numbers than large ones. Since in general, each type of organism tends to feed on organisms smaller than itself, the process of trophic transfer has the effect of smoothing out the rapid fluctuations. Conversely, predators may impose on their prey longer-term fluctuations which correspond with fluctuations in predator numbers.

PLAN OF THE BOOK

Part A begins by introducing turbulent motion and viscous boundary layers which determine the unusual feeding and locomotion techniques of the very small organisms. These plants and animals are the base of the food chain and account for about half the total biomass of the ocean. Their survival depends on a variety of physical processes outlined in Chapters 3 and 4. In the open ocean survival depends on the annual creation and destruction of the seasonal pycnocline. In shallow coastal waters the effects of freshwater run-off and tidal mixing can be the dominant processes.

In Part B, the consequences of winds near coasts and of the Coriolis force lead to the Ekman drift in the surface layers and coastal upwelling. This process is responsible for some of the most productive regions in the ocean. The enhanced biological activity near various types of fronts is covered in Chapter 6 and is followed by a discussion of tides including explanations of tidally-generated internal waves which transport nutrients onto the continental shelves.

Large-scale phenomena are treated in the next three chapters beginning with an explanation of the wind-driven circulation, the intense western boundary currents such as the Gulf Stream (Plate 2), and the warm- and cold-core rings that are generated by instabilities in the boundary currents. The unique biological properties of the rings and other circular circulation patterns such as gyres are then reviewed. The El Niño–southern oscillation story in Chapter 9 introduces the effect on biological productivity of changing circulation in the ocean. Chapter 10 reviews the greenhouse effect and the role of the oceans in this cycle emphasizing the biological pump which is the main mechanism transferring carbon dioxide from the upper layers to the bottom of the ocean.

In Chapter 11 we discuss questions for the future. There is a sense in which the whole book is an exploration of these questions, so we give them here:

1 Is there a common mechanism to account for the occurrence of high biological productivity in a variety of physical environments?

2 To what extent are events in marine ecosystems determined by physical processes? To paraphrase Hartline (1980), do physical factors feed fish?

3 How can we develop concepts and models that span the enormous range of scales in marine ecology, from the microscopic to the global and from seconds to geological ages?

We shall see that a tentative answer to the first question was provided by Legendre (1981). He said in effect that vertical mixing followed by stratification of the water column leads to a phytoplankton bloom, and that this can be seen to happen in a variety of habitats and at a range of temporal and spatial scales. Our review supports this, but are there other mechanisms?

One is tempted to respond to the second question by saying that physical factors obviously determine the course of biological events, and the converse rarely happens. In fact, if we take a long-term view we see that the greater part of the carbon dioxide released into the atmosphere during the life of the earth has been fixed by phytoplankton and deposited in marine sediments as carbonates or organic matter. Without these processes the carbon dioxide content of the atmosphere would be much higher, the earth would be much hotter, and the circulation of the oceans would be totally different. Even on the short time scale there are examples of phytoplankton altering the penetration of light and heat into the water column and hence the functioning of the ecosystem. Interactions between physics and biology are not entirely, or even mainly, in one direction.

The third question has been much discussed without any real resolution. It is a problem for ecologists generally, for we do not understand how to include bacterial processes on scales of millimetres and seconds in the same models that deal with animals that live for decades and may range over thousands of kilometres. Marine ecologists have the added difficulty that the biological events take place in a physical medium that exhibits processes on the same range of scales, thus compounding the difficulties.

We have found it useful to keep these questions in mind as we review the developments of marine ecology as an integrated physical, chemical and biological discipline.

PART A
PROCESSES ON A SCALE OF LESS THAN 1 KILOMETRE

2

Biology and Boundary Layers

2.1 INTRODUCTION

In this chapter we shall explore the intimate relationships between the small-scale processes in sea water and the lives of plants and animals. In order to do so, we shall have to shed many of the concepts that are ingrained in our way of thinking simply because we inhabit bodies of a particular size. To take one example, it seems natural for us to think that if we are in the sea and use our arms to push water backwards, we shall move forwards, coasting for many seconds or minutes before the viscosity of the water brings us to a halt. For a micro-organism this is not so.

Viscosity is all-important. A picoplankton cell of about $1\,\mu m$ diameter swimming at about $30\,\mu m\ s^{-1}$ and then stopping would come to halt in about $0.6\,\mu s$, having travelled only about $10^{-4}\,\mu m$ (Purcell 1977). Or again, a small crustacean which extended a pair of stiff limbs at right angles to the body and attempted to 'row' itself forwards would rock forwards and back staying in exactly the same place. Hence, traditional ideas about the locomotion of small organisms have to be drastically modified. In order to do so we have to understand that motion through the water is a function of two key variables, momentum and viscosity, and that the relative proportions of these change according to the scale of events being studied.

Consider the situation of a planktonic larva that is approaching the sea bed and is about to choose a site for settlement and metamorphosis. Interesting laboratory studies have been made, showing for example how certain larvae respond to chemical cues. In the real world, most areas of the sea bottom are exposed to one or two daily cycles of tidal currents. As Simpson (1981) put it, 'in stress terms, these tidal streams are equivalent to hurricane-force winds in the atmosphere blowing regularly twice per day.' Careful analysis shows that the only place that larvae can find water quiet enough for them to swim about and explore the bottom is a thin layer about $100\,\mu m$ deep immediately adjacent to the sea floor. In this thin layer they barely have room to manoeuvre, and it turns out that they use their swimming powers only to descend, sample the sea floor and rise again if it is unsuitable. This has been called the 'balloonist technique'. An understanding of the situation requires familiarity with the fundamental properties of boundary layers which form around objects when the water is in motion relative to the object.

Turning now to a consideration of the small-scale boundary layers of the surfaces of phytoplankton or seaweeds, viscosity causes the average speed of the flow to decrease from its value in the open water to zero at the boundary. The size of the turbulent eddies in the water also decreases to zero at the boundary. This creates problems for organisms which require the transport of nutrients towards their surfaces and waste substances away from them. Turbulent eddies transport nutrients efficiently in the open water but are too weak to transport nutrients through the boundary layer. Unless some special action is taken, the organism's metabolism is restricted by the lack of turbulent transport which is replaced by very slow transport due to molecular diffusion. This is often known as diffusion limitation of metabolism.

The thickness of a boundary layer is reduced in proportion to the speed of the water moving past it. For large plants such as seaweeds, thinning of the boundary layer is achieved by attaching themselves to a solid surface in a zone where tidal currents and wave action cause vigorous water movement. This technique has its dangers, for if the water movement is excessive, the drag on the plant may tear it from its attachment. Many seaweeds are capable of making changes in their shape during growth, to reflect the trade-off between the need to maximize turbulence close to the plant surface and the need to reduce drag.

The conventional understanding of a planktonic organism is that it moves passively with the water. However, the need to overcome diffusion limitation is just as real for a phytoplankton cell as it is for a seaweed. There are two main techniques available. One is to have a heavy cell wall which causes the cell to sink through the water column. This technique is employed by diatoms and works best in mixed layers where the stirring tendency of the turbulent flow counteracts the sinking of the organisms. Without the turbulence the diatoms would all end up in deep water where there is insufficient light for photosynthesis. Investigation of the efficacy of the technique requires an understanding of the physics governing the sinking rate of particles in a fluid, as well as the physics of small-scale boundary layers.

The second technique adopted by phytoplankton is to perform locomotory movements. This is the solution adopted by the flagellates, but it is a far from simple process because of the problems associated with overcoming viscosity at small scales.

When we take all these physical aspects of life in the sea into consideration, we find that many of our existing concepts, based on experiments carried out in the laboratory in still water, are in need of drastic revision. Grappling with the physics of turbulent flow is hard labour for many biologists, but it is absolutely essential for understanding contemporary marine ecology.

2.2 PHYTOPLANKTON AND BOUNDARY LAYERS

Phytoplankton productivity in the world ocean is now a major concern, because of the role it is thought to play in modifying the carbon dioxide content of the atmosphere and hence the scenario for global climate change. One of the major themes running through this book is the need to understand how phytoplankton productivity is influenced by the physics of the ocean. This influence operates at many scales from ocean basin circulation, through localized areas of upwelling down to the smallest scales of turbulence that affect individual cells. Kinetic energy is imparted to the world ocean by sun, wind and tides, and the energy of large-scale motions is transmitted progressively to smaller and smaller scales of motion until, at very small scales, the motion is resisted by the molecular viscosity of the water and is eventually dissipated as heat. In the sections that follow, we shall make a fairly long detour into the physics of turbulence, viscosity, molecular and turbulent diffusion and the structure of boundary layers, before returning to the physiology of phytoplankton in section 2.2.10.

2.2.1 Turbulent motion

To begin, consider the hypothetical record of velocity in the ocean shown in Fig. 2.01. The signal has a mean or average, $\bar{u}$, over the record but at most times the velocity deviates from the mean by an amount u', called the fluctuating part of the flow. The sum of the two at any instant gives the total velocity.

The fluctuations in the motion indicate the presence of turbulence. When $u' = 0$, turbulence is absent and the flow is said to be smooth, or laminar. For turbulent flow, u' is a function of time and is made up of fluctuations of many periods randomly mixed together. The most rapid fluctuations may have a period of about a second and are the smallest turbulent motions with scales of a few millimetres. The longest fluctuation in the record may represent motions that are a few metres in size with periods of tens of minutes.

Turbulent motions of these small scales are usually assumed to be three dimensional and statistically similar in all directions; that is, the turbulence is said to be homogeneous or isotropic. This convention arose because many major advances in the theoretical understanding of turbulent motion associated with important practical problems were only possible by using this assumption which allowed a great simplification to the equations governing turbulent motion. This idealized state is a fairly good assumption for scales between the viscous and buoyancy scales, calculated below, but for scales as large or larger than the depth of water, the motions lose homogeneity and become two-dimensional eddies which are sometimes called geostrophic turbulence to denote the fact that the eddies are random, large, and adjusted to the influence of the rotation of the earth.

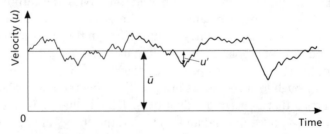

Fig. 2.01 A hypothetical record of water velocity in turbulent flow illustrating the difference between the average value $\bar{u}$ and the fluctuating component u'.

2.2.2 Sources of turbulent energy

The energy in the turbulent eddies is extracted from larger-scale motions via many different instability mechanisms. The most common and widely known instability is the breaking of surface waves that occurs when the waves get too steep. The breaking converts the regular and predictable motion of the wave into random turbulent motion. Deeper down in the ocean, internal waves propagate on and through the vertical density gradients. These waves also can become unstable and break up into turbulence. In the upper layer of the ocean, the wind, besides generating waves, forces the water to move relative to the layers below. This relative movement, or shear, can also lead to unstable motions that break the flow up into turbulent motions. Finally the large permanent currents, such as the Gulf Stream, develop meanders that create the large two-dimensional eddies of the geostrophic turbulence which eventually break up into smaller scales of motion.

2.2.3 Viscosity

The energy in turbulent motion is continually being transferred from large scales of motion to small scales. The little eddies that are 5 cm across get their energy from larger eddies which in turn get their energy from still larger ones. This process, called the energy cascade, does not change the total amount of energy in the turbulence nor does it convert the kinetic energy of the turbulent motion to another form of energy.

With the decrease in the size of the turbulent eddies comes an increase in the velocity gradient across the eddies. When the eddies are small enough and the velocity shear great enough then molecular viscosity, the internal resistance of the water, acts to resist and smooth out the gradients in velocity. This smoothing of the flow by viscosity is the way the energy in the turbulence is finally converted to heat and dissipated. The stress generated by the viscous forces is discussed in Box 2.01.

2.2.4 Comparing forces, the Reynolds number

The importance of the forces due to viscosity is often quantified by calculating the Reynolds number, which is the ratio of the inertial force to the viscous force acting on the body of interest, be it an animal or fluid element. The inertial force is the force that was necessary to accelerate the body to the velocity it now possesses, or to stop the body now travelling at a constant speed under its own inertia. The

Box 2.01 Calculating the stress due to viscosity

Viscous stresses for most oceanic phenomena are negligible, partly because the viscosity of water is so small, in fact one of the lowest found in naturally occurring liquids, and partly because only at the smallest scales are the velocity gradients large enough to make the viscous stresses significant when compared to the other forces present. A simple calculation can show the magnitude of the viscous forces in a specific situation. The molecular viscous stress τ that one layer such as A in Fig. 2.02 exerts upon layer B is

$$\tau = \rho v \, d\bar{u}/dz, \tag{2.01}$$

where $\rho \simeq 10^3$ kg m^{-3} is the density, $v \simeq 10^{-6}$ m^2 s^{-1} is the coefficient of kinematic viscosity and $d\bar{u}/dz$ is the gradient of the average velocity perpendicular to the flow. A typical change in mean velocity in the ocean of 1.0 m s^{-1} over 1000 m gives a velocity gradient of 10^{-3} s^{-1} and leads to a minute viscous stress of 10^{-6} N m^{-2}, but where velocity changes by 0.001 m s^{-1} over 0.01 m, as it may in small eddies, the viscous stress is a significant 10^{-4} N m^{-2}.

ratio, a dimensionless number, works out to be the velocity u times a typical dimension d divided by the kinematic viscosity v,

$$\text{Re} = ud/v. \tag{2.02}$$

The number was originally used as a criterion of whether flow in pipes would be smooth or turbulent. The typical dimension d was the pipe

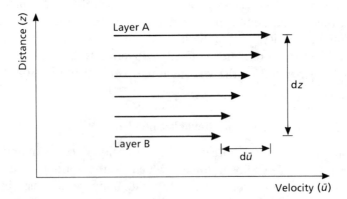

Fig. 2.02 The mean velocity $\bar{u}$, parallel to the x axis, increases by an amount $d\bar{u}$ in the distance dz, but viscosity creates a stress across the gradient that tends to retard the faster moving water at level A, and speed up the slower moving water at level B.

diameter, but it may equally be a typical dimension, such as length or width, of an organism around which flow is being considered.

The most useful feature of dimensionless numbers, such as the Reynolds number, is that they apply, universally, to all similar situations. For pipes of any size, for example, the division between smooth and turbulent motion occurs when $Re \simeq 2500$, which suggests that the viscous forces keep the flow smooth by suppressing small irregularities even though the inertial forces are 2500 times as large as the viscous forces. Since Reynolds first introduced the ratio it has been applied usefully in many quite different situations, but all for the same purpose, to compare the inertial and viscous forces.

Sometimes the ratio is applied to bodies of water. For example, if a mass of water, 1 km across, is moving with a uniform speed of 10 cm s^{-1}, $Re \simeq 10^8$ indicating a region where the viscous forces are too small to suppress the small perturbations that grow into turbulent eddies. On the other hand if the water mass is 1 cm across and moving at 1 cm s^{-1}, $Re \simeq 10^2$ indicating a flow where viscous forces are getting to be important in suppressing small perturbations in the flow.

It is also common to see calculations of the Reynolds number of solid bodies in the water such as grains of sand and animals. A 0.1 m fish, for example swimming at 1.0 m s^{-1} has a Reynolds number of 10^5, and obviously inertial forces dominate its life and viscosity can be ignored when considering the fish as a whole. A microscopic animal $50 \,\mu\text{m}$ long swimming at $10 \,\mu\text{m s}^{-1}$, on the other hand, exhibits the minuscule Reynolds number of 5×10^{-4} indicating that inertial forces can be ignored in this animal's world which is dominated by viscous forces. The Reynolds number is, then, a useful guide in assessing the relative strength of the inertial and viscous forces and for comparing similar situations.

The enormous range of Reynolds numbers associated with living organisms is illustrated in Table 2.1 compiled by Vogel (1981). The

Table 2.1 Approximations to the magnitude of the Reynolds number of various organisms. From Vogel (1981).

	Re
A large whale swimming at 10 m s^{-1}	300,000,000
A tuna swimming at the same speed	30,000,000
A duck flying at 20 m s^{-1}	300,000
A large dragonfly going 7 m s^{-1}	30,000
A copepod in a pulse of 20 cm s^{-1}	300
Flight of the smallest flying insects	30
An invertebrate larva, 0.3 mm long moving at 1 mm s^{-1}	0.3
A sea urchin sperm advancing the species at 0.2 mm s^{-1}	0.03

Reynolds number of an even wider range of organisms was calculated by Okubo (1987) using a characteristic dimension, d, and a typical swimming speed, u, for animals and bacteria, or sinking speed for phytoplankton. He did this for the whole range from bacteria to whales, then for good measure added the point for humans with a height of 2 m and a swimming speed of 1 m s^{-1}. The resulting plot, Fig. 2.03, shows that the Reynolds number increases systematically with the size of the organism according to the regression

$$\text{Re} = ud/\nu \simeq 1.4 \times 10^6 \times d^{1.86} \;, \tag{2.03}$$

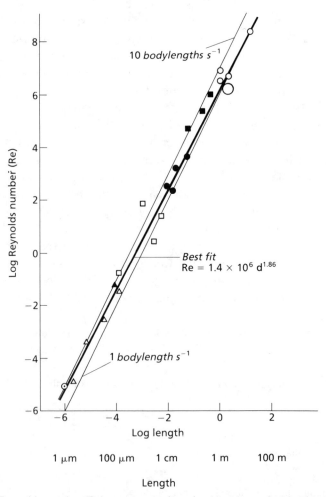

Fig. 2.03 Reynolds number (Re) versus organism size (d). Mammals (○), Fish (■), Amphipods (●), Zooplankton (□), Protozoa (▲), Phytoplankton (△), Bacteria (⊙), Human (◯). The heavy line is the best fit to the data. Thin lines illustrate the relationships for swimming at 1 and 10 bodylengths per second. Adapted from Okubo (1987).

where d is measured in metres. Substituting $v = 10^{-6} \text{ m}^2 \text{ s}^{-1}$ we find the relationship between the characteristic length and swimming speed,

$$u(\text{m s}^{-1}) = 1.4 \times d^{0.86} \ . \tag{2.04}$$

These equations provide a useful quantification of the everyday observation that large animals swim faster than small ones. For additional clarification, we have added two lines to Fig. 2.03 to indicate what the relationship would look like if the organisms moved at 1 and 10 body lengths per second. Okubo's best fit line lies at an angle between these two and suggests that small organisms can move about 10 bodylengths in a second but larger animals swim at nearer 1 bodylength per second. Such statements must, however, be treated with caution as the values of swimming speed are quite scattered, especially for zooplankton, and the relationship may only be accurate to within a factor of ten. It is also worth remembering that many of the data, especially those for the smallest organisms, were obtained under a microscope in a laboratory and not in the natural environment.

2.2.5 Molecular diffusion

As indicated above, turbulent energy is passed from large to small eddies and the viscosity of the water limits the size of the smallest eddies to a few millimetres diameter (see below). Such turbulence is ineffective in transporting nutrients and wastes for organisms less than 1 mm in diameter (which make up more than half the total biomass of the oceans, Sheldon *et al.* 1972). Small organisms must rely on the flux due to molecular diffusion, that is, the slow mixing caused by the random motion of molecules. As is shown in Box 2.02, this is a very slow process, requiring about 10 s to produce an effect over a distance of 100 μm.

2.2.6 Scales of turbulent structures

When dealing with small organisms living in turbulent water, it is sometimes important to estimate the distributions of velocity, temperature and nutrients near the organism. For example, we might want to know the sizes of the smallest turbulent eddies or the sizes of the smallest fluctuations in temperature, salinity and nitrate, etc.

The size of the smallest velocity fluctuation is determined by the strength of two competing forces. The force due to viscosity works to remove variations in velocity while the inertial force associated with the

Box 2.02 Fick's law and the diffusion time scale

Molecular diffusion is the slow mixing caused by the random motion of molecules. The flux of some constituent through the water due to molecular diffusion is given by Fick's first law of diffusion. This law states that if the concentration of some constituent C changes by an amount dC over a short distance dz, the flow of C down the concentration gradient of molecular motion is

$$F = - D \, dC/dz \ . \tag{2.05}$$

If C is in kg m^{-3}, F is the flux of C in kg m^{-2}s^{-1} and D is the coefficient of molecular diffusion which for large molecules, i.e. salt, in water is about 1.5×10^{-9} m^2 s^{-1}.

The coefficient, D, contains information related to how fast the molecules of the diffusing substance move through the fluid. To use this information, the definition of D is converted into a formula,

$$D = 10^{-9} \, \text{m}^2 \, \text{s}^{-1} = L^2/t \ , \tag{2.06}$$

where L and t are characteristic values of length and time. Turning the formula around yields $t = L^2/D = 10^9 L^2$, giving an estimate of the time it takes molecular diffusion to go the distance L. Using this relation, the time it would take molecular diffusion to cause an effect over $100 \, \mu$m, is about $(100 \times 10^{-6})^2 \times 10^9 \simeq 10$ s.

turbulent motions tends to create velocity fluctuations. The size of the eddies where these opposing forces are in balance is normally taken as the limiting size of the velocity fluctuations. If the viscous, or smoothing, force is represented by the kinematic viscosity v, while the twisting or shearing force in the turbulence is represented by the rate of turbulent-energy dissipation ε W kg^{-1}, the distance across the smallest eddies, known as the viscous or the Kolmogoroff length scale, L_v, is estimated from (Gill 1982),

$$L_v = (v^3/\varepsilon)^{1/4} \ . \tag{2.07}$$

In much of the oceanographic literature, however, this length is written

$$L_v = 2\pi(v^3/\varepsilon)^{1/4} \ , \tag{2.08}$$

where the factor of 2π has been added for mathematical convenience as it simiplifies the manipulation of equations. As discussed by Lazier and Mann (1989) the factor of 2π also results in a more realistic value of the length scale. Measurements have shown that there is virtually no energy in the turbulent eddies at the scale defined by Eq. 2.07.

The variable, ε, representing the dissipation of turbulent energy, is estimated from measurements of the finest scales of the velocity gradient. By doing this, Oakey and Elliott (1980) calculate that ε in the top 50 m layer over the Scotian shelf off Nova Scotia varies from $10^{-6}\,\mathrm{W\,kg^{-1}}$ when the wind is $15\,\mathrm{m\,s^{-1}}$, to $10^{-8}\,\mathrm{W\,kg^{-1}}$ when the wind is less than $5\,\mathrm{m\,s^{-1}}$. In these situations the smallest scales L_v, using Eq. 2.08, would vary roughly from 6 to 20 mm. At greater depths or when the wind is light, the level of turbulent energy decreases to roughly $10^{-9}\,\mathrm{W\,kg^{-1}}$, corresponding to a smallest eddy size of 35 mm (Osborn 1978). The smallest fluctuations of variables such as temperature and salinity are smaller than they are for velocity as is shown in Box 2.03.

At the other end of the size range are the largest turbulent eddies which are important because they determine the vertical excursion of the small passive organisms being moved about by the turbulent flow. Near the surface these large eddies determine how much time the organism spends in the euphotic zone. The largest eddies occur when the inertial forces associated with the turbulence, which tend to stir the water, are about equal to the buoyancy forces, which tend to keep the water stratified. This size is estimated from the turbulent-energy dissipation rate ε and the buoyancy or Brunt–Väisälä frequency N

Box 2.03 The smallest scale of different variables

We saw in section 2.2.6 that viscosity limits the size of the smallest turbulent eddies to about 6 mm, in a highly energetic environment. Viscosity may be thought of as the molecular diffusion of momentum with a value of $\simeq 10^{-6}\,\mathrm{m^2\,s^{-1}}$ while the molecular diffusivity of heat is about $1.5 \times 10^{-7}\,\mathrm{m^2\,s^{-1}}$ or about one-tenth the value for viscosity. The diffusivity of salt and nitrate are two orders of magnitude lower again, $1.5 \times 10^{-9}\,\mathrm{m^2\,s^{-1}}$. The effect of these lower diffusivities is to permit smaller fluctuations to persist longer, before being smoothed out by diffusion.

The length scale of the smallest fluctuation of any property of diffusion constant D is given by

$$L_d = 2\pi(\nu D^2/\varepsilon)^{1/4} , \qquad (2.09)$$

which is called the Batchelor scale. If ε varies from 10^{-6} to 10^{-9} as suggested earlier, the smallest scale for temperature fluctuations is 2–13 mm and the smallest scale for salt or nitrate fluctuations is 0.2–1.0 mm.

(section 3.2.4) which is proportional to the density stratification. The size of the largest turbulent eddies, sometimes called the buoyancy length scale, is estimated by Gargett *et al.* (1984) to be

$$L_b = (\varepsilon/N^3)^{1/2} \ . \tag{2.10}$$

In the mixed layer if $\varepsilon \simeq 10^{-7}\,\mathrm{W\,kg^{-1}}$ and $N \simeq 10^{-3}$ rad $\mathrm{s^{-1}}$, $L_b \simeq 10$ m. In the deep ocean, where ε is small, or in stratified regions, where N is high, the buoyancy length scale works out to about one-tenth of the value found in the mixed layer.

2.2.7 Turbulent or eddy diffusion

Across distances greater than a few millimetres the eddies of the turbulence mix the water much more effectively than does molecular motion. The eddy-caused diffusion works the same way as molecular diffusion except that the random movement of the eddies is much larger than the molecular motion. Eddy diffusion is also different from molecular diffusion in that it is the same for all properties, that is heat, salt and nitrate will have the same eddy diffusion constant. A typical eddy diffusivity for horizontal diffusion in the ocean is $\simeq 500\,\mathrm{m^2\,s^{-1}}$ which is about 10^9 times the molecular diffusivity for heat. More details about turbulent diffusion are contained in Box 2.04.

2.2.8 Boundary layers

It was mentioned in the introduction that solid boundaries, such as the surfaces of organisms, have associated with them a boundary layer in which water movement is reduced. Since all organisms in the sea have a need to exchange molecules of O_2, CO_2, NH_3, etc., with the surrounding medium, the boundary layer is liable to reduce that rate of exchange. There is also a boundary layer associated with the water above the sea floor, and this can affect the process of exchange of essential substances between the benthic community of organisms and the overlying water. This section examines in detail the properties of boundary layers.

The fundamental property to be considered is the 'no-slip condition'. Water molecules in contact with a solid surface stick to that surface and are therefore stationary with respect to it. If we plot the average velocity near the boundary as in Fig. 2.04(a) we see that it increases as we move away from the boundary until we come to water moving with the 'free-stream velocity', i.e. the velocity which it would have if the solid surface were not present. However, Fig. 2.04(a) is simplified, as if the water in a

Box 2.04 Eddy fluxes and time scales

Ideally the turbulent flux of a constituent, C, can be determined by measuring the turbulent velocity fluctuations along with the fluctuations in the concentration of C. Such measurements are, however, difficult to make and are only done in special situations. Usually it is assumed, by analogy with the molecular case, that the flux is dependent on the gradient of C and a diffusion constant, except in this case the diffusion constant is a diffusion due to eddies rather than a molecular diffusion. Thus the flux F of the constituent C is written

$$F = K_e \, dC/dz, \tag{2.11}$$

where K_e is the eddy diffusivity and dC/dz is the gradient in C. The value of K_e varies greatly throughout the ocean, depending partly on the level of turbulence and partly on the stratification. Also, because of stratification, eddy diffusion is not the same vertically as horizontally. One recent estimate quoted by Gargett *et al.* (1984) gives, deep in the ocean, a horizontal eddy diffusivity $K_h \simeq 500 \text{ m}^2 \text{ s}^{-1}$ and a vertical eddy diffusivity $K_v \simeq 0.6 \times 10^{-4} \text{ m}^2 \text{ s}^{-1}$.

As with the molecular case (Eq. 2.06), the eddy diffusivity contains information on the rate of diffusion. For the case of turbulent diffusion the approximate length of time the diffusion takes over a distance L is calculated from L^2/K_e and if L is 1 km, horizontal eddy diffusion will show an effect in $L^2/K_e = 10^6/500 \simeq \frac{1}{2}$ h. In the deep stratified ocean, eddy diffusion will transport an effect vertically through 10 m in $100/0.6 \times 10^{-4} \simeq 20$ days. In the mixed layer where the eddy diffusion rate is high in all directions the time taken to mix properties is obviously less than the time taken in the deep ocean.

particular layer is moving at a constant velocity and always in the same direction. In real life flows are turbulent and exhibit fluctuations perpendicular to the direction of the mean flow as illustrated in Fig. 2.04(b). The amplitude of these fluctuations decreases toward the boundary as there can be no flow into or out of the boundary.

The decrease in the magnitude of the mean flow is caused by the stress, τ, or drag exerted on the water by the solid boundary. The transmission of this stress across the layer of water is accomplished, in turbulent flow, by eddy diffusion. As the boundary is approached the flow decreases until a level is reached where viscosity smooths out almost all turbulence and the stress between the water and the solid is transmitted by viscous stresses.

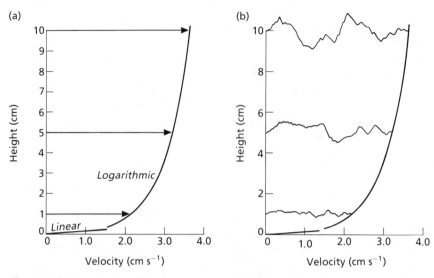

Fig. 2.04 (a) A vertical profile of mean water velocity through the boundary layer above a smooth surface showing the linear sublayer where viscous stresses dominate the stress between the water and the surface and the logarithmic layer where turbulent or Reynolds stresses dominate. (b) The same profile as in (a) including time series measurements of velocity at three levels to illustrate the increase in the size of the turbulent fluctuations with height above the boundary.

It is therefore possible to recognize two distinct layers within the boundary layer, the part in which turbulent fluctuations transmit the stress (Reynolds stresses) and the viscous sublayer adjacent to the boundary. If the velocity profile in Fig. 2.04 is plotted on a semilogarithmic scale, as in Fig. 2.05 (lower line), the different layers are easily distinguished. From 10 cm down to about 0.2 cm from the boundary the profile is represented by a straight line, commonly called a 'log-linear' relationship. Adjacent to the boundary, in the viscous sublayer, the profile is no longer log-linear. In fact, in this thin, viscous sublayer the velocity decreases linearly which appears as a curve on the logarithmic plot. The convention is to call the outer region the logarithmic layer, 'log-layer', and the region adjacent to the boundary the linear or viscous sublayer. The equations for these layers are presented in Box. 2.05.

Since boundary layers are regions of reduced turbulent transport, which may reduce the exchange of essential substances between an organism and the water, it is important to know the thickness of the layers. As we move from the surface of an organism out into the water the speed of flow increases as the effects of the no-slip condition decreases, but just where the speed reaches the free-stream value is not easily defined. The equation introduced in Box 2.05 to describe the

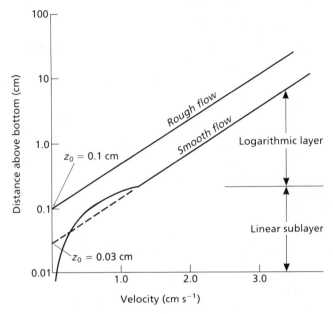

Fig. 2.05 Logarithmic profiles of mean velocity through a boundary layer for the cases of smooth ($z_0 = 0.03$ cm) and slightly rough ($z_0 = 0.1$ cm) flow. Both profiles have $u_* = 0.25$ cm s^{-1}. The linear sublayer where viscous forces are important is only a prominant feature of the smooth flow profile.

mean flow in the boundary layer (Eq. 2.13) predicts an ever increasing velocity with distance from the boundary. This is because the equation only describes the logarithmic layer without any information about the free-stream velocity. In this situation, the practical solution is to define the thickness of the boundary layer as that distance from the boundary at which the velocity is 99% of the free-stream velocity.

The thickness of the boundary layer surrounding an object moving through water increases with increasing distance from the leading edge. A familiar analogy is that of an aircraft wing. At the front of the wing the boundary layer is very thin, but increases in thickness with distance from the leading edge according to the expression (Prandtl 1969):

$$\delta = (xv/u)^{1/2}, \tag{2.14}$$

where x is the distance from the leading edge, v is the viscosity of the medium and u is the speed of the flow. When applied to an organism moving through water, two conclusions may be reached. The first is that the larger the organism, the thicker the boundary layer, since there are more places where the value of x is large. Secondly, the boundary layer becomes thinner as the velocity of the water relative to the organism increases. For sedentary organisms, rapidly moving water will cause a

Box 2.05 Equations for the boundary layers

Two equations are required to describe these two layers (Chriss and Caldwell 1984). For the viscous boundary layer, as we have already seen (Eq. 2.01), the viscous stress between adjacent layers moving at different velocities is given by

$$\tau = \rho v \, du/dz \, .$$

Because τ and ρ are constant through the boundary layer the ratio $(\tau/\rho)^{1/2}$ is also constant, and is conventionally replaced by u_* which is called the 'friction velocity'. After integrating and introducing u_* the velocity profile becomes

$$u(z) = u_*^2 z/v \, . \tag{2.12}$$

From this we see that in the viscous boundary layer the velocity, u, is a linear function of z, the distance from the boundary. Hence the straight line close to the boundary in Fig. 2.04.

For the log layer, the mean velocity is given by

$$u(z) = (u_*/k) \ln(z/z_0) \, , \tag{2.13}$$

where k is the Von Karman constant, $\simeq 0.4$. In this layer the velocity increases logarithmically with distance from the boundary.

In these examples we have assumed that the boundary is a smooth surface. In the real world few surfaces are perfectly smooth and the irregularities, or roughness elements, in the surface are liable to generate turbulent wakes in the sublayer. If these wakes are not smoothed out by viscosity the whole sublayer becomes turbulent and the flow is called 'rough'. This has two consequences. First, the boundary exerts more stress on the water because there is more friction. The velocity at any given distance from the boundary is therefore less than before as indicated by the upper line in Fig. 2.05. Second, the relationship between velocity and distance from the boundary becomes logarithmic throughout the boundary layer and the linear sublayer controlled by viscosity disappears.

As indicated in Fig. 2.05 the height, z_0, where the mean velocity is predicted to go to zero is greater in rough flow than in smooth flow. This height can be used as an estimate of the size of the roughness of the surface, and is called the roughness length. In practice it is about one-tenth the height of the roughness elements. The logarithmic profile is reasonably accurate above the level of the roughness elements but in amongst the roughness elements the flow is a complicated function of position and can not be predicted by the equations.

The variable z_0, representing the size of the roughness elements, along with the friction velocity u_*, are sometimes used as the typical length and velocity scales in the Reynolds number (Eq. 2.02). When this is done it is called the roughness Reynolds number and is denoted by Re_*.

thinning of the boundary layer, and can be expected to assist exchange of substances. For motile organisms, rapid swimming can be expected to have the same effect.

2.2.9 Drag

When a machine or an animal moves through water, two forces arise to oppose the motion. The first of these drag forces originates in the no-slip condition in association with viscosity. Because of the no-slip condition the moving object is stuck to the immediately adjacent layer of fluid which in turn pulls along, because of viscosity, some of the layers of fluid further out. The resultant shear in the water is maintained by a stress, or force per unit area (Eq. 2.01), which must be supplied by the moving object if it is to maintain its motion. Because the magnitude of this drag force is directly dependent on the surface area of the object it is often called skin friction instead of viscous drag. In cases of low Reynolds number flow, as encountered with small plants or animals falling or swimming slowly, this is the only appreciable drag force encountered and is an important feature of the feeding strategies as will be explored in section 2.3.2.

The other drag force, called form or pressure drag, arises because the water, which has mass, must be pushed out of the way of the moving object, and then move back into place behind. Discussion of this force is thoroughly presented by Vogel (1981) who shows that it may be represented by the formula

$$F_D \simeq \frac{1}{2} C_D \rho u^2 A , \tag{2.15}$$

where ρ is the density of the water, u the speed of the object and A is some measure of the cross-sectional area of the object, such as the area projected onto the plane perpendicular to the direction of motion. The term C_D, the drag coefficient, is a complicated function of velocity which must be determined empirically for each object. C_D is so complicated because of the way water flows around the moving object especially around the stern or back end. Objects which are designed to have as little drag as possible, like aeroplanes and fish, are shaped in a streamlined way so that the fluid moves in behind in a smooth and orderly way. If the object is not streamlined the motion creates vortices and turbulent wakes at the back which lead to a reduction in pressure on the back side. Lower pressure at the back than at the front generates the pressure drag force in the direction opposite to the motion. The speed at which a particular object creates the turbulent wake or the place on the object where the wake will be produced cannot be deduced *a priori*, and the form or pressure drag can only be determined by experiment for each item. For some simple shapes, such as spheres and cylinders, etc., values of C_D have been determined once and for all as functions of Reynolds number and these are available in the reference literature. Plants, however, change shape as the speed of the flow changes and it is not yet possible to determine the form drag of such objects without direct measurements.

2.2.10 The problem for phytoplankton

The problem of boundary layers around phytoplankton cells was briefly referred to in previous sections. The generally accepted story, first worked out quantitatively in a classic paper by Munk and Riley (1952) is that phytoplankton cells suspended motionless in the water tend to use up the nutrients in the water around them. Their nutrient-uptake rate is then limited to the rate at which the nutrients can diffuse towards them. One way of overcoming this limitation is to generate movement relative to the water, so that the zone of nutrient-depleted water is periodically renewed. They may do this either by sinking passively or by generating their own locomotion. As discussed earlier, there is inevitably a region around the organism in which the velocity of the water relative to the cell is less than the absolute rate of movement of the cell through the water. The cause of this is the no-slip condition, which states that molecules in contact with the cell surface adhere to that surface and cannot slip past it. In the boundary layer around the cell the water velocity relative to the cell changes from zero at the cell surface to the 'free-stream velocity' beyond the boundary layer. A sinking or swimming phytoplankton cell drags its boundary layer with it, causing fluid in the depleted region to be sheared away. New fluid in which the nutrients have not been depleted takes its place. This decreases the effective size of the depleted region and increases the rate of uptake by the cell. The magnitude of the increase depends on the velocity of the cell relative to the water.

2.2.11 Sinking of phytoplankton

Most phytoplankton cells are more dense than water. The density of sea water varies from about 1.021 to $1.028\,\mathrm{g\,cm}^{-3}$, but the density of cytoplasm is from 1.03 to $1.10\,\mathrm{g\,cm}^{-3}$. Diatoms have a cell wall of hydrated silicon dioxide with a density of about $2.6\,\mathrm{g\,cm}^{-3}$ and coccolithophorids have plates of calcite, aragonite or valerite with a density of

$2.70–2.95\,\mathrm{g\,cm^{-3}}$. Thus there is plenty of evidence that many phytoplankton cells are substantially more dense than water; which means they will sink except where an upward movement of the water prevents it. The rate at which the cells fall can be estimated (Box 2.06) if the Reynolds number is $\ll 1.0$ and if the cells are approximately spherical.

For a given excess density and viscosity, a sphere will increase its sinking speed in proportion to the square of the radius (Eq. 2.17 presented in Box 2.06). Few phytoplankton cells are spherical; nevertheless for $Re \leq 0.5$, Stokes' law may be applied to non-spherical bodies without important errors (Hutchinson 1967). Roughly speaking, phytoplankton up to $500\,\mu m$ in diameter fall within this category. Empirical evidence for increasing sinking speed with increasing size has been reviewed by Smayda (1970) (Fig. 2.06). Although the data fall within the $1–500\,\mu m$ range, the slopes of the lines are considerably less than the value of 2 predicted by Stokes' law. In considering the reasons for this, shape can probably be ruled out. According to the theoretical analysis of Munk and Riley (1952), the effect of geometry is different in different size classes. In order of decreasing sinking rate, the pattern for $5\,\mu m$ particles is plate $>$ cylinder $>$ sphere. For $50\,\mu m$ particles it is cylinder $\simeq$ plate $>$ sphere, and for $500\,\mu m$ particles it is cylinder $>$

Box 2.06 Falling speed of small spheres

The formula used to calculate the terminal velocity of small spheres at low Reynolds number was devised by Sir George Stokes in 1850. He first found that the fluid slows the falling sphere by exerting a resisting force equal to $6\pi\rho vVr$, where ρ and v represent the density and the kinematic viscosity of the fluid, and V and r represent the velocity and radius of the sphere. This force must be equal to the weight of the sphere less the upward push due to buoyancy. The weight of a sphere is equal to its mass times the acceleration due to gravity, g, and the mass is the volume times the density, ρ', thus the weight is $\frac{4}{3}\pi r^3 \times \rho' \times g$. The upward force of buoyancy is equal to the weight of the water displaced by the sphere, $\frac{4}{3}\pi r^3 \times \rho \times g$. Equating the resisting force to the net weight gives

$$6\pi\rho vVr = \frac{4}{3}\pi r^3 g(\rho' - \rho),\qquad(2.16)$$

and after rearranging, the rate of fall is given by

$$V = \frac{2}{9}gr^2(\rho' - \rho)/\rho v.\qquad(2.17)$$

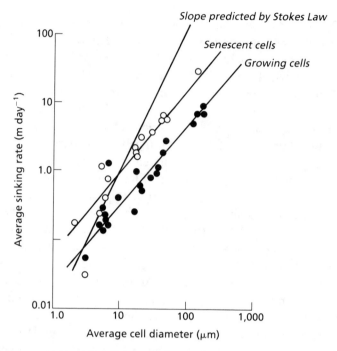

Fig. 2.06 Average sinking rate of phytoplankton cells grown in culture and measured in the laboratory and line predicted by Stokes' law. After Smayda (1970).

sphere > plate. Moreover, the sinking rates of different shapes tend to equalize with increasing size. The data in Fig. 2.06 have not been corrected for changes in density and viscosity with temperature. At the natural extremes this is not negligible: a change from 0 °C to 25 °C results in lowered density and viscosity so that the sinking rate of 20 μm particles approximately doubles. There is no reason to think that the data displayed in Fig. 2.06 were carried out at such an extreme range of temperature difference.

It is seen from Fig. 2.06 that senescent cells sink faster than actively growing cells. It has also been found that preserved cells sink faster than living cells, and that natural blooms of phytoplankton tend to sink in the water column as they approach senescence. Factors that are thought to contribute to this effect include: variation in the cell content of low density oils and fats, variation in the amount of light ions such as NH_4^+ and possible effects of excreted products on the viscosity of the fluid surrounding the cells. The effectiveness of sinking as a means of combating diffusion limitation is considered in section 2.2.13.

2.2.12 Swimming by phytoplankton

Swimming is the alternative to sinking, as a means of reducing the diffusion limitation on nutrient uptake. The locomotion of small organisms is a very different proposition from what we are familiar with from our own attempts at swimming or from watching macroscopic organisms such as fish or insects (Purcell 1977). For small organisms which perform movements relatively slowly, the Reynolds number is low and viscous forces predominate. A picoplankton organism of about $1\,\mu m$ radius moving at about $30\,\mu m\,s^{-1}$ generates inertial forces that are so small that after the swimming effort ceases it will come to a stop in about $0.6 \times 10^{-6}\,s$, travelling only $10^{-4}\,\mu m$. It is rather like a human swimming in a pool of molasses and forbidden to move any part of the body faster than $1\,cm\,min^{-1}$. One consequence of this situation is that jet propulsion just does not work: it relies on inertial forces. A scallop-like animal at this scale would just expel the liquid as it closed its shell and take it in again on opening, getting nowhere in the process. The mechanisms that work best under these conditions are the 'flexible oar' (cilium) and the 'corkscrew' (flagellum pointing forward) (Purcell 1977). Of the two, flagella are the most commonly used by phytoplankton.

2.2.13 The effectiveness of swimming or sinking

A stationary phytoplankton cell cannot take up nutrients faster than they are transported towards it by diffusion (section 2.2.10). There are some types of phytoplankton that take up nutrients so slowly that diffusion is not a limitation on their metabolism. This is not true for most phytoplankters which must move relative to the water by sinking or swimming in order to increase their nutrient uptake (Pasciak and Gavis 1974). The critical question is: how effective is relative movement between the cell and the water in increasing the rate of nutrient uptake?

We see in what follows that the conclusions of various studies are quantitatively the same. For small organisms in the $1-10\,\mu m$ range, viscosity dominates their world and diffusion is faster than water movement in supplying nutrients through their boundary layer. As far as we can tell, these small organisms move in order to find better concentrations of nutrients in the environment, not to reduce their diffusion limitation.

On the other hand, flagellates above $10\,\mu m$, many of which are able to swim more than 10 times their own body length per second, may achieve a significant increase in nutrient uptake by swimming. The relative advantages are greater still for cells of $50\,\mu m$ or more, not because they move faster but because they have a smaller ratio of surface area to volume and have a stronger nutrient limitation in the first place.

Most diatoms are unable to swim, but create water movement past their surfaces by sinking. Large cells sink faster than small cells, and it is only the larger cells that sink fast enough to gain a significant advantage from it in terms of nutrient uptake. In no case can these cells abolish nutrient limitation by sinking, but the differences between cells are thought to be great enough to influence interspecific competition. The evidence for these findings is as follows.

Pasciak and Gavis (1974) considered the question of diffusion limitation in conjunction with conventional Michaelis–Menten nutrient dynamics. They concluded that when the concentration of nutrients in the environment is not many times greater than the half-saturation constant of the cell, and when the cell is small, diffusion can be a severe limitation on cell metabolism. Gavis (1976) then introduced motion relative to the water into the calculations. Taking published rates of sinking of phytoplankton cells, he concluded that an appreciable increase in nutrient uptake rate would occur only if the cells are relatively large (100μm) and the concentration of nutrients is low. In the optimum case, sinking would alleviate only about 30% of diffusion limitation.

A similar result was obtained using published swimming rates. The effect was most marked with large (100μm) cells swimming at about 3 bodylengths per second. There was an appreciable increase in nutrient uptake but it was still less than one third what it would have been in the absence of diffusion limitation. Hence, the conclusion was that swimming or sinking are not very effective methods of alleviating diffusion limitation, but that they might make enough difference to affect the outcome of competition between species.

Berg and Purcell (1977) and Purcell (1977) did not take into account the biological properties of the cells but considered the case of a spherical cell of radius r, propelled at a constant velocity u through water containing molecules 'for which the cell is a perfect sink'. They calculated the relationship between velocity and the fractional increase in the transport of nutrients to the cell. They showed that this increase was a complex function of ru/D, where D is the diffusion constant of the molecules absorbed. More recently Sommer (1988) took this formulation by Berg and Purcell (1977) and plotted the results for 19 species of marine flagellates for which the size and swimming velocity were known (Fig. 2.07). Swimming speeds ranged from about 2 to a maximum of 100 bodylengths per second, with the smaller cells having the higher relative velocities. Even so, the conclusion (Fig. 2.07) was that cells of 5μm, or less, increased their nutrient flux by only 5–30% when swimming. On the other hand, flagellates of 50μm diameter had the potential to double or even treble their nutrient-uptake rate by swimming.

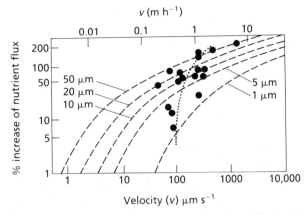

Fig. 2.07 Fractional increase of nutrient flux to the cell surface as a function of swimming velocity and cell diameter, calculated for spherical cells, after Berg and Purcell (1977). The circles represent observations and the finely dotted line is a regression through the observations. From Sommer (1988).

We may also apply this diagram to a consideration of sinking, treating it as a special case of locomotion. From Fig. 2.06 we see that a $50\,\mu m$ diameter cell sinks at approximately $1\,m\,d^{-1}$ or $12\,\mu m\,s^{-1}$. According to Fig. 2.07 this might increase the nutrient flux by 30–40%. On the other hand, a cell of $10\,\mu m$ would sink at only about $3\,\mu m\,s^{-1}$ which would give it no increase in nutrient-uptake rate.

There is yet another way of trying to understand that creating movement of water relative to a very small cell does not benefit nutrient uptake. Berg and Purcell (1977) discussed it in relation to 'stirring', defined as the creation of water currents by means of an appendage, but the concept applies equally well to water movement induced by sinking or swimming. The time required for transporting a nutrient distance L by water movement is L/u, where L is the distance transported and u is the water velocity. On the other hand, the time required for transport by diffusion is L^2/D, where D is the diffusion constant (see section 2.2.5). The effectiveness of water movement versus diffusion, for any given distance and diffusion constant (the Sherwood number) is given by:

$$\frac{\text{time for transport by diffusion}}{\text{time for transport by movement}} = \frac{L^2/D}{L/u} = Lu/D.$$

For scales of the order of $1\,\mu m$ the ratio works out at about 10^{-2}. Diffusion is about 100 times faster than movement. Hence, in this world of low Reynolds numbers nothing is gained by trying to reduce the diffusion barrier by generating turbulent advection. In this context, the only possible advantage to the organism of undertaking locomotion is that it might encounter nutrients in a higher concentration. Purcell (1977) summarized it by saying that the organism does not move like a cow that is grazing pasture, it moves to find greener pastures.

2.2.14 The role of turbulence

In the foregoing discussion on diffusion limitation and the possible effects of swimming or sinking on the rate of uptake of nutrients, no mention was made of the effects of turbulent motion. Turbulence, we know, is a ubiquitous feature of the ocean but it is not immediately clear how the turbulent motion affects the flow around very small organisms or if that flow will alter the rate of molecular diffusion towards or away from a cell. Since the interaction between the small organisms and the turbulence cannot be observed directly in the ocean we must rely on theoretical constructions and indirect observations.

The first discussion of the effect of turbulence on the molecular diffusion in the vicinity of small organisms is due to Munk and Riley (1952). They suggest that the turbulent-pressure fluctuations in the ocean create a small relative motion between the organisms and the water because the two are of slightly different densities. They conclude however that the effect of this relative motion on the diffusive flux near an organism would be negligible. The question was re-examined by Lazier and Mann (1989) who argued that turbulence is manifest across even the smallest distances by a linear shear whose magnitude depends on the strength of the turbulence. The shear is a property of the fluid and exists whether there are organisms present or not. The effect that this shear has on the diffusion towards or away from a perfectly absorbing sphere is calculated using the experimental results of Purcell (1978).

A summary of the increase in the diffusive flux due to both relative motion through the water and turbulence is given in Fig. 2.08. The two curves constructed from the results of Berg and Purcell (1977), B&P, show the change in the diffusive flux as a function of diameter for relative motion through the water at 1 and 10 diameters per second $(d\,s^{-1})$. The calculations of Munk and Riley (1952) for spheres are recast in the curves labelled M&R and on the right are two curves indicating the response to weak and strong turbulence.

The B&P results indicate that for cells less than $\simeq 5\,\mu m$ in diameter sinking or swimming at even $10\,d\,s^{-1}$ will have no effect on the rate of molecular diffusion in the neighbourhood of the cell. For cells of larger diameter relative motion causes a significant increase in the diffusive flux. A $40\,\mu m$ cell, for example, will increase the flux by about 100% if

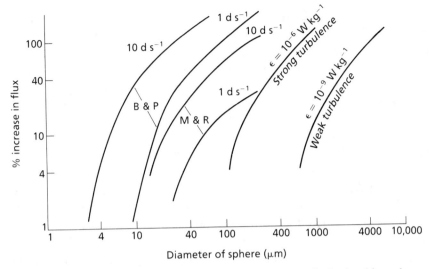

Fig. 2.08 Percentage increase in the diffusive flux towards a perfectly absorbing sphere of diameter d caused by the relative motion or the shear created by turbulence, adapted from Lazier and Mann (1989). The curves for relative motion are for sinking or swimming at 1 and 10 diameters per second ($d\,s^{-1}$) based on the work of Munk and Riley (1952) (M&R) and Berg and Purcell (1977) (B&P). The data for the effect of turbulence are derived from the work of Purcell (1978) for dissipation rates of 10^{-6} and $10^{-9}\,W\,kg^{-1}$. For interpretation, see text.

it moves through the water at $10\,s^{-1}$. Similar increases in flux are noted for relative motion at $1\,d\,s^{-1}$ but larger cell diameters are required. The M&R curves are roughly the same as the B&P which is encouraging considering the completely different methods of analysis. The curves representing the effect of turbulence show that even strong turbulence will not have an appreciable effect compared to that due to relative motion through the water. But for stationary cells the turbulence will start to have an effect on cells greater than $100\,\mu m$ in diameter. In weak turbulence, only cells larger than $1\,mm$ will be affected.

2.2.15 The paradox of cell growth in low-nutrient environments

Phytosynthesis has been shown to occur in waters where the concentration of nitrogenous nutrients is below the limit of detection (McCarthy and Goldman 1979). How can phytoplankton be physiologically active and grow enough to maintain their populations when nutrients are at such a low level? Perhaps the phytoplankton are able to take up the nutrients from small local concentrations such as might be produced by the excretion of a single zooplankton organism or the regeneration of nutrients from the bacterial decomposition of a particle of organic matter. It was demonstrated in the laboratory that phytoplankton are able to take up enough nutrients for a cell doubling during an exposure to high nutrients for less than 4% of a doubling period. If the model is correct, and phytoplankton cells are frequently encountering very small patches of high nutrient concentrations, this could explain the anomaly of phytoplankton growth at nutrient concentrations which, on average over larger scales of sampling, are below the limit of detection by present day methods.

Jackson (1980) argued against this hypothesis. Using data from the North Pacific Gyre, he showed that a balanced budget can be obtained when the phytoplankton undergo only 0.1 doublings per day. By extrapolation from the observation that *Thalassiosira pseudonana* can grow at 2.6 doublings a day at an ammonia concentration of 0.1 μmol, he suggested that it could maintain 0.1 doublings per day at a concentration of only 0.8 nmol, a value well below detectable limits. He further calculated that the nutrient plumes generated by the most important herbivores, the ciliates, would disperse by molecular diffusion much too fast for them to be useful to a phytoplankton cell, under most circumstances.

An even more detailed model was constructed by Williams and Muir (1981) using excretion rates appropriate to *Oithona* and *Clausocalanus*. They used the same diffusion coefficient for $N-NH_3$ as previous authors, but considered both the persistence time of the patches of ammonia and the probability of them being encountered by phytoplankton that were floating passively or actively sinking. They concluded that there cannot be enough of these patches of ammonia, nor can they persist long enough to be of importance in maintaining primary productivity in the oligotrophic ocean.

A further contribution to the debate came from freshwater ecology. Lehman and Scavia (1982a) labelled 2-mm-long female *Daphnia pulex* with ^{32}P so that they excreted labelled phosphorus. These were then added to cultures of the alga *Chlamydomonas*. Autoradiography was used to determine the distribution of the label among the algae after a period of exposure to the labelled *Daphnia*. In a stirred vessel the algae became labelled randomly according to an expected Poisson distribution, but in an unstirred vessel the cells became labelled in a non-random manner, presumably because some encountered micro-patches of nutrients while others did not. This certainly seemed to indicate that organisms of 2 mm (larger than most marine copepods) could produce patches of nutrients that could be utilized by phytoplankton. The authors argued that the unstirred experiments were representative of conditions in nature because at a scale of millimetres diffusion processes predominate over turbulent processes.

In a second paper Lehman and Scavia (1982b) produced a model of the nutrient plume released by a 2-mm-long *Daphnia*, using the assumption that only diffusion is operative at this scale. The postulated excretion rate was 0.0618 pmol of P per animal per second; the rate of swimming 0.015 cm s^{-1} and diffusion coefficient 10^{-5} cm^2 s^{-1}. Figure 2.09(a) shows the concentration ranging between 625 nM close to the animal and 5 nM 60 cm behind it. Figure 2.09(b) shows how the modelled distribution was expected to change when *Chlamydomonas* was present. Knowing the densities of cells in the experiment the authors were able to show that the algae would indeed take up significant amounts of nutrients, leading to the modified nutrient distribution shown. The model was used to predict the frequency distribution of P uptake in a population of *Chlamydomonas* and the model results were

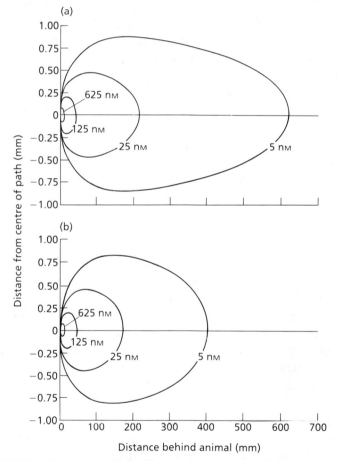

Fig. 2.09 (a) Modelled contour diagram for steady-state plume of phosphorus behind a swimming *Daphnia*. (b) Contour diagram with algal uptake of phosphorus. From Lehman and Scavia (1982b).

statistically indistinguishable from the experimental results, supporting the view that excretory plumes of zooplankton are an important source of nutrients for phytoplankton.

Currie (1984) used Lehman and Scavia's (1982a,b) data to calculate that a stationary phytoplankton cell which experienced the most favourable exposure to a patch by being in the centre line as the *Daphnia* swam by, would obtain from it only about 10% of the P required for one cell division. To support a doubling rate of 1.0 per day there would have to be an unrealistically high concentration of animals. Currie (1984) concluded that the results of Lehman and Scavia were artifacts resulting from: (i) high concentrations of phytoplankton, producing unrealistically high encounter rates; and (ii) absence of a uniformly distributed 'background' level of labelled P. Currie (1984) suggested, as Jackson (1980) had done, that slow continuous uptake of P from low concentrations is sufficient to sustain most phytoplankton populations, and absence of any way to compare rapid and slow uptake routes invalidated the statistical approach used by Lehman and Scavia. The latter authors, in a response to Currie's (1984) paper, pointed out that they merely claimed that phytoplankton were *able* to take up nutrients from plumes created by zooplankton. Neither they nor anyone else had satisfactorily demonstrated the ecological significance of the phenomenon. This amounts to an admission that Currie (1984) was probably right.

As was mentioned earlier, two mechanisms have been suggested for the production of localized high concentrations of nutrients in oligotrophic waters. One is zooplankton excretion, the other is bacterial breakdown of organic matter. For the latter to be effective, there must be a mechanism to bring together the bacteria, their organic food and the phytoplankton cells that are to benefit from the nutrients released by the bacteria. Attention has recently focused on various types of organic aggregates that are abundant and widespread in the sea. One type is microscopic, formed by the precipitation of dissolved organic matter at interfaces such as the air–water interface of bubbles and at the sea surface (Riley 1970). Another type is macroscopic and is believed to be formed from the mucus structures and remains of gelatinous zooplankton. Known collectively as 'marine snow', these macroscopic aggregates are highly amorphous and very fragile. The best way to collect samples for study is to scuba dive and collect the specimens by hand. They are found to have concentrations of phytoplankton, bacteria and protozoa (Silver *et al.* 1978) and they occur both in surface waters and at great depth (Silver and Alldredge 1981).

Both the microscopic and the macroscopic aggregates have been postulated as sites where bacterial action would permit the accumulation of nutrients, thus facilitating phytoplankton growth in nutrient-deficient waters. For the microscopic aggregates, Goldman (1984b) put forward his 'aggregate-spinning wheel' hypothesis (Fig. 2.10) according to which phytoplankton, micro-grazers and bacteria live attached to the surfaces of the aggregates and are involved in the rapid recycling of nutrients and hence a high turnover of phytoplankton biomass. As he himself admitted, the hypothesis is untestable by contemporary methods.

On the other hand, some progress has been made in obtaining evidence that the macroscopic marine snow forms sites of high nutrient concentration. Alldredge and Cohen (1987) used microelectrodes to investigate the small-scale distribution of oxygen and pH around marine snow particles. In the light, they found clear evidence of photosynthetic oxygen production in a zone about 1.5 mm thick at the surface of the aggregates and a zone of enhanced oxygen concentration extending about 1 mm beyond the aggregate. In the dark, they found oxygen depletion within the aggregates, and this also extended about 1 mm beyond the boundary of the particle. Thus there was evidence of both photosynthesis and respiration. Water was moved past the aggregates at the time of measurement, at a rate thought to roughly correspond with the rate of sinking in nature. The authors therefore concluded that in spite of water movement, a boundary layer forming a diffusion barrier exists around the aggregates. Since it had already been shown (Shanks and Trent 1979) that the aggregates are sites of elevated phosphate, ammonium and nitrate, they concluded that the boundary layer will

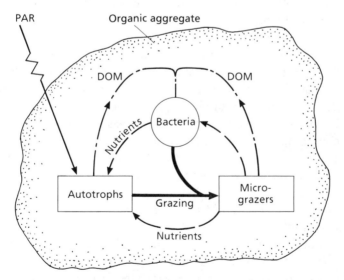

Fig. 2.10 Conceptual scheme of a microbial food chain within a discrete organic aggregate. From Goldman (1984).

tend to maintain high nutrient concentrations within the aggregates, thus providing phytoplankton with the opportunity for rapid turnover of biomass.

2.2.16 Conclusions

We have seen that a phytoplankton cell that remained neutrally buoyant in still water would soon deplete the nutrients in the water surrounding it, and create a gradient. Under these conditions, the only mechanism that would tend to replace the nutrients would be molecular diffusion. This may be adequate for very small cells of the order of $1\,\mu m$, but it tends to be extremely limiting of the metabolism of phytoplankton of the scale $10–100\,\mu m$. It has been proposed that they overcome this difficulty by either sinking or swimming, rather than floating neutrally buoyant. A detailed study of the dynamics suggests that the observed rates of sinking go some way towards reducing diffusion limitation, perhaps enough to account for one species having a competitive edge over another, but even then the uptake rate of nutrients is likely to be no more than one-third of what the rate would be in the absence of diffusion limitation.

Swimming is found in all classes of organisms from bacteria-sized upwards. For bacteria and picoplankton it seems that the role of swimming is to move the organisms to regions of higher nutrient concentration. The relative importance of viscosity is so great for these organisms that their locomotion does not materially change their diffusion limitation. For cells of about $100\,\mu m$ the effect of viscosity is less drastic, and it has been calculated that at observed rates of swimming they have the possibility of reducing diffusion limitation by about one-third.

Finally, there has been some study and much speculation about the possibility that the excretion of microbes and animals creates micro-zones of high nutrient concentration in the sea, and that phytoplankton are capable of taking advantage of them. It has been shown under laboratory conditions that zooplankton as large as *Daphnia* are able to create nutrient-rich zones from which *Chlamydomonas* can benefit, but the ecological applicability of these results is in doubt. Most calculations from field conditions suggest that the nutrient plumes generated by zooplankton are too few, and disperse too fast to be of great significance to phytoplankton.

An idea that is still largely untested holds that, especially in oligotrophic seas, phytoplankton cells live in close proximity to bacteria on the surfaces of amorphous organic aggregates formed by the precipitation of dissolved organic matter, or on the marine snow thought to be derived from large gelatinous zooplankton. The bacteria may be consumed by micro-grazers such as ciliates and these may be responsible for the excretion of the nutrients needed by the phytoplankton. These ideas are intriguing but await further investigation.

2.3 ZOOPLANKTON

2.3.1 Life in a viscous environment

For particles of the size of phytoplankton, inertial forces are very small and by comparison viscous forces are strong and predominate. As a consequence, the relationship between a zooplankter and its food is quite different from anything in our own experience. Viscous flow is reversible and a limb moving backwards and forwards creates a symmetrical pattern of flow. A copepod may scratch a bacterium from its surface, but if the motion is symmetrical the bacterium will return to the place of origin. A complex asymmetrical movement must be made if the bacterium is to be removed from the cuticle (Strickler 1984). On the other hand, viscous properties can be used to advantage when feeding. Algae captured by copepods and assembled between the setae of the mouthparts will remain there so long as the mouthparts make only symmetrical movements.

When it comes to swimming, zooplankton occupy an interesting transitional range of Reynolds number, between 0.1 and 500 (Fig. 2.03). At the scale of the setae on the feeding appendages, viscous forces are important and inertial forces are trivial. However, some copepods are capable of accelerating their mass to 35 cm s^{-1} and inertial forces can then be exploited (Strickler 1984).

2.3.2 Feeding in a viscous environment

The effect of boundary layers on the action of mouthparts, when viscous forces are predominant, is graphically illustrated by a scaled-up model (Strickler 1984). The medium is honey, the food particle to be captured is a grain of rice, and the mouthparts are a pair of knife blades. Figure 2.11(a) shows that if a food particle is close to the surface of the mouthpart and that mouthpart is moved in a direction parallel to its flat surface, the particle will move with it, because the medium closest to the blade does not slip, and layers of medium progressively further from the blade are drawn along by viscous stress. Figure 2.11(b) shows that if the two mouthparts are brought together in the vicinity of the food particle (a clap) the particle escapes because its inertia is negligible compared with the viscous forces carrying it away in the fluid (this contrasts

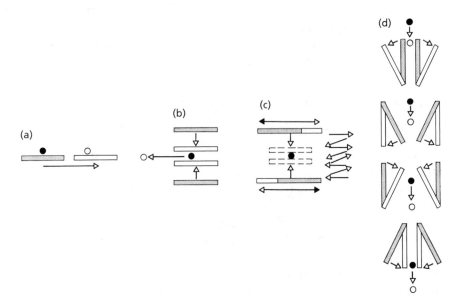

Fig. 2.11 Diagrams of experiments on low Reynolds number flow. Starting positions shaded. For details see section 2.3.2. Modified from Strickler (1984).

with the situation in which we successfully capture a mosquito in air by clapping). Figure 2.11(c) shows that if the mouthparts move in opposite directions simultaneously they can exert equal and opposite viscous drag forces on the particle so that it remains approximately stationary and is captured. Finally, Fig. 2.11(d) shows a technique that has come to be known as 'fling and clap'. The two mouthparts are flung apart, drawing in a parcel of medium containing the food particle. The edges closest to the inflow then converge behind the particle, effectively trapping it.

The older view of the way in which copepods feed was that the appendages pushed water postero-laterally, creating a large swirl on each side of the animal (Fig. 2.12). The maxillipeds swung outwards then inwards, sucking some of the swirling water into the mid-line of the animal and forwards through a filter formed by the second maxillae. The captured food was transferred to the mouth by the endites of the first maxillae. There were several studies of the efficiency of this filtering mechanism for the retention of particles of different sizes. Koehl and Strickler (1981) were at a loss to understand how the water would be forced through the sieve formed by the setae, when it could take the path of least resistance and go around it. They also noted that the feeding mechanism had been studied by holding animals in drops of

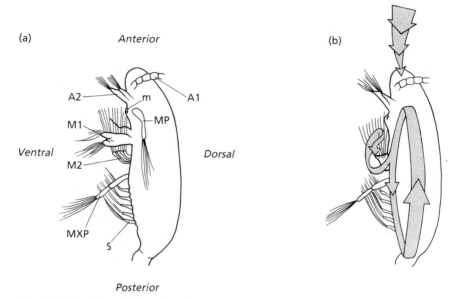

Fig. 2.12 *Eucalanus pileatus.* (a) Diagram of an animal in typical feeding position. Only left appendage of each pair is shown. A1, A2 — first and second antennae; M1, M2 — first and second maxillae; MXP — maxilliped; S — swimming leg. (b) Diagram of feeding currents observed by earlier workers when animal was imprisoned in a drop of water. From Koehl and Strickler (1981).

water under the microscope and suspected that the large lateral swirls were artifacts resulting from the confinement.

Koehl and Strickler (1981) reinvestigated the feeding mechanism by tethering specimens of *Eucalanus pileatus* in an optical glass cuvette holding 120 ml of water. They took high-speed cinematographic movies of the feeding process while India ink was released from a micro-pipette close to the animal. They found that second antennae, mandibular palps, first maxillae and maxillipeds combined in action to produce a pulsing stream of water past the ventral side of the animal (Fig. 2.13). When a food particle came within range, the appendages began to beat asymmetrically, simultaneously drawing the particle towards the mid-ventral line and turning the animal towards the particle. At the critical moment, the second maxillae performed a fling and clap movement that drew the alga and a small volume of water into the space between them, then closed in to retain the alga while forcing the surrounding water out through the spaces between the setae. There are several important points to note. The first is that forcing water continuously through a fine sieve is an energetically costly activity, and the copepods avoid much of this cost by scanning the water moving past them and forcing through the second maxilla only that small volume of water surrounding the

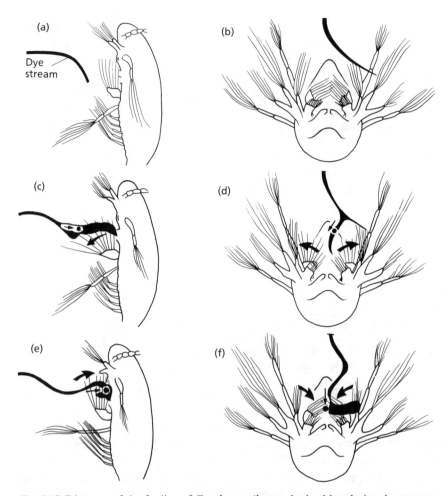

Fig. 2.13 Diagrams of the feeding of *Eucalanus pileatus* obtained by placing dye streams from a micro-pipette close to the animal and using high speed cine-camera. Heavy arrow indicates movement of second maxilla. Light arrows and circles show movement of a particle during food capture. (a), (c), (e) — view from left side of animal; (b), (d), (f) — view from head end. From Koehl and Strickler (1981).

food particle. Secondly, the flow of water around and between the setae of the second maxillae is very different on the small scale and low Re from what we expect from observing events on the larger scale, and it seems that algae are often directed into the mouth without actually making contact with the setae. When they do make contact with the setae, they stick, as mentioned earlier, and they are combed off by the endites of the first maxillae. The large-scale analogue of this operation is using a fork to comb crumbs off another fork when both are immersed in honey (Koehl 1984), and this is clearly not a trivial task.

Finally, the observations that the feeding appendages change the symmetry of their beat when an alga is in the proximity shows that there is a mechanism for sensing the algal presence before it makes any contact with the copepod, and that the animal is performing a 'detect and capture' operation rather than behaving as a passive filter feeder.

2.3.3 Detection of food

It seems probable that detection of food particles is by chemoreception or mechanoreception. In his review, Strickler (1984) pointed out that structures have been identified which look like chemoreceptors, but there is no proof that they function as such. However, when he offered copepods glass particles that were too large to ingest, they captured them provided that they were simultaneously exposed to an appropriate chemical stimulus delivered by means of a micro-pipette, but ceased capture and discarded any they had already captured as soon as the chemical stimulus was interrupted. Clearly, chemical stimuli do affect the feeding behaviour of copepods. Both taste and smell types of chemosensors have been identified.

Wong (1980) showed that cyclopoid and calanoid copepods can sense moving bodies with low Re flow fields, at 3–4 bodylengths away. This suggests the existence of mechanoreceptors. However, algae swim weakly or not at all and are unlikely to produce signals that would be detected by mechanoreceptors more than 1 mm away. Furthermore, Friedman and Strickler (1975) searched for ultrastructural evidence of mechanoreceptors in calanoid copepods and failed to find them, but they speculated that there really had to be mechanoreceptors to facilitate the handling of food after capture.

2.3.4 Possible modes of selective feeding

The methods copepods use to feed selectively on different sorts of particles are not yet well understood, but various mechanisms can be proposed in the light of our new understanding of the feeding process in a viscous environment (Koehl 1984). It is now no longer possible to regard the second maxillae with their array of setae as simple sieves through which water is forced while particles are retained. The no-slip condition, to which we have repeatedly referred, means that water moving between two closely-spaced setae has zero movement at the point of contact with the setae and is subjected to shear deformation in the boundary layer surrounding each seta. Hence there is great resistance to the movement of water between closely spaced setae. A copepod appendage with an array of setae behaves more like a solid paddle than an open rake. Only when water is trapped and has no other route of escape can any quantity be forced between the setae.

Copepod second maxillae have been found to retain particles that are smaller than the spaces between the setae. Rubenstein and Koehl (1977), using filtration theory, drew attention to five possible mechanisms (Fig. 2.14). The first is conventional sieving, in which a particle is retained because it is larger than the space through which the water is passing. The second is direct interception, in which the water containing a particle passes so close to a seta that the particle collides and, because of the low Re, adheres. The third is inertial impact, in which a stream of water is diverted round a seta, but a particle has sufficient momentum that it carries on in more or less a straight line and collides with the seta. The fourth is diffusion, or motile-particle deposition, in which a particle has a random movement in relation to the stream of water carrying it, and is brought into contact with the seta. Finally, there is gravitational deposition, in which the weight of a particle produces a small sinking motion that brings it into contact with the seta. As particle size increases, the filtering element's ability to capture by inertial impaction, gravitational deposition and direct interception is improved. As particle size is reduced, diffusion is enhanced.

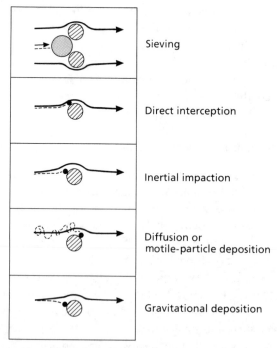

Fig. 2.14 Diagram illustrating five mechanisms of particle retention, as described in section 2.3.4. From Rubenstein and Koehl (1977).

Centropages typicus has second maxillae with widely spaced setae and these move almost 20 times faster than those of *Eucalanus pileatus*. Hence the Reynolds number is very different in the two cases, and the operation of the various mechanisms of particle interception will be very different. Many copepods move their second maxillae in a different way when feeding on small algal cells than they do when feeding on large ones (Price and Paffenhofer 1980). The precise consequences of these various adaptations have yet to be investigated.

2.4 BENTHIC PLANTS

2.4.1 The problem

The concept applied to phytoplankton applies also to attached benthic plants. If a plant is surrounded by completely still water it will soon produce a zone around itself in which the nutrients are depleted and metabolism is limited by the rate at which nutrients are able to diffuse along the concentration gradient (Box 2.02). The problem is alleviated if there is a movement of water relative to the plant, for the zone of nutrient depletion is partially replaced by less depleted water. We saw that phytoplankton achieve a movement relative to the water by passively sinking or by actively swimming. They may also preferentially occupy zones of locally high-nutrient concentration. For benthic plants, the key to their productivity is the movement of water over them while they are held in one place by their attachment to the substrate.

Benthic plants are limited to areas where the sea floor is within the zone where there is sufficient light to support photosynthesis and it is usual to find a fringe of marine macrophytes associated with the various types of coastline. Seaweeds, i.e. attached algae, predominate in eroding areas where there is a rocky bottom for attachment of the holdfast, while flowering plants such as seagrasses, marsh grasses or mangroves predominate in sedimenting areas where their roots can take hold and obtain nutrients. In the section that follows we shall explore the question of how dependent benthic plants are on the movement of water around them.

2.4.2 Water movement and productivity

The earliest work on this question was with freshwater macrophytes. It was shown experimentally that increasing the movement of water over various freshwater species increased the rate of uptake of ^{32}P and the rate of photosynthesis (Whitford and Schumacher 1961; Schumacher and Whitford 1965; Westlake 1967). Koehl (1986) has given a good summary of the water flow encountered by benthic marine algae. Giant kelps that colonize water of about 10 m depth and have buoyant fronds near the surface encounter orbital water movement as a result of surface wave motion. Closer to shore, where the water depth is less than half the crest-to-crest distance of the waves, the algae experience a back-and-forth oscillation of the water, parallel with the bottom, as the waves pass overhead. This movement is accentuated in the shallow intertidal zone, where waves break and water rushes up the shore, to recede again after each wave.

Within these broad zones of water movement algae are able to experience flow micro-habitats, according to their growth habit. Very small forms, or encrusting species may be so close to the substrate that they are in the boundary layer of slowly moving water; either the viscous sublayer or the turbulent zone described in section 2.2.8. Because the boundary layer builds up as water moves across a surface, as described by Eq. 2.14, algae further from the leading edge of a structure such as a rock will experience a thicker boundary layer than those living near the edge. Again, algae growing close to one another may be compressed into one large clump at high water velocities, so that water currents are directed around, rather than through the clump, and current speed within the clump is markedly reduced. Full quantification of these processes has not yet been achieved, but there are interesting studies of components of the problem.

Wheeler (1978, 1980) showed for blades of the giant kelp *Macrocystis* that both photosynthesis rate and uptake of nutrients increased with increasing current speed up to an asymptotic value of about 5 cm s^{-1} (Fig. 2.15). He showed that the mean flow of currents in a *Macrocystis* bed are commonly below this value and suggested that a useful management technique might be to harvest kelp in such a way that current flows through the beds might be increased.

Gerard (1982) approached the question by making indirect measurements of turbulence rather than direct measures of mean flow. Using a technique pioneered by Doty (1971) she used the rate of dissolution of small hemispherical plaster buttons as an index of turbulence. She calibrated them against flow rates and temperature in the laboratory, and she attached them to the surfaces of *Macrocystis* blades in the sea. She also measured NO_3 uptake in the laboratory, at differing levels of turbulence as measured by the dissolution rate of the plaster hemispheres. From these experiments she concluded that the level of turbulence in the natural environment was normally sufficient to saturate the NO_3

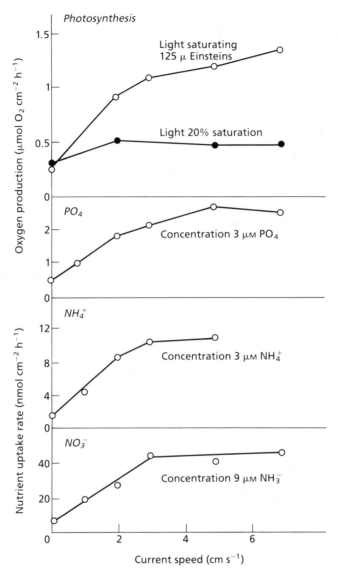

Fig. 2.15 Photosynthesis and nutrient uptake as a function of current speed, for discs taken from the blades of *Macrocystis*. From Wheeler (1978).

uptake mechanism of the plants. She pointed out that in addition to the mean flow, there were turbulent movements adjacent to the blades produced by wave and swell action. Each blade has a hollow float and the rise and fall of the blades, which are attached to the bottom by their long stipes, resulted in a marked horizontal dragging of the blades through the water.

Gerard and Mann (1979) reasoned that if water movement increased photosynthesis in kelps, the productivity of *Laminaria* at sites exposed to Atlantic swells ought to be higher than at sheltered sites. When comparing plants from two such sites in Nova Scotia, Canada, they noted that the plants from the exposed site had narrower, thicker blades than those from the sheltered site. They also had solid stipes, whereas those at the sheltered site had hollow stipes with gas-filled bladders near the junction of the stipe and blade. Furthermore, the plants at the sheltered site developed corrugations on their blades. The results of the study showed that the plants at the sheltered site had a higher annual productivity per plant and a higher annual productivity per unit area of kelp bed. When plants from the sheltered site were transferred to the exposed site they were torn to pieces by the wave action. When the reverse transplant experiment was carried out, the plants from the exposed site, living in sheltered conditions, began to grow wider, frilly blades, showing that the morphological differences were phenotypic responses. The conclusion was that the level of turbulence at the exposed site was well in excess of the optimum for the species. Its narrow, thick blade was streamlined to reduce drag but at the same time had the effect of reducing the photosynthetic area. At the sheltered site, the wide blades had an increased area for both photosynthesis and for nutrient uptake. The hollow stipes had the effect of holding them closer to the surface, thus maximizing light capture both absolutely and relative to competing species. The authors speculated that the frilly edges of the blades had the effect of generating small-scale turbulence, thus maximizing water movement near the boundary layer, even though the mean flow was lower at the sheltered site.

In a subsequent study Gerard (1987) subjected *Laminaria* blades to constant longitudinal tension by the use of weights and pulleys and found that this caused the blades to grow longer and narrower, but not thicker. She speculated that the wave action to which the plants were exposed in nature created a longitudinal drag which caused the elongation. She suggested that the thickening was a response to frequent bending.

Much more quantitative studies were carried out by Koehl and Alberte (1988), working on the Pacific coast of North America with *Nereocystis*. This kelp also has narrow, flat blades in exposed situations and wider, undulate blades in more sheltered localities. They used electronic current meters to produce continuous records from the vicinity of the plants and from these were able to describe not only the current velocities but also the turbulence intensities. They pointed out that while the mean velocity gives a measure of the advective transport of nutrients, turbulence intensity indicates the importance of small-scale motions that lead to enhanced turbulent diffusion as outlined in section 2.2.7.

Turbulence intensity is usually estimated with the rms or root mean square velocity, $(\overline{u'^2})^{1/2}$, where u' represents the velocity fluctuations or differences $(u - \overline{u})$, between the measured velocity and the mean velocity and the overbar means that the average of u'^2 is calculated. Koehl and Alberte (1988) estimated this turbulent intensity for various positions in the kelp bed. Representative values in Table 2.2, show that the ratio of the rms velocity to the mean velocity is 0.2–0.3, or the mean is about five times the rms velocity, when the mean is between 0.2 and 0.3 m s^{-1}. During slack water the mean velocity drops to near zero, but there is enough residual turbulence to give a relative turbulence intensity of 10.6 even though the rms velocity has also decreased to the lowest value measured. We shall return later to the importance of small-scale local turbulence.

Koehl and Alberte (1988) also measured the velocity gradient at the surface of *Nereocystis* blades. They used a measure analogous to the boundary friction velocity u_* discussed in section 2.2. Water velocity profiles along strips of the blades were determined by placing them in a tank through which particles moved while being illuminated by strobe flashes. The distances between the particles as seen on a photographic image gave a measure of water velocity at different positions. u_* was calculated from Eq. 2.13, with $z_0 = 0$ for the case of a smooth strip of algal blade. For blades of different velocities they characterized the average friction velocity by a technique similar to Gerard's (1982) use of plaster buttons. They sewed the candies known as 'Pep-o-Mint Life Savers' to the blades and used weight loss during 10 minutes as an index of friction velocity. They calibrated the technique against the

Table 2.2 *Nereocystis luetkeana.* Examples of mean velocities and turbulence intensities* encountered on windy days. From Koehl and Alberte (1988).

Location and time	Velocity (m s^{-1})		Turbulence intensity* (m s^{-1})
	Maximum	Mean	
Current-swept site edge of bed, ~1 h after slack tide†	0.49	0.33	0.21
middle of bed, ~1 h after slack tide†	0.39	0.22	0.29
middle of bed, slack tide†	0.12	0.004	10.64
Protected site peak tidal current	0.57	0.36	0.22

* Turbulence intensity was calculated from velocities taken at 0.5 s intervals ($n = 256$ per record) on records of flow like those shown in Fig. 2.
† Low, low water.

measurement of u_* in a tank. From this they were able to clearly demonstrate that the effect of current on photosynthetic rate was through the increase of the velocity gradient. They also found that there were conditions of low flow in nature which led to gradients that corresponded to a definite limitation on photosynthetic rate.

From measurements made in a flow tank they determined that frilly wide blades 'flapped' with a greater amplitude than narrow blades in response to the current, but caused greater drag on the plant. Narrow flat blades, on the other hand, collapsed into more streamlined bundles in a strong current, but in this position they were subjected to more self-shading. They showed that photosynthetic rate decreased in slow flow, but that blade flapping enhanced it. Hence, the frilly blade morphology, which caused increased flapping, enhanced photosynthesis. There was a trade-off between the photosynthetic advantages of wide, frilly blade morphology and the risk of the plant breaking, due to increased drag.

Anderson and Charters (1982) placed the bushy alga *Gelidium nudifrons* in a water tunnel in the laboratory and observed turbulence by a dye stream technique. Any turbulence in the water entering the network of filaments comprising the thallus was strongly suppressed, and below a critical velocity the flow leaving the plant was smooth. However, above a critical velocity of $6-12$ cm s^{-1} (depending on the diameters and spatial density of the branches) the plant created its own micro-turbulence. The authors speculated as follows: 'The transition in the flow induced by the branches . . . is probably a phenomenon of considerable adaptive significance, because the turbulence generated by the plant itself or by neighbouring plants may be the only turbulence in water motion past the plant that is of the right scale to enhance nutrient uptake and effect exchange of gases and solutes.' This may be an overstatement, considering the spatial heterogeneity of the average rocky shore, which seems certain to create micro-turbulence, but it is an interesting idea, which deserves further study.

2.4.3 Water movement and drag

As we have seen from several examples, seaweeds often have the ability to respond phenotypically to the ambient level of water movement. Those in quiet waters may develop structures variously described as undulant, frilly, or ruffled, while those in exposed situations may be much more streamlined. Since the undulant form has been shown to have a higher rate of photosynthesis in moderate currents, but a greater drag, and since plants that are torn from their holdfasts are often cast up on the beach and die, there is an evolutionary trade-off between photosynthetic capability and streamlining. For this discussion of drag on algae we are mainly indebted to Koehl (1986) and the references therein. The magnitude of the form drag force, F_D, on a macroscopic organism is calculated with Eq. 2.15:

$$F_D \simeq \frac{1}{2} C_D \rho u^2 A .$$

Note that drag, for high Reynolds number flows, is proportional to velocity squared, so a technique for minimizing exposure to high velocity, as by growing in the shelter of another object, or growing close to the substratum will have the greatest effect. Koehl and Alberte (1988) measured the drag force on different morphological types of *Nereocystis* by towing them stipe-first through the water at different speeds and expressing the drag in Newtons per m^2 of blade area (Fig. 2.16). At a given current speed, all parameters were comparable except the width and degree of ruffling of the blade, so this has a clear effect on the drag coefficient. An obvious adaptation to minimizing drag in an oscillating current is the ability of a plant to bend in the

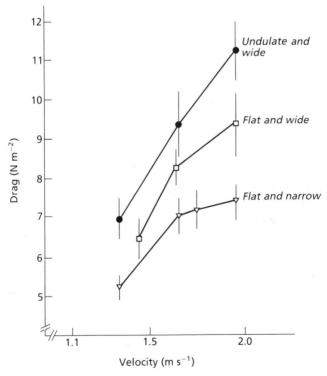

Fig. 2.16 Effect of blade morphology in *Nereocystis* on the relationship between water velocity and drag. From Koehl and Alberte (1988).

direction of the flow at a given time thus minimizing S, the area presented to the current. However, as Koehl (1986) pointed out, flexibility may have disadvantages too. A stiff plant may be able to hold its blades above those of its more flexible neighbours, out-competing them for light. On the other hand, a flexible plant, by its lashing about, may shake off epiphytes, or herbivorous animals. Some seaweeds, as they bend in the wave-induced currents, sweep the area of sea floor surrounding each plant keeping it free from competitors. What degree of flexibility is optimal for this?

Koehl's (1986) review gives a great deal of information about the biomechanics of seaweed structures, and interested readers may care to go more deeply into this subject.

2.5 BENTHIC ANIMALS

2.5.1 Filter feeding in the benthic boundary layer

As we have seen, benthic plants take up nutrients more rapidly and increase their metabolism as the current flows faster, because there is an increase in the supply of limiting substances, through increased turbulent diffusion. Benthic animals behave in a similar way. Current speed affects the growth rate of such benthic suspension-feeders as barnacles and bivalves (Crisp 1960, Richardson *et al.* 1980, Wildish and Peer 1983). Wolff (1977) reviewed the evidence for particularly high benthic secondary production in estuaries and suggested that the prevalence of tidal currents was an important factor. He remarked 'In the Grevelingen estuary, for example, tidal currents in the range $0.1–1.0 \text{ m s}^{-1}$ would cause the vertical component of the turbulent diffusion coefficient to be approximately $50–500 \text{ cm}^2 \text{ s}^{-1}$, implying that surface production becomes available to benthic filter-feeders at a few metres depth within an hour or less.'

Wildish and Kristmanson (1979) extended the idea still further, producing evidence that benthic communities in areas affected by tidal currents formed a spectrum. In waters with very low tidal currents, deposit-ingesting animals such as polychaete worms, nematodes and certain kinds of bivalves predominated, while in areas with more water movement suspension-feeding bivalves and holothurians were much more abundant. Finally, in areas where the mean tidal currents, measured $1–2$ m above the bottom, exceeded 30 cm s^{-1}, they showed that the communities were stressed and the fauna impoverished. Conditions at the two extremes can be understood without much difficulty. In very quiet waters there is an accumulation of nutritious organic matter on the bottom and this is most easily utilized by animals that ingest sediment directly. At the other extreme, high currents erode the sediments, leaving little organic matter and making suspension-feeding difficult because there is so much inorganic matter in suspension. For the intermediate stages, the authors found a gradient in abundance of suspension-feeding organisms which they thought could be explained by interactions between the filter-feeders and the tidal currents. They suggested that the suspension-feeding organisms depleted the concentration of food particles near the sediment–water interface, and moderate tidal currents generated enough turbulence in the bottom boundary layer to increase the supply of food to the suspension-feeders by turbulent diffusion.

In later work the same authors (Wildish and Kristmanson 1984) confirmed in a laboratory flume tank that blue mussels, *Mytilus*, and horse mussels, *Modiolus*, were able to cause a reduction in the living organic matter in suspension at a height of 1 cm above the bottom, compared with 30 cm above the bottom.

A more detailed study of the dynamics of a benthic community in relation to turbulent diffusion in the water column was made by Frechette and Bourget (1985a,b) and Frechette *et al.* (1989). For their study area they chose a dense intertidal bed of the mussel *Mytilus edulis* in the St Lawrence estuary, Canada. They measured phytoplankton concentration from chlorophyll and phaeopigments, at 100, 50 and <5 cm above the mussel bed, following the semi-diurnal tidal cycle at intervals through the fortnightly tidal cycle. They found that in most cases there was a depletion of the particulate organic matter in the layer above the mussel bed, and this was negatively correlated with current speed, i.e. was most marked at low current speeds. In the cases where there was not depletion above the mussels, there was strong wave action which appeared to be enough to cause complete mixing in the water column.

To test the hypothesis that the mussels were food-limited in their growth, Frechette and Bourget (1985a) compared the growth of mussels at 0.05 m and at 1.0 m above the bottom, taking care to arrange, through clever use of the local topography, that the two sets of mussels were immersed for the same proportion of each tidal cycle. They carried out the comparison at both low and mid-shore intertidal levels and found that the growth was influenced both by position on the shore (through the amount of time submerged on each tidal cycle) and by height above the bottom. Since they had already shown that the waters close to the bottom had a reduced food concentration most of the time, they inferred that mussels sited 1.0 m above the bottom had an improved food supply and hence an improved growth rate.

The study was carried further, in an attempt to show that turbulent diffusion in the benthic boundary layer is an important factor in the supply of food to the mussels. Frechette *et al.* (1989) estimated biomass from chlorophyll at distances of 1.0, 0.5 and <0.05 m above the mussel bed, over semi-diurnal tidal cycles at intervals through the fortnightly tidal cycle. In addition, current was measured at 1.0 and 0.5 m above the mussels, and complete current profiles above the bed were determined on several occasions. Phytoplankton consumption of the mussels was measured in the field, using filtration chambers and standard flow-through techniques.

On every occasion except one, the phytoplankton concentration was higher at 50 cm than immediately above the mussels. Hence there was a vertical gradient down which biomass could be transported by turbulent diffusion to the mussel bed. They constructed a steady, two-dimensional, finite difference model to represent the balance between horizontal advection and vertical diffusion of phytoplankton. The model showed a marked reduction in phytoplankton concentration close to the mussel bed and demonstrated the dependence of consumption rate on flow speed. Enhanced vertical diffusive transport at higher current speeds resulted in a higher rate of replenishment of phytoplankton in the waters close to the mussels, enabling them to increase their feeding rate.

Thus, the evidence from community structure, single-species populations and from physiological experiments on individual animals supports the view that food consumption of benthic filter-feeders is a function of current speed in the benthic boundary layer, as well as of the more traditional parameters such as food concentration, temperature, etc., that have been investigated in the past.

In the case of the mussels discussed above, the animals normally live attached to solid substrates in areas where water movement is sufficiently vigorous to prevent the accumulation of sediment. There are also many kinds of benthic animals that live in areas where the intensity of turbulence is sufficiently low to permit the accumulation of sediment. Some of these animals feed exclusively by ingesting the sediment, but many ingest food from the suspension of particles that is found just above the sediment. In the next section we look at the vertical distribution of these particles and its effect on the feeding strategies of animals.

2.5.2 Suspension-feeding benthos

If we envisage a current just above the bottom carrying relatively heavy mineral particles and lighter organic particles, we might guess that the heavy particles would be concentrated close to the bottom and the lighter particles higher up. Muschenheim (1987a) studied this question in theory and in practice. The theoretical argument goes as follows.

The mean flow in the boundary layer above the ocean bottom shown in Figs 2.04 and 2.05 and described by Eq. 2.13

$$u(z) = (u_*/k)\ln(z/z_o)$$

corresponds to a logarithmic decrease in velocity as one approaches the sediment surface. The equation

$$C(z) = C(a)(a/z)^p \qquad (2.19)$$

gives $C(z)$, the concentration of suspended material at height z, relative to the concentration $C(a)$ at height a. Particles of different sizes are differentiated by their settling velocity w_s in the Rouse ratio p given by:

$$p = w_s/ku_*, \qquad (2.20)$$

where u_* is the friction velocity and k is von Karman's constant. When $p = 1$ the particles are heavy and the settling velocity w_s is high relative to the friction velocity. These particles tend to be concentrated near the bottom. For lower values of p the particles are light and easily stirred up by the turbulence so they are distributed rather higher in the water column.

The horizontal flux, F, of particles at any height z above the bottom is given by multiplying the horizontal velocity u at the height z by the concentration at z, i.e.

$$F(z) = u(z)C(z). \qquad (2.21)$$

The horizontal flux of the heavy particles ($p = 1$) is a maximum quite close to the boundary. For lighter particles ($p<1$) the maximum horizontal flux is found higher up in the water column. Thus heavy inorganic particles will have their maximum horizontal flux close to the bottom, while light organic particles will have their maximum flux a few centimetres above the bottom. This leads to the hypothesis that it is to the biological advantage of benthic animals to place their feeding organs a few centimetres above the bottom.

Muschenheim (1987b) studied a polychaete worm, *Spio setosa*, which builds a sand tube that allows it to feed 4–6 cm above the sediment. Worms taken from the field were found to have gut contents that were enriched in organic-mineral aggregates, relative to the sediment surface. In a flume tank study, half of the worm tubes were cut down close to bed level so that the worms were forced to feed at a height of 0–2 cm, while the remainder were left at 4 cm length so that the worms fed at 4–6 cm. Those feeding at the lower level were found to collect particles with a significantly lower organic content. Measurement of the settling velocity of surficial sediment collected from the field site showed that it had a Rouse number consistent with it being carried in suspension several centimetres above the sediment surface.

These results therefore support the view that in a boundary layer above a sediment surface, the differential settling velocities of particles of different densities have a natural sorting effect, such that the maximum flux of organically-rich particles occurs at a higher level above the sediment than the maximum flux of inorganic particles. This explains why a number of different organisms, especially polychaete worms, build tubes that protrude above the sediment surface.

2.5.3 The boundary layer and larval settlement

In this section we consider the problems encountered by small planktonic larvae as they descend from the water column into the benthic boundary layer prior to settling on the bottom to begin their adult life. Tidal currents above the bottom often greatly exceed the swimming speeds of the larvae, but they decrease in a logarithmic manner as the bottom is approached. There is a viscous layer close to the bottom in which turbulence is minimal. Can the larvae enter this layer and then swim strongly enough to be able to select a site for settlement, or are they forced to locate a settlement site by some more haphazard approach?

Butman (1986) studied the question at a site where the tidal currents 0.5 m from the bottom reached $5–10\,\mathrm{cm\,s^{-1}}$ on each ebb and flow. A vertical profile of the boundary layer was constructed by deploying current meters at 30, 50, 100 and 200 cm above the sea floor, and using the data to calculate u_* and Re_* (Box 2.05). Calculations were made for smooth-turbulent and rough-turbulent velocity profiles and these were compared with the size and swimming speed of polychaete larvae (Figs 2.17 and 2.18). In smooth-turbulent flows, larvae of $300\,\mu$m would not have room to manœuvre by swimming if they found themselves below the horizontal dotted line, and would not be able to swim against the current if they were in the region to the right of the vertical dotted line. When rough-turbulent profiles were calculated it was concluded that eddies would reach so close to the sea floor that there would be no region in which the larvae could effectively move by swimming.

Hence it seemed that during about 40% of the tidal cycle it is physically possible for the larvae to swim around near the bed, exploring sites available for settlement. They would be confined to a region within $100\,\mu$m of the bed. However, in the real world there is the additional complication that wind-driven and density-driven currents are superimposed on the tidal cycle. The author concluded that the larvae probably do not search for preferred habitats by active horizontal swimming near the bed. It seems more likely that they swim vertically in smooth-turbulent flows, going down to test the substrate and up to be advected to another site downstream. This has been called the 'balloonist technique'.

From this study we see how consideration of the physics of the boundary layer creates a totally new perspective on the settlement of planktonic larvae. Results from classical studies conducted in the laboratory are in need of reinterpretation.

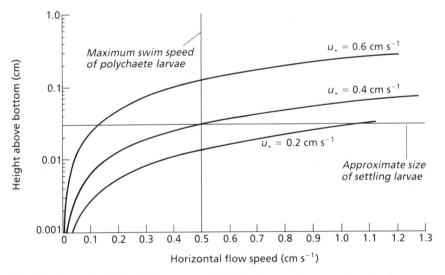

Fig. 2.17 Smooth-turbulent velocity profiles in the viscous sublayer, calculated for a range of near-bottom flow speeds measured at a site in Buzzards Bay, Mass., USA. Light lines indicate size and swimming speed of settling worm larvae. From Butman (1986).

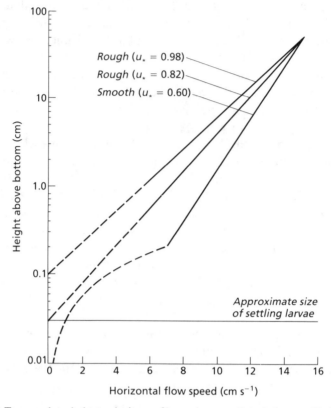

Fig. 2.18 Two rough-turbulent velocity profiles and a smooth-turbulent profile from a site in Buzzards Bay, Mass., USA. From Butman (1986).

2.6 SUMMARY: LIFE IN BOUNDARY LAYERS

In this chapter we have explored the influence of the small-scale properties of sea water on the life of plants and animals. Two phenomena crop up most frequently: the existence of a boundary layer around all organisms and the importance of the viscosity of the water at scales of millimetres. We saw that large-scale turbulence generated by ocean currents, tides and waves cascades down through various scales of turbulence until its energy is dissipated by viscosity in small-scale eddies. This mixing transports essential substances such as carbon dioxide, plant nutrients, or food particles into the general vicinity of the organisms, but the last stage of the journey into the organisms is inhibited by the boundary layer. Water molecules stick to the surface of organisms creating a zone of zero water movement. Beyond this zone viscosity creates a zone of reduced water movement known as the viscous boundary layer, into which turbulent eddies cannot penetrate so that the passage of essential substances is limited by the very slow rates of molecular diffusion. The thickness of this boundary layer decreases with the speed of the water moving beyond it.

For large plants, such as seaweeds, thinning of the boundary layer is achieved by attaching themselves to a solid surface in a zone where tidal currents and wave action cause vigorous water movement. It has been shown that if the drag on them is excessive the plants are in danger of being broken free and cast up on the shore to die. Many of them are capable of elegant changes in shape during growth, reflecting the trade-off between the need to maximize turbulence close to the plant surface and the need to reduce drag.

For medium- to large-sized phytoplankton, thinning of the boundary layer is achieved either by passive sinking or by active swimming. The ability to sink in the water column varies from species to species according to the amount of skeletal material present, and a careful analysis shows that the limitations caused by the boundary layer can be alleviated by up to one third by either passive sinking or active swimming. Differences in these properties may account for differing growth rates among species, and account in part for the succession of dominant species observed to occur with the progression of the seasons.

Very small phytoplankton and heterotrophic bacteria live in a world where viscous forces predominate and any swimming mechanism that relies on the inertial properties of water is simply unworkable. These organisms must make complex asymmetrical flagellar movements to travel through water. When they do so, they do not reduce their boundary layers significantly. The main advantage of moving about appears to be the increased possibility of encountering a higher concentration of nutrients.

On the sea floor there is a boundary layer of reduced water movement extending for tens of centimetres above the bottom. Animals living on the sea floor benefit from the turbulence created by tidal currents, which have the effect of increasing downward transport of food material through this boundary layer. In areas of low turbulent energy soft sediments containing both organic and inorganic matter accumulate on the sea floor. When this is transported laterally by tidal or wind-generated currents a natural sorting mechanism ensures that the denser inorganic materials are in a higher concentration close to the bottom, while the lighter organic materials are concentrated a few centimetres above the bottom. Many animals such as tube worms are adapted to take their food from the upper of these two layers. Microscopic larvae are unable to hold their position above the bottom during the ebb and flow of tidal currents, unless they are in the viscous layer very close to the bottom. This layer is often too thin for them to move about in, and they are forced to select their sites for settlement by making vertical excursions in and out of the viscous layer.

3

Vertical Structure of the Open Ocean: Biology of the Mixed Layer

3.1 INTRODUCTION

One of the problems confronting phytoplankton in the ocean is that they need light and nutrients for growth and reproduction, but the source of light is from above, while the source of nutrients is at depth. The sun's energy that reaches the surface waters is absorbed as it passes downwards, decreasing exponentially with depth. There is a finite layer, the euphotic zone, in which there is enough light for photosynthesis and growth to take place. In a water column with no turbulence, the euphotic zone would become depleted of nutrients as a result of uptake by the phytoplankton.

The reserve of nutrients in deeper water is constantly replenished by the decomposition of organisms from the euphotic zone which sink and decay. In the situation of zero turbulence that we have envisaged, there

61

would be a very low level of nutrients in surface waters, a high level at depth, and the only mechanism for transfer from one to the other would be molecular diffusion, which as we have seen, is extremely slow. In the real world, the ocean is filled with turbulent motion, generated by wind stress at the surface, internal waves, and so on. Phytoplankton depend absolutely on this turbulence to bring nutrients into the euphotic zone where they can be utilized in photosynthesis. This chapter is therefore concerned with some of the physical processes that affect the vertical distributions of light, heat and nutrients so as to better understand the dynamics of phytoplankton production.

On average, the situation in the open ocean is less complex than in coastal waters where the influence of fresh water run-off from the land, of tidal currents and of coastal topography lead to a high degree of complexity. This chapter therefore deals with the open ocean, while coastal waters are deferred until Chapter 4.

3.2 VERTICAL STRUCTURE AND PHYTOPLANKTON PRODUCTION: TROPICAL WATERS

Figure 3.01 shows a vertical profile of temperature, chlorophyll, primary production and nitrate from the eastern tropical Atlantic. It may be taken as representative of large areas of tropical ocean, and has been

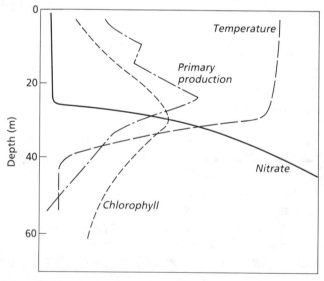

Fig. 3.01 Schematic diagram showing typical vertical structure of the water column in tropical latitudes ('typical tropical structure' — TTS). Note that the thermocline and the nutricline are at the same depth. The peak of primary production is more shallow than the peak of chlorophyll (= phytoplankton biomass).

referred to as 'typical tropical structure', TTS (Herbland and Voituriez 1979). The water column is clearly divided into a warmer, lighter, upper mixed layer, and a cooler, heavier lower layer, separated from the upper layer by a region of rapid change of temperature and density, the thermocline or pycnocline. The mixed layer has a very low level of nitrate, the lower layer has a higher level, and the zone of rapid change is referred to as the nutricline. The difference in nutrient concentration between upper and lower layers was noted in the introduction, but it is not immediately obvious why there is a sharp gradient in properties such as nitrate and a maximum concentration of chlorophyll and primary production at a depth of 20–30 m.

Several explanations have been offered for the subsurface chlorophyll and production maxima. Some, especially the early theories, depend on biological factors such as the sinking of diatoms or the alteration of the sinking rate according to light or nutrient levels. The grazing impact of the zooplankton has also been suggested as a significant process. Recently, the decrease with depth in the vertical mixing rate of the ocean has likewise been demonstrated to cause a subsurface chlorophyll maximum in certain circumstances. Before treating these ideas in detail, we first explore the physics of mixed layers and thermoclines, beginning with the way heat is gained and lost through the surface of the ocean.

3.2.1 Heat gain and loss

The heat that warms the upper ocean comes, of course, from the sun and the amount that arrives each day at the outer surface of the atmosphere varies in predictable ways with latitude and season (Budyko 1974). The amount that reaches and penetrates the sea surface, however, depends on unpredictable variables such as the amount of cloud. Measurements of the radiation reaching the sea can be made directly but they are difficult and expensive. Usually estimates are made by using empirical relationships derived from ships observations of cloud cover. Examples of these calculations are given in the climate atlas of Isemer and Hasse (1987).

The sun's radiation that reaches the sea surface, often called the short-wave radiation, contains energy over a wide range of wavelengths from the blue light at 0.3 μm to the far infrared at 4.0 μm. The amount of energy at each wavelength is as shown in Fig. 3.02, with minimal amounts at the shortest and longest wavelengths and with a maximum in the red at a wavelength of about 0.5 μm.

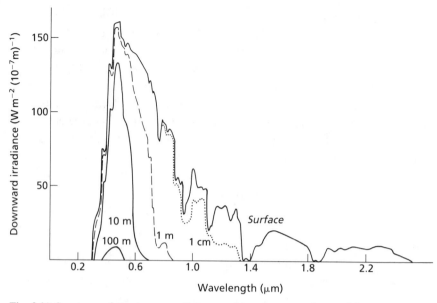

Axis labels — y-axis: Downward irradiance (W m^{-2} (10^{-7} m)$^{-1}$); x-axis: Wavelength (μm). Curves labelled: Surface, 10 m, 100 m, 1 m, 1 cm.

Fig. 3.02 Spectrum of short-wave radiation reaching the sea surface and four depths. Note the progressive elimination of longer wavelengths as depth increases. From Jerlov (1976).

The short-wave radiation is absorbed as it passes through water and the intensity at each wavelength decreases exponentially with depth, but the rate of decrease is different for each wavelength (Fig. 3.02). The intensity of the light of wavelength λ, at depth z, can be calculated from the intensity of the light at the sea surface, I_o, as:

$$I = I_o \exp(-\alpha_\lambda z), \tag{3.01}$$

where the absorption coefficient α_λ is a function of wavelength, increasing from 0.004 m^{-1} for blue light to 1000 m^{-1} in the far infrared at a wavelength of 1.5 μm. Thus blue light is absorbed much less rapidly than infrared, and at any particular depth the intensity of blue light is a larger fraction of its surface value than is the longer wavelength light.

Figure 3.03 shows three different exponential rates of attenuation with depth, calculated using Eq. 3.01. For purposes of comparison, the intensity at the surface is taken to be 100% for all wavelengths even though Fig. 3.02 shows this not to be the case. The curves are also drawn on the assumption that the water is clear open ocean water. In other locations, especially coastal areas, the water may be less clear, with the result that α_λ values are much higher and the vertical absorption profiles are very different (see, for example, Jerlov 1976).

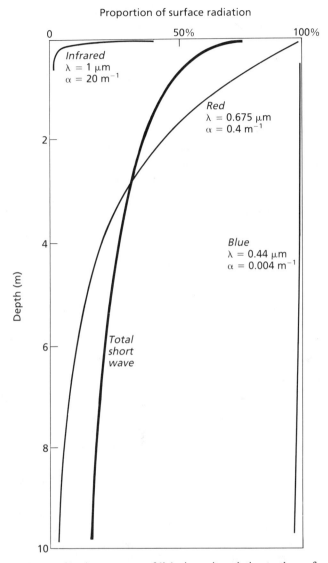

Fig. 3.03 Vertical profiles in sea water of light intensity relative to the surface for three wavelengths and for the total incoming short-wave radiation. For explanation, see text.

The variation with depth of the total amount of short-wave radiation is also shown in Fig. 3.03 constructed from data published by Ivanoff (1977). The curve represents the cumulative effect of all the different wavelengths being absorbed exponentially with depth at different rates. Although it is made up of many exponential curves, it is not itself an exponential function. In some mathematical thermocline models, such as that of Denman (1973), the rate of attenuation of the total incoming radiation is assumed to be exponential, partly for mathematical convenience and partly because it does not introduce large errors when the mixed layer is tens of metres deep. Woods and Barkmann (1986) on the other hand use an empirical absorption profile similar to the curve of total absorption in Fig. 3.03 because it is significantly more accurate when dealing with shallow mixed layers.

Although the absorption of radiation and the resultant gain of heat occurs through the top few metres of the ocean, the loss of heat is almost entirely from the top centimetre. Losses occur mainly through evaporation, infrared (long-wave) radiation, and conduction. Like the estimates of short-wave radiation, estimates of heat losses are calculated from empirical formulae derived from the common meteorological observations obtained from ships at sea. These formulae and their applications are discussed in Smith and Dobson (1984) and Isemer and Hasse (1987) (see Box 3.01).

Box 3.01 Calculating the temperature increase in the mixed layer

The temperature increase which results from the absorption of short-wave radiation can be estimated with the equation

$$\Delta T = \Delta Q / mc, \tag{3.02}$$

where ΔT is the temperature increase, ΔQ is the energy absorbed in kJ, m is the mass of the water, and c, the specific heat (i.e. the amount of heat required to raise the temperature of 1 kg of water 1 °C) is $\simeq 4.2$ kJ $kg^{-1}°C^{-1}$. Over short periods of time, the amount of energy absorbed is estimated from the empirical formulae mentioned above but for periods of a month or more, the data may be extracted from climate atlases.

To give an example, Isemer and Hasse (1987) indicate that over the course of an average July day in the North Atlantic Ocean, at 40° N 40° W, roughly 22,500 kJ m^{-2} of energy are absorbed by the ocean. From the same publication the heat loss, from evaporation, etc., over the same period is about 10,400 kJ m^{-2}. If the mixed layer is 5 m deep, Fig. 3.03 indicates 76% or 17,100 kJ m^{-2} of the incoming energy is absorbed in the 5 m mixed layer while the remainder is absorbed below 5 m depth. Since 10,400 kJ m^{-2} is lost from the surface of the water, the net energy gain or ΔQ in the mixed layer over the day is 6700 kJ m^{-2}. To get the temperature increase this ΔQ must be divided by the mass, which for the 5 m mixed layer is 5000 kg, and the specific heat (4.2 kJ $kg^{-1}°C^{-1}$). This works out to about 0.3 °C.

3.2.2 Temperature increases and mixed layers

Absorption of short-wave radiation causes the temperature of the water to increase by an amount directly proportional to the amount of energy absorbed. Thus the temperature rise after a given length of time will be very similar to the curve in Fig. 3.03 representing the total absorption of short-wave radiation. The greatest increase will be in the surface layers where the absorption is greatest and will decrease rapidly with depth. Observations of temperature with depth in the ocean, however, do not normally resemble this absorption curve because the upper layer of the ocean is usually stirred up by wind waves or by convection which is generated by the loss of heat at the surface. This stirred layer where the temperature remains constant with depth is of course called the mixed layer. Sometimes it is called the surface mixed layer to distinguish it from homogeneous layers in the interior or at the bottom. A method for estimating the increase in temperature in the mixed layer is outlined in Box 3.01.

3.2.3 Deepening the mixed layer

The depth of the wind-mixed layer in the previous section was assumed to stay constant with time. If the wind continues to blow, the turbulence generated by the waves will work to deepen the mixed layer. Imagine a homogeneous 5 m thick mixed layer with a temperature of 2 °C lying on top of a deep homogeneous layer at 0 °C. Now suppose that the stirring of the wind waves deepens the mixed layer from 5 to 10 m. Since the amount of heat does not change, the temperature of the layer is the result of mixing 5 m of 0 °C and 5 m of 2 °C water. Thus the temperature of the mixed layer will be the average of the two, or 1 °C. Obviously energy must be used to accomplish the mixing since the colder and therefore denser water must be lifted upwards against gravity, while the warmer and less dense water is carried downwards, against buoyancy. The energy used to mix the water appears as a change in the distribution of mass or, in terms of energy, as a change in the potential energy of the water column. This energy change is discussed more fully in Box 3.02.

3.2.4 The pycnocline barrier

At the base of the mixed layer a density gradient or pycnocline separates the lighter water of the mixed layer from the denser water below, Fig. 3.04. If a particle of density ρ_2 from below the pycnocline gets displaced

Box 3.02 Calculating potential energy increase of a deepened mixed layer

The potential energy of a mass m is defined as the product mgh, where g is the gravitational constant, $10\ \text{m}\ \text{s}^{-2}$, and h is the height of the centre of the mass m, above a reference level. For convenience we assume that the potential energy of a mixed layer is zero when the temperature is $0\ °C$. As heat is absorbed the water expands which raises its centre of mass and consequently its potential energy. The amount that a column of height h increases by is

$$\Delta h = \alpha h \Delta T, \tag{3.03}$$

where α is the coefficient of thermal expansion for water ($\simeq 10^{-4}\ °C^{-1}$) and ΔT is the increase in temperature. If h is 5 m before it absorbs the heat and it warms up by 2 °C, Δh is $\simeq 10^{-4} \times 5 \times 2 = 10^{-3}$ m (1 mm). The centre of mass rises by half of this or 0.5×10^{-3} m. Thus the increase in potential energy is $mg\Delta h/2$ or $5000 \times 10 \times 0.5 \times 10^{-3} = 25$ J m^{-2}.

After mixing, the layer is 10 m deep but, because heat is conserved, the temperature is 1 °C rather than 2 °C and the potential energy referred to a constant temperature of 0 °C is 50 J m^{-2}. Thus the wind mixing has increased the potential energy of the water column by 25 J m^{-2}. This is an important calculation when considering changes in mixed layer depths but it is only a tiny amount of energy compared with the 4.2×10^4 kJ m^{-2} of heat associated with the temperature of 1 °C through the top 10 m.

According to Denman and Miyake (1973) the rate at which the energy in the wind becomes available for increasing the mixed layer thickness is

$$E_m \simeq 0.0015\, \rho_a C_{10} U_{10}^3, \tag{3.04}$$

where ρ_a is the air density ($\simeq 1$ kg m^{-3}), U_{10} is the mean wind speed measured 10 m above the sea surface, and C_{10} is the drag coefficient to be used with 10 m winds ($C_{10} \simeq 0.001$, Smith 1980). If $U_{10} \simeq 10$ m s^{-1}, then $E_m \simeq 1.5 \times 10^{-3}$ W m^{-2}. This is the rate in joules per second at which the potential energy of the water column is changed by the mixing. In the example, the potential energy increased by 25 J m^{-2}, which would be supplied by the 10 m s^{-1} wind in $25(\text{J m}^{-2})/1.5 \times 10^{-3}(\text{J m}^{-2}\ \text{s}^{-1}) \simeq 16.7 \times 10^3$ or 4.6 h.

The turbulent motions arising from convection rather than wind mixing also act to deepen the mixed layer. Deardorff et al. (1969), in a laboratory experiment, found the effect to be small and it is ignored in the thermocline models of Denman (1973) and Woods and Barkmann (1986).

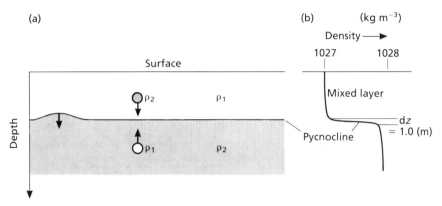

Fig. 3.04 (a) A cross-section through two layers of homogeneous water having unequal density separated by a thin pycnocline. Displacements above or below the pycnocline give rise to the buoyancy forces indicated by the arrows. (b) A vertical profile of the density through the two layers.

above the pycnocline, as shown in the diagram, it will be surrounded by water of density ρ_1. The denser particle will be heavier than the surrounding water and will fall back towards the interface. Similarly, a particle of water from the upper layer displaced below the interface will be lighter than the surrounding water and will move up to the interface. The buoyancy force per unit volume which pushes the displaced particles back to the interface is (Gill 1982)

$$(\rho_2 - \rho_1)\, g, \tag{3.05}$$

where the constant g is the gravitational acceleration ($\simeq 10$ m s^{-2}). The restoring force increases as the density difference $(\rho_2 - \rho_1)$ increases and is the same as the force required to displace the particle in the first place. Since displacing particles from one side of the pycnocline to the other is equivalent to stirring, it follows that it takes more energy to stir water across a pycnocline with a large density difference than one with a small density difference. The fact that energy must be supplied by the turbulent motions in order for vertical mixing to occur in the pycnocline presents a barrier to vertical transport.

The effectiveness of the barrier increases with increasing density difference $(\rho_2 - \rho_1)$. The reduced vertical transport means that the layers on either side of the pycnocline are somewhat isolated from each other. This, as we discuss in the next section has important biological consequences: the phytoplankton grow mostly in the upper layer but the supply of nutrients from the lower layer into the upper layer is, to some degree, blocked by the pycnocline.

A variable defining the strength of the density gradients in the ocean is often required. It could be specified by the density gradient $d\rho/dz$ but it is more usual to use the Brunt–Väisälä or buoyancy frequency

$$N = (g/\rho \, d\rho/dz)^{1/2} , \tag{3.06}$$

where g is again the acceleration due to gravity ($\simeq 10$ m s^{-2}), ρ is the density ($\simeq 1000$ kg m^{-3}) and z is the depth. Physically, N is the frequency of the oscillation that results when the pycnocline is displaced, then left to return to its rest position. For example, in Fig. 3.04(a) the pycnocline has been pushed above its position of rest. Buoyancy forces then act to push it back but because of inertia the interface overshoots the rest point and moves below the original position. Buoyancy forces then act in the opposite direction and push the interface back up. This process continues and results in an oscillation of the pycnocline which spreads out as a moving wave. The process is just the same as when waves are made at the surface of the water. The buoyancy frequency is the most popular variable representing the strength of the vertical density gradient because physicists use it as a reference in studies of other types of waves in the ocean and it is calculated by the method shown in Box 3.03.

3.2.5 Phytoplankton production in tropical and subtropical oceans

As we saw in section 3.2, large areas of the world's ocean are permanently stratified, with a high concentration of nutrients below the pycnocline, a low concentration above, and the pycnocline acting as a barrier

---

Box 3.03 Estimating the Brunt–Väisälä frequency

The value of the Brunt–Väisälä frequency is easily found using a profile of the vertical density gradient, such as that shown in Fig. 3.04(b). Here the density difference across the pycnocline is 1 kg m^{-3} and the thickness of the pycnocline is about 1 m. These give N = $[10(g) \times 1.0(\Delta\rho)/1000.0(\rho) \times 1.0 \, (\Delta z)]^{1/2} \simeq 0.1$ rad s^{-1}. The radian frequency is 2π times the cycle frequency f, that is, $N = 2\pi f$ or $2\pi/\tau$ where τ is the period of the waves, thus $\tau = 2\pi/N \simeq 2\pi/0.1 \simeq 1$ min. This is an atypically high gradient and fast wave for the open ocean. The minimum period in the upper ocean, Phillips (1977), is about 10 min while in the deep ocean or in a well mixed surface layer the period may be as high as 2–4 h.

to turbulent diffusion of nutrients from one to the other. This situation does not completely inhibit photosynthesis, because the organisms that graze upon the phytoplankton excrete nutrient substances such as ammonia, so that phytoplankton growth and reproduction can be maintained at a certain level with these recycled nutrients. However, the situation is potentially unstable, because any removal of organisms and their contained nutrients will lead to a running down of the production based on recycling. For example, the sinking of dead organisms, or excretion by zooplankton which migrate down through the thermocline, both deplete the nutrients in the mixed layer. If the food web in the mixed layer of a particular locality consists of phytoplankton, zooplankton, and small forage fish which feed upon the zooplankton, and a shoal of tuna comes through the area, consumes large quantities of forage fish and moves away, the amount of nutrients available for recycling is diminished. On the other hand, if some physical event occurs that permits the vertical transport of a significant quantity of nutrients from below the thermocline, the nutrient pool in the mixed layer is increased. It is clear that in the long term, the persistence of a phytoplankton community in the mixed layer depends on the losses from the pool of recycling nutrients being balanced by vertical transport from below. There might, of course, be horizontal advection from one part of the mixed layer to another, but it remains true for the mixed layer as a whole that primary production ultimately depends on the vertical transport of nutrients.

Dugdale and Goering (1967) working with nitrogen metabolism, noted that recycled nitrogen is usually in the form of ammonia, while nitrogen transported from below is usually in the form of nitrate. They therefore introduced the term 'regenerated' production for that based on ammonia and 'new' production for that based on nitrate. More recently, the ratio of new production to total production, the f ratio (Eppley and Peterson 1979), has come to be widely discussed in the context of the global carbon budget, see section 10.3.1.

In tropical waters vertical distributions of nitrate, chlorophyll and primary production are more or less the same in all the seasons of the year, as in Fig. 3.01. This constancy suggests that the processes maintaining the low nitrate levels in the surface mixed layer and the maxima of chlorophyll and primary production near the pycnocline must be in balance. This steady-state condition is quite unlike the situation found in temperate waters where the vertical structure changes from season to season. The tropical structure, however, is very similar in appearance to the late summer structure observed in the temperate zones, and the following tentative explanation of the tropical profiles is extrapolated from the more complete body of knowledge that exists for the temperate areas. This work is described more fully in section 3.3.4.

To begin with the simplest case, consider the idealized vertical profile of density in Fig. 3.05(a). Here a homogeneous mixed layer lies over a thin pycnocline and a homogeneous deep layer. The level of vertical mixing as symbolized by the column of circles is vigorous enough to keep the upper layer homogeneous and the pycnocline thin. The light intensity, Fig. 3.05(b), allows phytoplankton growth to a depth somewhat below the pycnocline. Any nitrate diffusing up through the pycnocline is quickly spread by the turbulence throughout the mixed layer and is consumed by the phytoplankton which is also distributed evenly through the layer. If the nitrate level is low in the mixed layer, the phytoplankton production will be low and limited by the vertical flux of nitrate. Below the pycnocline, the phytoplankton are mixed into the lightless deep layer by the turbulence, and lost. This arrangement, then, is not capable of exhibiting chlorophyll maxima.

In the more realistic situation illustrated in Fig. 3.05(c), it is possible to create the maxima because the intensity of vertical mixing decreases with depth gradually rather than suddenly as in the simpler model. Nitrate diffusing up is not quickly spread through the mixed layer but moves slowly up through the pycnocline. At the depth where the light intensity is high enough, photosynthesis begins. The nitrate flux is sufficient to support photosynthesis through a few metres of the water column but there is not enough nitrate to keep phytoplankton growing through the whole of the upper layer. The vertically diffusing nitrate is consumed in a layer of production bounded on the bottom by low light and on the top by low nitrate.

This explanation of the subsurface chlorophyll and productivity maxima in the tropics must be viewed sceptically. It is not, as we show in section 3.3.4, the only possible mechanism and it is neither quantitative nor rigorous. It has been extrapolated from work carried out in temperate oceans and may not apply in the tropics, however it is a promising starting point from which improvements can be devised. The most important factor in the argument, the vertical variation in the turbulent diffusion coefficient, K_v, is very difficult to measure directly and has only recently been done in a few special locations. The main thrust of investigations of vertical profiles in the tropics has been to determine: (i) how much of the primary production is new production and how much is based on recycled nutrients; (ii) the rate of nitrate consumption in the euphotic zone; and (iii) the vertical flux of nitrate through the pycnocline. These are the efforts we now describe.

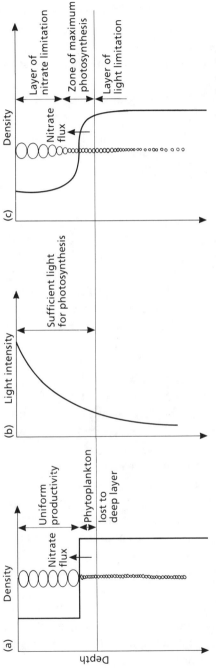

Fig. 3.05 (a) A vertical density profile through a thin pycnocline separating two homogeneous layers. The rate of vertical mixing, symbolized by the circles, changes abruptly through the pycnocline and is high above and low below the pycnocline. (b) A vertical profile of the average light intensity with an indication of the depth where phytoplankton growth is light limited. (c) A vertical density profile similar to (a) except that the density increases and the vertical mixing decreases gradually through the pycnocline.

One group of studies established that in the oligotrophic ocean gyres and some coastal waters the vertical eddy diffusivity is normally low, of the order of 0.3×10^{-4} m^2 s^{-1}. From this we may infer that the normal rate of delivery of nitrate to the mixed layer from below is extremely small, and the tropical oceans probably do have a low rate of primary production, in spite of claims to the contrary (see sections 8.5.1 and 10.3.1).

King and Devol (1979), working in the eastern tropical Pacific, used nitrate gradients at the top of the nutricline (which coincided with the thermocline) along with phytoplankton nitrate uptake rates in the mixed layer to calculate vertical eddy diffusion coefficients. Following Okubo (1971) they used Eq. 2.11, the eddy diffusion equation:

$$F = K_v dN/dz,\qquad\qquad(3.07)$$

where F is the flux of the property N, and K_v is the vertical eddy diffusivity. Putting values of the integrated nitrate-uptake rate by phytoplankton in the mixed layer as the vertical flux, F, and the vertical gradient of nitrate dN/dz in the nutricline into the equation enabled them to calculate eddy diffusion coefficients ranging from 0.05 to 1.10×10^{-4} m^2 s^{-1}. They found that these were in good agreement with coefficients calculated by others on the basis of heat flux through the thermocline, and with those using radiochemical tracers. They are also similar to those given in section 2.2.7.

Eppley *et al.* (1979) made similar calculations for coastal waters of southern California and came up with eddy diffusion coefficients of $0.01-0.3 \times 10^{-4}$ m^2 s^{-1} in the thermocline. The agreement between the vertical eddy diffusion coefficients calculated on the basis of the nitrate uptake of the phytoplankton and those calculated from non-biological considerations lend support to the view that in these conditions vertical eddy diffusion is the responsible agent in the transport of nitrate to the mixed layer and is a controlling factor in primary production. The agreement also supports the expectation that the turbulent or eddy diffusion constant, in Eq. 3.07, is the same for all properties such as heat, salt and nitrate. This is unlike molecular diffusion in which different properties have different molecular diffusivities, as we discussed in section 2.2.6.

Eppley *et al.* (1979) also found that at the various stations studied, the rate of regenerated production was proportional to new production. This implies that increased new production leads to increased production of consumers, and hence to increased recycled nitrogen. In other words, vertical transport of nitrate is a controlling factor for total primary production.

A detailed study of vertical nitrate fluxes at a site in the oligotrophic eastern Atlantic approximately 1000 km southeast of the Azores was made by Lewis *et al.* (1986). They determined the turbulent kinetic energy dissipation rate (ε) from repeated measurements of micro-scale velocity shear and found low values with a mean of 1.7×10^{-9} W kg^{-1}. When combined with the buoyancy frequency, as in Eq. 3.12, they were able to calculate profiles of eddy diffusivity K_v. The mean value was 0.37×10^{-4} m^2 s^{-1}. Using Eq. 3.07 they obtained a vertical flux value of 0.14 mmol N m^{-2} d^{-1} (95% confidence interval 0.002 to 0.89), which is stoichiometrically equivalent to a carbon fixation rate of 0.9 mmol C m^{-2} d^{-1}. This extremely low figure was used to contribute to the continuing debate about the level of new production in the oligotrophic oceans and its importance in the global atmospheric carbon budget (see section 10.3.1).

During the same cruise Lewis *et al.* (1986) made direct measurements of nitrate uptake and obtained an integrated value for the water column of 0.807 mmol N m^{-2} d^{-1}. This value, though higher, is not significantly different at the 95% confidence interval from the nitrogen flux calculated from physical parameters.

The conclusion to be drawn from this work and that of others in the Pacific is that in the absence of special upwelling situations the level of new production in oligotrophic ocean waters is extremely low, and is limited by the stable vertical structure with its pycnocline barrier and the low level of vertical transport of nitrate and other nutrients. The results also do not contradict the simple explanation given above for the chlorophyll and productivity maxima being located in the pycnocline rather than at the surface.

On the other hand, Dandonneau (1988) showed that seasonal and longer-term changes in the climate of various tropical regions, and the advection of water masses from coastal upwelling areas, cause relatively frequent deviations from the general picture of an oligotrophic mixed layer in tropical waters. This matter will be discussed more fully in connection with large-scale ocean circulation.

3.2.6 Equatorial upwelling and domes

There are many parts of the tropical oceans where the concepts of a stable mixed layer and a low coefficient of vertical eddy diffusivity in the thermocline do not apply. The first of these is at the equator, where extensive upwelling occurs. The detailed physics of the process is best presented in connection with major ocean currents, but a general description is needed here.

At the equator, westward-blowing trade winds give rise to a westward flow along the equator in the mixed layer (Gill 1982; Leetmaa *et al.* 1981). In the Pacific (Fig. 3.06), this westward current is the south equatorial current found roughly between 5° S and 5° N. A short distance away from the equator the Coriolis force (section 4.2) causes the current north of the equator to be deflected north (i.e. to the right in the northern hemisphere), while the current south of the equator is deflected south (i.e. to the left in the southern hemisphere). A divergence or spreading of the water is thus set up in the surface layers at the equator. The lost water is replaced by water which upwells from beneath the equator.

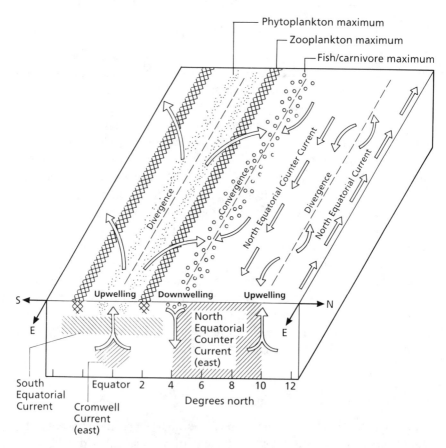

Fig. 3.06 Schematic diagram of equatorial upwelling system. Based on information in Vinogradov (1981). Water upwelling at the equator diverges north and south. Water travelling north eventually converges with water from the north equatorial counter-current, and sinks. For explanation of the biological zones, see text.

Because of the continuous flow of surface water from east to west along the equator, the mixed layer is deeper in the west than in the east. At 155° W it is 80 m or more while at 100° W it is only 10–20 m thick. This means that the water upwelled along the equator comes from the nutrient-rich waters below the shallow pycnocline in the east, but further west, where the mixed layer is deep, the upwelled water comes partly from the nutrient-depleted water in the mixed layer. As Sorokin *et al.* (1975) have pointed out, this leads to greater biological productivity in the east (Table 3.1).

The trade winds blow from the northeast and generate the westward flowing north equatorial current between 10 and 20° N. In the band between this current and the south equatorial current, from 5 to 10° N, is the north equatorial counter-current, so called because it is counter to the prevailing winds. Right at the equator at the top of the thermocline is another counter- or eastward-flowing current. This is the Cromwell in the Pacific and the Lomonosov in the Atlantic. They are about 100 m thick, 200 km wide and have top speeds of $\simeq 1$ m s^{-1}.

Table 3.1 Biomass of organisms in equatorial upwelling systems at various longitudes. From Vinogradov (1981).

Group of organisms	155° W	140° W	122° W	97° W
Phytoplankton	4.8	4.5	5.1	46.5
Bacteria	2.9	5.2	2.8	16.5
Flagellates	0.51	0.91	0.83	1.7
Ciliates	2.1	0.54	1.7	0.24
Other protozoans (radiolarians, foraminifera)	0.05	0.12	0.25	0.04
Total protozoans	2.6	1.6	2.8	2.0
Fine filterers, metazoans (Appendicularia, Doliolidae, small calanoids)	1.7	2.0	1.1	2.7
Copepod nauplii	0.5	0.3	0.4	1.6
Coarse filterers (calanoids, juv. euphausids)	1.1	1.1	0.87	3.3
Total non-carnivorous metazoans	3.3	3.4	2.37	7.6
Cyclopoids	0.39	0.91	1.4	2.3
Predatory calanoids	1.4	0.49	0.38	0.83
Other predators (Chaetognatha, Polychaeta, Amphipoda, etc.)	0.69	0.52	0.38	1.5
Total mainly carnivorous metazoans	2.5	1.9	2.1	4.6
Total zooplankton (including protozoans)	8.4	6.9	7.27	14.2
Total plankton	16.1	16.6	15.17	77.2

*Biomass (wet weight g) of the components of plankton communities along the equator in the Pacific Ocean in the 0–120 m layer in January, 1974.

Equatorial upwelling leads to a biological community with a striking pattern that runs parallel with the lines of latitude (Vinogradov 1981). Consider the fate of a parcel of water upwelling at the equator. It is rich in nutrients but not in phytoplankton. As time passes it moves away from the equator and the phytoplankton population grows rapidly. A zone of maximum biomass is found between latitudes 0.5 and 1.0° from the equator. After a further period of time the water has moved further from the equator and a population of herbivorous zooplankton has developed. Still later, and further from the equator, a population of carnivorous zooplankton reaches its maximum. Finally, by the time the parcel of water is close to the boundary with the north and south equatorial counter-currents, a mature community including macroplankton and young fishes has developed. While these biological events are in progress, the water mass also travels westward with the surface drift. Vinogradov (1981) estimated that water containing a fully developed community has been displaced 250–450 km away from the equator and 1800–2500 km west since being upwelled. Predatory fish such as tuna congregate near the convergence of the two currents to feed on the products of the equatorial upwelling system. A line of high phytoplankton biomass along the equator can easily be seen in satellite images (Plate 1).

A seasonal study of the equatorial upwelling system has been carried out at 5° W (Vinogradov 1981). Between February and May there is a warm season characterized by a slackening of the southeast trades. The surface layers reach their highest seasonal temperature and upwelling is at a minimum. This coincides with the period of minimum phytoplankton productivity, the equatorial area showing little difference from the surrounding oligotrophic ocean. In June–July the southeast trades strengthen to force 4–5 and upwelling becomes very vigorous. After a bloom of phytoplankton the process of development of a large biomass of zooplankton follows, reaching its peak during September to November. As a result, the production in these systems may show as much seasonal variability as in temperate latitudes.

As the equatorial counter-currents approach the coast on the eastern boundaries of the major oceans, part of the water turns to the left and part to the right, eventually joining the north and south equatorial currents, respectively. In doing so, they create cyclonic gyres which have ascending water masses in their centres. A well-studied example is the Costa Rica dome, so named for the tendency of the mixed layer to be thin and the thermal structure below it to be dome-like (Wyrtki 1964). The ascending velocity within the dome is estimated at 10^{-6} m s^{-1}, with the upwelling water originating between 75 and 200 m. Physically, chemically and biologically the waters in the domes are similar to the water in the equatorial divergence (Blackburn 1981), with a rich development of phytoplankton downstream of the upwelled water.

At the eastern end of the Atlantic equatorial counter-current is found the Guinea dome. Unlike the Costa Rica dome, it is probably seasonal. A similar feature, again seasonal, exists in the Indian Ocean south of Java (Wyrtki 1962).

3.2.7 Magnitude of equatorial phytoplankton production

The eastern half of the equatorial region of the Pacific is particularly productive. As the satellite picture (Plate 1) shows, there is a triangular region which clearly has higher chlorophyll biomass than the central gyres. Chavez and Barber (1987) made an estimate of its contribution to global phytoplankton productivity. They recognized a 'cold tongue' area of equatorial upwelling extending from 90 to 180° W, extending 10° in meridional width, with a total area of 1.3×10^{10} km^3. They found that the mean primary productivity was 0.54 g C m^{-2}d^{-1} or 197 g C m^{-2}y^{-1}. For the total area they estimated annual production at 1.9×10^{15} g C y^{-1}. Using the model of Eppley and Peterson (1979) they estimated that 0.85×10^{15} g C y^{-1} is new production. As will be seen in Chapter 5, the Peru upwelling system, which has been widely regarded as a major contributor to global fish production, has a total primary production that is an order of magnitude less than the eastern Pacific equatorial upwelling system. The rate of production per unit area in the coastal upwelling system is much higher, but the area is only a minute fraction of that of the equatorial upwelling system. Taken together, the two contribute about 10% of the global primary production as estimated by Shushkina (1985) or 18% of that estimated by Platt and Subba Rao (1975). The old idea that Pacific waters are mostly nutrient-deficient and unproductive is no longer tenable.

3.3 VERTICAL STRUCTURE AND PHYTOPLANKTON PRODUCTION: TEMPERATE AMD POLAR WATERS

We discussed the mixed layer of the open tropical ocean as if it were relatively constant and unchanging, except in special upwelling areas. This is a reasonable first approximation. However, in temperate and polar waters seasonal change is very evident, and it is now being realized that diurnal changes are also important.

As we move poleward from the stable tropical and subtropical waters we find that in the winter season downward mixing by convection of cooled surface water and wind-driven turbulence combine to cause a progressive deepening of the mixed layer. This means that turbulence penetrates deeper and deeper into the zone of high nutrient concentration and brings nutrients up into the euphotic zone. On the other hand, it means that phytoplankton cells are carried deeper and deeper by the turbulent mixing, and spend a larger proportion of their time below the euphotic zone, where photosynthesis is no longer possible. At the end of winter, when surface warming begins, the reverse process occurs. The mixed layer becomes shallower and phytoplankton cells trapped above the pycnocline spend more and more of their time in the euphotic zone. The net result is a great burst of phytoplankton activity which we call the spring bloom (see section 3.3.2). Reference to Fig. 3.07 shows that at latitude 41° N deepening of the mixed layer continued until late March, but there was a rapid shallowing during the early part of April. The figure also shows that from January onwards there was a diurnal oscillation in the depth of the mixed layer. In the next section we consider the physics of these processes.

3.3.1 Diurnal and seasonal changes in mixed layer depth

The discussions in sections 3.2.2 and 3.2.3, concerning changes of temperature and depth of mixed layers, assumed the inputs of heat and mixing energy to be constant in time. These factors however vary over the day and the year giving rise to diurnal and seasonal changes in the temperature and depth of mixed layers.

To illustrate diurnal changes, the variation of net heat gain through the ocean surface at 40° N 40° W, mid-Atlantic site, through the course of an average day in both July and December are shown in Fig. 3.08. In July the sea loses heat at the rate of $\simeq 120 \text{ W m}^{-2}$, mostly by evaporation and long-wave radiation. The losses continue at much the same level throughout the day because the controlling factors, such as wind speed, air–sea temperature difference and cloud amount, tend to remain fairly constant.

Heat gain on the other hand depends mostly on the altitude of the sun which varies greatly throughout the day. Heat input varies from zero during the night to almost 900 W m^{-2} at noon. Under these circumstances a mixed layer of constant depth becomes warmer during the day and cooler at night but because there is a net gain of energy over the day there will be a net increase in temperature over the day. If the wind decreases, the mixed layer becomes thinner and the temperature increase in the layer will be greater. This is because most of the radiant energy is absorbed in the top few metres as indicated in Figs 3.02 and 3.03, and because weaker winds allow the heat to be concentrated in a

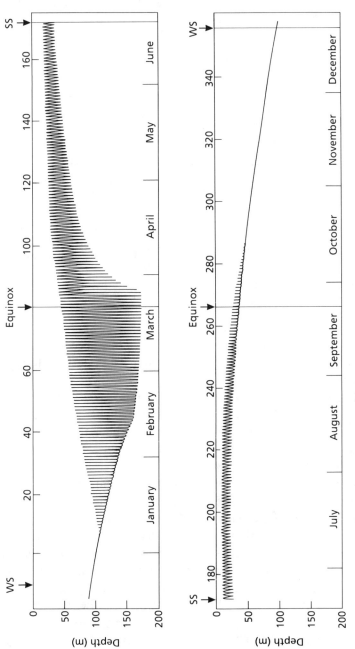

Fig. 3.07 Model simulation of the diurnal variation of the mixed layer depth over the year. Note that there is a diurnal movement of the thermocline from January to October. WS, winter solstice; SS, summer solstice, From Woods and Barkmann (1986).

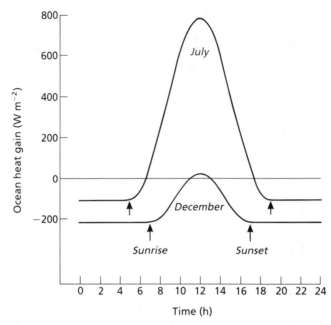

Fig. 3.08 Diurnal changes in the net heat gain during an average day in July and December at 40° N 40° W.

thinner layer. If the wind increases, the mixed layer becomes deeper and the heat input is spread throughout a greater depth, leading to a smaller temperature rise over the day.

In December the sun is above the horizon each day for only 10 h at 40° N 40° W and delivers a total of only 4300 kJ m^{-2} over the day, while the heat losses total 19,000 kJ m^{-2}. Under these circumstances there is not enough heat gain during the day to form a warm surface layer for even part of the day.

The annual variations of the heat budget components at the site of ocean weather ship Echo, 35° N 48° W, are illustrated in Fig. 3.09. Included are the heat input from short-wave radiation, and the losses from radiation, evaporation and conduction. These data, extracted from the atlas of Isemer and Hasse (1987), show that short-wave radiation is the only source of heat and that the latent heat of evaporation is responsible for most of the heat loss in all months. The sum of all the factors indicates that the sea is only gaining heat through the sea surface at this location from April to August, and over the year more heat is lost than is gained, which indicates that heat must be advected into this region to keep the mean temperature the same from year to year.

The curve of total gains and losses suggests that all the heat gained through the surface in the summer would likely be lost by the following

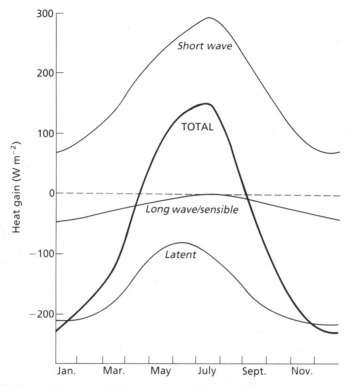

Fig. 3.09 Seasonal changes in net short-wave radiation, net long-wave radiation, sensible heat exchange, latent heat exchange and the total at 35° N 48° W.

January or February. By that time no seasonal thermocline would remain in the upper layers of the ocean.

Woods and Barkmann (1986) have created a numerical model to predict the depth of the mixed layer which incorporates both the diurnal and seasonal variations in the heat gains and losses. They used input parameters from 41° N 27° W (north of the Azores) and a result of their calculations over a year is shown in Fig. 3.07. In mid-December at the winter solstice the model indicates there is no diurnal thermocline, but that the depth of the seasonal thermocline is approaching 100 m. In early January a diurnal thermocline begins to be formed. Its minimum depth decreases from about 100 m in January to 10 m in July. The maximum depth of the diurnal thermocline, which is the depth of the seasonal thermocline, continues to increase from December through to late March. It then quite suddenly decreases when the heat gained during the day is neither lost at night nor mixed down to great depths by the wind.

Through the spring and early summer the upper mixed layer becomes more stable. The depth of the seasonal thermocline decreases as does the difference between the minimum and maximum depths of the diurnal thermocline. From late summer through to early winter the amount of heat lost to the atmosphere from the sea steadily increases as does the mixing energy from the wind. The diurnal thermocline tends to disappear as the seasonal one moves to greater depths.

3.3.2 The mechanism of the spring bloom

Early biological oceanographers were concerned to explain the onset of the spring bloom. Atkins (1928) suggested that it was triggered by the spring increase in solar radiation, but Marshall and Orr (1928) found that there was enough light for vigorous phytoplankton growth at the latitude of the British Isles even in winter. In the course of an investigation into the productivity of the Bay of Fundy, Gran and Braarud (1935) suggested that stabilization of the water column by thermal stratification could have the effect of greatly stimulating primary production. The idea was that in winter time in temperate latitudes the phytoplankton cells are being circulated to the full depth of the mixed layer and hence spend a large proportion of their time in regions where there is not enough light for growth. With the onset of surface warming and the formation of a much shallower mixed layer, phytoplankton cells are held for longer periods in the euphotic zone, i.e. the zone in which there is sufficient light for growth and cell division.

It was pointed out in Mills (1989) that the ideas contained in Gran and Braarud (1935) had been incubating in the mind of Professor Gran, a Professor of Botany in the University of Oslo, for several years. In 1931, commenting on results obtained in the Weddell Sea, he wrote:

> My explanation is that the diatoms move with the vertical movements of the water, and that therefore no accumulation is found in the illuminated zone, with the effect that the whole production is retarded, because too many of the diatoms sink below the balance depth of the photosynthesis, in this area probably a depth of about 50 m. [After the water column has become stratified . . .] It has the effect that the vertical circulation is stopped and the sinking of the diatoms retarded, and now it is clearly seen how the phosphates (and nitrates) are consumed in the surface layers as far down as the photosynthesis is effective. (Gran 1931).

Next year he wrote:

> All these examples show the importance of the stabilization of the surface layers. A marked stratification excludes circulation of the nutrient salts with the result that the surface layers

become depleted of plant nourishment and the diatoms with their high requirements sink and disappear. On the other hand continuous vertical mixing prevents them from accumulating in the lighted zone and from utilizing the nutrient salts present to such an extent as might have been expected.

We shall see in later chapters that this mechanism, by which turbulent mixing of the entire water column brings nutrients to the surface waters, after which stratification sets in and the phytoplankton are in held in the euphotic zone so that they can multiply rapidly, is the key to understanding phytoplankton productivity in a variety of situations including estuaries, upwelling areas and tidal fronts. We shall henceforth refer to it as the Gran effect.

The idea was further developed by G. A. Riley and used to explain the spring bloom on Georges Bank (Riley 1942). It was also incorporated into a major model of the phytoplankton dynamics of the North Atlantic (Riley *et al.* 1949). However, the classic quantitative exposition of the theory which we present here in detail is by Sverdrup (1953) and is entitled 'On conditions for the vernal blooming of phytoplankton'.

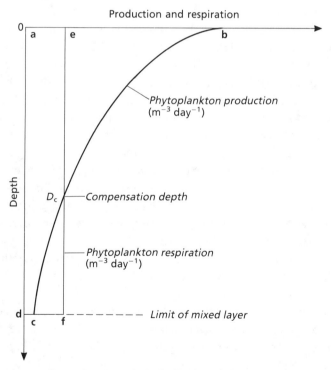

Fig. 3.10 Diagram illustrating theoretical distribution of phytoplankton production and phytoplankton respiration. After Sverdrup (1953).

Consider the situation (Fig. 3.10) of a water column in which the mixed layer extends to depth d, the turbulence is sufficiently strong to distribute the phytoplankton cells uniformly through it, nutrients are not limiting and the extinction coefficient of light (α in Eq. 3.01), is a constant. Since photosynthesis is proportional to light intensity, and assuming that the light intensity at the surface is not high enough to inhibit photosynthesis, the expected distribution of the daily rate of phytoplankton production will be as shown, with a maximum at the surface, decreasing logarithmically to a low level at depth. The rate of respiration, on the other hand, being more or less independent of light, will be constant throughout the mixed layer. At some depth D_c the daily rate of photosynthesis is just balanced by the daily rate of respiration. This is the compensation depth. Experiments have shown that for commonly occurring diatoms the light energy flux I_c at the compensation depth is $\simeq 1.5-1.7$ W m^{-2}.

This information, together with the daily irradiance at the surface and the extinction coefficient permits calculation of the compensation depth, D_c. Taking the daily mean surface irradiance I_0 as 200 W m^{-2} and the extinction coefficient α as 0.4 m^{-1} we have, from Eq. 3.01, $I_c/I_0 = 1.5/200 = \exp(-0.4D_c)$. Taking the natural logarithm of each side gives $-4.9 = -0.4D_c$ from which $D_c \simeq 12$ m.

Now consider a phytoplankton cell circulating in a random manner throughout the mixed layer. When it is near the surface the rate of photosynthetic production will greatly exceed respiration. When it is near the bottom of the mixed layer the reverse will be true. The condition for net positive population growth is that the integrated production, represented by the area *abcd*, be greater than the integrated respiration, represented by the area *aefd*. As the depth of the mixed layer increases, respiration increases proportionately but production increases by a small amount, or not at all. There is therefore a critical depth of the mixed layer at which integrated production just equals respiration. At shallower mixed layer depths the phytoplankton population has an excess of production, and therefore grows. At greater depths of the mixed layer it fails to grow. Note that the effect of grazers has not been considered up to this point.

Sverdrup (1953) showed that the critical depth of the mixed layer D_{cr} can be calculated by algebraically equating the areas representing production and respiration. Assuming that daily photosynthetic production at any depth is proportional to the mean daily light energy at that depth, we may regard the line *bc* in Fig. 3.10 as the distribution with depth of the mean daily light energy or the mean daily photosynthetic production. The length of the line *df* or *ae* is equal to the light intensity at the compensation depth, I_c, and is proportional to the rate of respiration at all depths in the mixed layer. From the discussion above, the intensity of radiation I_c, or the length of *df* or *ae* , at the compensation depth D_c can be expressed following Eq. 3.01 as:

$$I_c = I_0\exp(-\alpha D_c). \tag{3.08}$$

When the depth of the mixed layer is equal to the critical depth D_{cr}, the area *aefd* representing the total respiration is given by $I_c D_{cr}$. The area *abcd* representing the total photosynthesis is given by:

$$I_0\int_{D_{cr}}^{o}\exp(\alpha z)\mathrm{d}z = I_0[1 - \exp(\alpha D_{cr})]/\alpha. \tag{3.09}$$

Equating the two areas gives

$$I_c D_{cr} = I_0[1 - \exp(\alpha D_{cr})]/\alpha , \tag{3.10}$$

which can be solved graphically for D_{cr}.

Sverdrup then proceeded to calculate the values of D_{cr} for ocean weather ship M in the Norwegian Sea in the spring of 1949. When plotted with data for the actual depth of the mixed layer and the numbers of phytoplankton and zooplankton (Fig. 3.11), it became clear that the critical depth increased from about 30 m at the beginning of March to nearly 300 m at the end of May, and that the major increase of populations of both phytoplankton and zooplankton occurred when the depth of the mixed layer was much less then the critical depth.

Detailed examination of Fig. 3.11 reveals the following: (i) through March the amounts of phytoplankton and zooplankton were very small, and this was the period during which the depth of the mixed layer was much greater than the critical depth; (ii) the mixed layer thickness changed from more than 300 m on 2 April to 50 m on 4 April. This probably indicated advection of a different water mass containing abundant phytoplankton, rather than stabilization by warming; (iii) the development of zooplankton populations in April undoubtedly reduced the size of the phytoplankton populations; and (iv) during May, when the depth of the mixed layer was much less than the critical depth, phytoplankton populations reached high densities in spite of the presence of sizeable zooplankton populations. The overall picture was in accordance with the theory that the spring bloom is strongly dependent on the stability of the water column. We may note that this model takes no account of diurnal variation in the thickness of the mixed layer.

Sverdrup's (1953) paper was, then, a quantitative exposition and a field check of the phenomenon first described by Gran in 1931. In later chapters we shall find that enhancement of primary production by stabilization of the water column after a period of strong mixing is a

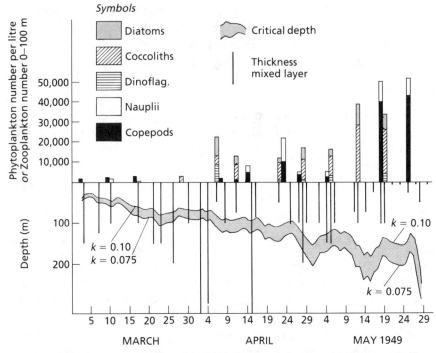

Fig. 3.11 Data from ocean weather ship M in the Norwegian Sea on plankton, mixed layer thickness and critical depth. Note that the mixed layer is always deeper than the critical depth until early April, when a large increase in phytoplankton biomass occurs. From Sverdrup (1953).

phenomenon that occurs in many situations, in estuaries, and in coastal upwelling areas, for example. Many oceanographers refer to it as the Sverdrup mechanism, but we prefer to call it the Gran effect.

It is important to remember that Sverdrup's model applies to temperate latitudes where, after the winter deepening of the mixed layer, surface waters contain a good supply of nutrients, and where spring warming leads to a shallowing of the mixed layer. We may now proceed to follow the fate of the phytoplankton as the year progresses. During the period of stratification the pycnocline presents a barrier to the vertical diffusion of nutrients and the phytoplankton progressively depletes those nutrients present in the mixed layer. At the same time, zooplankton are consuming phytoplankton and excreting nutrients. The budget for total nitrogen contained in the mixed layer has the following terms:

Depletion by:
1 uptake of nitrogen by phytoplankton;
2 sinking of phytoplankton out of the mixed layer; and

3 consumption of phytoplankton by zooplankton. These animals produce faecal pellets which tend to sink rapidly through the pycnocline. Zooplankton may also descend below the mixed layer at night and release nitrogen by excretion and defecation.

Addition by:
4 excretion by zooplankton and other heterotrophs; and
5 vertical diffusion.

A common result of these processes is a rapid depletion of the dissolved nitrogen content of the mixed layer and a fall in phytoplankton biomass, marking the end of the spring bloom.

During the summer period conditions may be similar to those described in section 3.2.5 for tropical oceans, i.e. there may be stable stratification and very limited diffusion of new nutrients into the mixed layer. Phytoplankton production is then confined to that which can be sustained by recycled nitrogen and the small amount of diffusion, and it is common to find that the phytoplankton community changes from one dominated by diatoms to one dominated by flagellates. (This is because the flagellates have a better surface : volume ratio and are able to perform locomotion out of micro-environments with low nutrient concentrations. See section 2.3.4 and Margalef (1978a).) However, as the days shorten and surface waters begin to cool, convective and wind-driven processes once more begin to cause a deepening of the mixed layer, so that new nutrients are entrained from below the thermocline, and an autumn bloom may occur before the depth of the mixed layer once again exceeds the critical depth. Figure 3.12 shows the seasonal changes in chlorophyll, nitrate and primary production in the mixed layer of coastal waters of Nova Scotia, at latitude 45° N (Platt and Irwin 1968). Deep winter mixing leads to increasing concentrations of nitrate, which reach a maximum in March. Spring warming and a shallowing of the thermocline lead to a spring bloom in late April and early May, as shown by the peak in chlorophyll concentration. Nitrate concentration declines rapidly during the spring bloom. During summer, primary production is moderately high, but the chlorophyll low, because zooplankton remove the phytoplankton biomass almost as fast as it is produced. Most of the primary production is made possible by the regeneration of ammonia (not shown), through excretion of the consumers.

The timing of spring and autumn blooms are dependent in part on latitude. At high latitudes, where the spring warming comes late and autumn cooling begins early, it is common to find just one bloom in mid-summer. This bloom is often more pronounced than in temperate waters and two

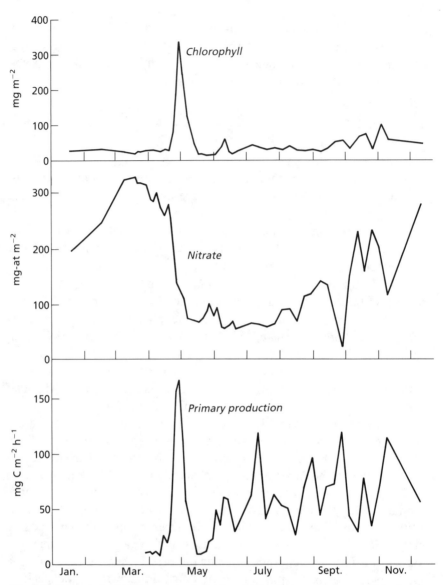

Fig. 3.12 Integrated mixed layer values of chlorophyll, nitrate and primary production in coastal waters of Nova Scotia. For interpretation, see text. From Platt and Irwin (1968).

factors undoubtedly contribute (Cushing 1975). The first is the longer daylight characteristic of high latitudes in summer, which permits the phytoplankton populations to grow for up to 24 h a day. The second is the low temperature, which slows the metabolism of the grazers and allows the phytoplankton populations to grow in spite of them. Photosynthesis is much less sensitive to temperature than is respiration and growth in zooplankton.

3.3.3 Large-scale turbulence and phytoplankton performance

In Chapter 2 we discussed the relationship between phytoplankton and small-scale turbulence. In this section we turn our attention to turbulence on the scale of metres or tens of metres that is characteristic of the mixed layer. The exponential decay of light intensity with depth means that as a phytoplankton cell is carried passively from near-surface to depth and back again it is exposed to a changing light environment. There has been a controversy over the question of whether phytoplankton cells in a fluctuating light regime are more productive than they would be if exposed to light of a constant average value. Marra (1978) thought that they were. He simply exposed a diatom culture to fluctuating light regimes on time scales from minutes to hours and concluded that the photosynthetic performance was superior to that found in a constant light regime. Gallegos and Platt (1982) came to the opposite conclusion. They set up a physical analogue model of the mixed layer in which cells were pumped from one incubation bottle to another along a light gradient and back again. They concluded that for phytoplankton from an Arctic station vertical mixing had little quantitative effect on water column primary production. Falkowski and Wirick (1981) set up a computer model in which phytoplankton were allowed to adapt to light or shade by altering carbon to chlorophyll ratios in response to varying light regimes. The vertical position as a function of time was modelled as the random walk

$$Z(t + \Delta t) = z(t) \pm (2\Delta t\, K_v)^{1/2}, \tag{3.11}$$

where K_v is the vertical eddy diffusivity, which remained constant throughout each experiment at values between zero and $10^{-3}\, m^2\, s^{-1}$. The plus or minus was chosen randomly but with equal probability at the end of each time step Δt. They concluded, like Gallegos and Platt, that vertical mixing had little effect on the integrated water column primary productivity.

Resolution of the controversy was provided by Lewis *et al.* (1984a) who pointed out that the metabolic machinery of a phytoplankton cell is complex and there is no reason to suppose that all of its component parts would adapt to changes in light level at the same rate. They proposed a model in which properties that adapted rapidly to the changing light levels would show marked differences between cells taken from the top and from the bottom of the mixed layer, while properties that adapt slowly would show minimal differences between surface and depth.

They suggested that this would explain the difference in the results of earlier investigations.

To test this model Lewis *et al.* (1984b) made a series of measurements of the turbulent energy dissipation rate (as a measure of vertical motion through the light gradient) in the mixed layer of a coastal inlet on different days, and simultaneously measured two photoadaptation properties of the phytoplankton on those same days. As an example of a relatively slowly adapting parameter they chose P_M^B, the maximum potential photosynthetic rate normalized to chlorophyll-a. The difference between surface samples and those from the base of the mixed layer was labelled ΔP_M^B. There was a marked decrease in ΔP_M^B with increase in turbulent energy dissipation rate. For their example of a property with a rapid photoadaptive response they chose *in vivo* fluorescence per unit chlorophyll. Its value was less at the surface than at depth on all days except one, which was the day of highest turbulence.

The results were in accordance with the model. At slow rates of turbulent mixing, and hence of light fluctuation, there was a marked difference in values between samples from the top and the bottom of the mixed layer for the slowly adapting parameter, as well as for the fast adapting parameter. As the rate of light fluctuation increased, the differences between top and bottom in the slowly adapting parameter decreased while those in the fast adapting parameter persisted.

Lewis *et al.* (1984b) then calculated the time scale of the turbulent mixing. We saw in Box 2.04 that the approximate time for vertical diffusion to occur over the distance L can be estimated using the vertical eddy diffusivity K_v from $\tau = L^2 K_v^{-1}$. Following Osborn (1980), the authors substituted $\varepsilon/4N^2$ for K_v (where ε is the energy dissipation and N is the Brunt–Väisälä frequency), so that they were able to calculate the time scale of turbulent mixing from the energy dissipation:

$$\tau = 4N^2 L^2 \varepsilon^{-1}. \tag{3.12}$$

They found that on the various days of observation the mixing time scale ranged from about 1 to > 10 h. Since it had previously been shown that the time scale for adaptation of the *in vivo* chlorophyll fluorescence is about 1 h, while the time scale for photoadaptation of P_M^B is 6–10 h, the results were in good agreement. Hence they concluded that the time scale of turbulence has strong effects on some of the parameters of photosynthetic performance, but is less strong for others. This important result will undoubtedly be incorporated in future models of phytoplankton productivity.

3.3.4 The oligotrophic phase in temperate waters

Much of the current controversy about the magnitude of primary production in the world's oceans revolves around the mechanisms controlling production in the subtropical oligotrophic basins that occupy a large proportion of the total area (see sections 8.3 and 10.5). Work on the transition from bloom conditions to oligotrophic conditions in temperate waters greatly assists in understanding the basins that are permanently oligotrophic.

We have seen that after the termination of the bloom in temperate oceans the water is stratified and the vertical diffusion of new nitrogen through the pycnocline is restricted. Production in the mixed layer is then largely confined to that which can be supported by nitrogen regenerated by the grazers. During this phase the profiles of nutrients, chlorophyll, and production resemble those of an oligotrophic tropical system, and the similarity extends to the presence of a subsurface chlorophyll maximum. In the course of the debate about what mechanism causes the formation of a subsurface chlorophyll maximum in temperate waters, much light has been thrown on the mechanisms that sustain production at the permanent chlorophyll maximum in tropical waters.

There are two processes that may produce the chlorophyll maximum. If the plants forming the spring bloom sink, they carry the chlorophyll maximum with them to greater depths. Variations on this basic model constituted the sinking rate hypotheses that were the first explanations of the existence of a chlorophyll maximum at some depth below the surface layer. A second group of theories argued that the chlorophyll maximum occurs because the cycle of phytoplankton growth and decay is faster in the upper mixed layer than in and below the pycnocline. At the top of the water column the intensity of light is high and growth will be rapid, but so will nutrient depletion and decay of the population. Deeper down the light level is lower than at the surface, and growth and decay of the phytoplankton population is slower, reaching a maximum after the growth at the surface has died out. Thus the chlorophyll maximum appears at ever increasing depths as the maximum in growth passes down the water column. This process also depends on the fact that turbulent mixing decreases in and below the pycnocline. This slows down exchange between the well-mixed surface layer and the layers below, and allows the growth and decay of the phytoplankton in the various layers to proceed relatively independently of one another. In this and the next section we examine some of these theories in more detail, starting with the sinking theories.

Riley *et al.* (1949) were among the first to present a theory based on the sinking of the phytoplankton. They assumed that the dense concentrations of phytoplankton formed in the spring bloom sank to greater depths and in their model of production in the western North Atlantic they assumed a constant sinking rate for the phytoplankton. Steele and Yentsch (1960) showed that an improved model of vertical distribution of chlorophyll and production resulted if the sinking rates were assumed to decrease with depth. They showed experimentally that *Skeletonema* decreased its sinking rate when exposed to dark, nutrient-rich waters, and proposed this as the explanation of the formation of the deep chlorophyll maximum. They suggested that cells reduced their sinking rate when they came close to the base of the euphotic zone and encountered the nutricline. More recently Bienfang *et al.* (1983) found that reduction in light alone produced a decrease in sinking rate. For *Thalassiosira,* reduction of light levels to those characteristic of the chlorophyll maximum was enough to cause the diatom to decrease its sinking rate. They inserted their data into a modified form of the model of Riley *et al.* (1949). Setting phytoplankton biomass constant with depth in the mixed layer, ignoring grazing, and allowing sinking rate to vary with light according to the experimental results, the authors found that phytoplankton biomass increased by 31–100 % between 40 and 86 m, the magnitude of the increase being negatively correlated with the value of K_v. This result was in good agreement with field data and supports the view that in temperate waters having a shallow mixed layer, a well-developed pycnocline, and phytoplankton populations characterized by the abundance of large-celled organisms, reduction in sinking rates is a prime factor in the formation of deep chlorophyll maxima. They emphasized that both modelling and experimental data indicate little need to invoke this mechanism to explain chlorophyll maxima in oligotrophic subtropical waters where the phytoplankton cells are smaller, and where the response to decreasing light is sufficient to account for the observed increase in chlorophyll with depth.

Jamart *et al.* (1977) produced a thorough model which includes all the conceivably important factors. They have terms representing the seasonal change in light intensity, the phytoplankton sinking rate, the rate of grazing by zooplankton, the phytoplankton respiration and a term representing the decrease with depth in the turbulent diffusion coefficient (K_v). Some of the changes in the vertical distributions of chlorophyll and nitrate through the spring and summer that are predicted by the model are shown in Figs 3.13(a) and (b).

At the start of the model run (day 75), the chlorophyll concentration is low throughout the top layer, decreasing from 0.25 mg m^{-3} at the

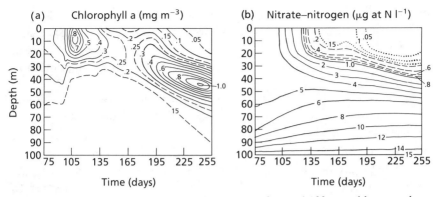

Fig. 3.13 (a) Chlorophyll distribution between the surface and 100 m and between day 75 and day 225 according to the model of Jamart *et al.* (1977). (b) Nitrate distribution over the same depth and time as in (a). Adapted from Jamart *et al.* (1977).

surface to 0.15 mg m^{-3} at 60 m. As the amount of light increases with the increasing day length the chlorophyll concentration increases slowly until day 96 when a pycnocline forms at about 25 m. This isolates the upper mixed layer from the deeper waters and the Gran effect begins to produce a rapid increase in phytoplankton concentration in the mixed layer. Simultaneous increases in the grazing by zooplankton limit the chlorophyll maximum to $\simeq 0.8$ mg m^{-3}.

Over the next 30–40 days the chlorophyll concentration stays high in the top 20 m but nitrate concentrations rapidly decrease to near zero by day 152 or the beginning of June. The phytoplankton die out in the nutrient-poor upper waters while production continues at the deeper levels. The depth of the chlorophyll maximum increases from 14 m at day 151 to 22 m on day 161. From mid-June onward the zooplankton grazing decreases and the concentration of chlorophyll at the maximum increases and appears at greater depths reaching a maximum of 1.0 mg m^{-3} at 45 m on day 240 (end of August).

This model then, exhibits the main features of the developing chlorophyll maximum through the spring and summer of the temperate zone. And although grazing by zooplankton helps to decrease the concentration of chlorophyll, the reason the chlorophyll maximum appears at greater and greater depths as the summer progresses is the decrease in the turbulent diffusion with depth. This tends to isolate the mixed layer from the pycnocline waters and allows the cycle of phytoplankton production, nutrient depletion and phytoplankton decay to proceed at a slower rate within the pycnocline than within the upper mixed layer. By the time the chlorophyll concentration reaches a maximum in the pycnocline it has decreased to near zero in the mixed layer. The importance of this process was emphasized in a subsequent paper (Jamart *et al.* 1979) where the authors showed by means of sensitivity analysis that a sinking rate which decreases with increasing nutrient concentration was not essential for the formation of a subsurface chlorophyll maximum in their model.

The simple explanation for the subsurface chlorophyll and productivity maxima in tropical waters, presented in section 3.2.5, is based primarily on this model by Jamart *et al.* (1977). We assumed that the decrease in the turbulent diffusion with depth combined with the decreasing light intensity are the two most important factors that the authors incorporated. The profile in Fig. 3.05(c) which is intended to represent the tropical situation is similar to that found in late summer in the model; for example near day 200 in Figs 3.13(a) and (b).

The effects of short-term light level changes on the production of chlorophyll were examined by Taylor *et al.* (1986). They modelled the vertical distribution of phytoplankton in temperate waters under stratification and showed that a marked reduction of incident light, as in very cloudy weather, may reduce the photosynthetic activity in the deep chlorophyll maximum to the point where upwelling nutrients are not utilized immediately but are allowed to diffuse upwards into the mixed layer. Banse (1987) considered whether the same mechanism would operate in a 'typical tropical structure' (TTS). Under these conditions the deep chlorophyll maximum is normally located deeper in the pycnocline (see Fig. 3.01) than it is in the temperate waters considered by Taylor *et al.* (1986). The conclusion was that severe cloud cover would be expected to lead to enhanced nutrient flux into the upper parts of the pycnocline, but not into the mixed layer.

Summarizing to this point, there are two processes that have been proposed to explain the existence of the sub-surface chlorophyll maximum in temperate waters. The first is mainly biological and depends on the sinking rates of phytoplankton. Increase of biomass is assumed to begin near the surface and to subsequently sink where it is observed as a subsurface maximum. The second process is physical in nature and depends on the gradual decrease in turbulent mixing through the pycnocline coupled with the decrease in light intensity with depth. As the light decreases, the time for phytoplankton to grow, deplete nutrients and decay increases. This, together with the smaller vertical mixing allows the maximum in production to occur at different times at different depths. It seems likely that both the biological and physical processes are important, but which dominates will depend on the circumstances.

3.3.5 The results of Lagrangian modelling

A new type of model of phytoplankton growth was introduced by Woods and Onken (1982). They pointed out that all previous models had studied a cloud of individual plankters relative to a continuous property of sea water, for example chlorophyll concentration. They called this the Eulerian-continuum method. This procedure suffers from the disadvantage that processes such as cell growth are assumed to proceed at the average rate. The authors suggested that it was important to understand the interactions of the physiology of each phytoplankton cell with the diurnal changes in mixed layer depth, light intensity and nutrient concentration. For this purpose they proposed to follow an ensemble of individual phytoplankters. At each time step of the model, the growth or decay of each cell was determined from the light and nutrient environments it was in. This incremental change was then added to or subtracted from the existing cell. They called this process of following the cells through their individual lives the Lagrangian-ensemble method.

The authors drew particular attention to the diurnal changes in the depth of the mixed layer. As we saw in section 3.3.1, increased cooling at night causes a progressive deepening of the mixed layer. This continues until about one hour after sunrise, when solar heating begins to exceed the rate of surface cooling and the mixed layer gets progressively shallower, continuing until noon. This diurnal cycle changes with the seasons (Fig. 3.07). The net result is that the water column can be divided into four zones that vary in thickness with the seasons. They are: (i) the mixed layer; (ii) the diurnal thermocline; (iii) the seasonal thermocline; and (iv) the interior of the ocean.

In the most recent version of the Lagrangian-ensemble model (Wolf and Woods 1988), which simulates a water column in the open ocean at 41° N 27° W, the seasonal variation in the mixed layer depth was as shown in Fig. 3.07. In the model it was assumed that the time scale of turbulence in the mixed layer was 30 min (Fig. 3.14). Since the model ran with time steps of 1 hour it was assumed that at the end of each period a particle in the mixed layer had equal probability of lying at any depth between the surface and the base of the mixed layer. Particles lying near the bottom were left isolated below the diurnal thermocline and were subject to only their normal sinking of 2 m d^{-1}. Most of them were re-entrained into the mixed layer when the thermocline descended next night, but at times when the diurnal mixed layer was becoming shallower, a proportion were not re-entrained and entered the seasonal thermocline where they remained until next spring.

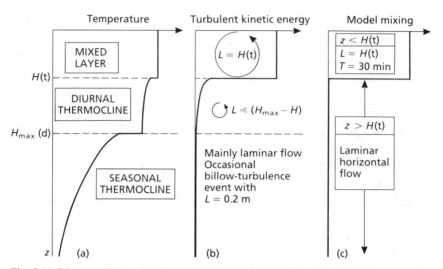

Fig. 3.14 Diagrams illustrating terms used in the model of Woods and Onken (1982): (a) the temperature profile; (b) the turbulent kinetic energy profile; and (c) the mixing regime. L, largest scale of mixing; H, thickness of mixed layer; T, time scale of turbulence; z, depth.

These physical conditions were used to determine the position of 5000 phytoplankters at the beginning of each hour. A simulated light field appropriate to each time and depth was used to determine the accumulation of energy, and an analogous process determined the accumulation of nutrients. When the accumulated energy (less respiration loss) and the accumulated nutrients both passed threshold values, the cell divided. Nutrient uptake was used to determine the nutrient profile. There was a standard run of the model with no upwelling, and there were subsequent runs with upwelling of various speeds and durations. The model simulated well the course of events in nature.

Wolf and Woods (1988) pointed out that phytoplankton cells in the model arrived at a particular depth and time with differing histories of exposure to light and to nutrients and hence differing degrees of adaptation to the prevailing conditions. This 'adaptation diversity' makes the model performance significantly different from that of Eulerian models. The broad conclusions were that the chlorophyll maximum formed as a consequence of the surface exhaustion of nutrients, with growth limited to cells in and below the nutricline. Cell division was nutrient-limited in the upper reaches of the nutricline and light-limited below. In these respects the model is similar to the one discussed in the previous section by Jamart *et al.* (1977). The chlorophyll maximum moves down through the water column because of the change in the turbulent mixing rates between the mixed layer and the pycnocline. Also like Jamart *et al.* (1979) the chlorophyll maximum in the model descends through the seasons without the invocation of a change in phytoplankton sinking rate.

In the standard model with no upwelling the nutricline moved deeper and deeper below the bottom of the mixed layer throughout the summer period. With moderate upwelling, at a speed less than the sinking rate of the phytoplankton, production increased but the nutricline and chlorophyll maximum stayed well below the mixed layer. Upwelling rates greater than the sinking speed caused the nutricline and chlorophyll maximum to move upwards until they entered the mixed layer and produced an explosive burst of phytoplankton growth. The authors proposed that episodic upwelling events are responsible for the transient patches of high surface chlorophyll ('hot spots') seen in satellite images of ocean colour (see section 3.2.6).

Considering this and the previous sections, it appears that the nutricline is formed as a result of progressive depletion of nutrients in the mixed layer during the oligotrophic phase in temperate waters, and that the subsurface chlorophyll maximum forms in and around the nutricline, with a magnitude of production that depends on the intensity of upwelling. While there is experimental evidence to show that reduction in light intensity and/or increase in nutrient concentrations brings about a reduction in phytoplankton sinking rate, this mechanism appears not to be essential to the formation of a subsurface chlorophyll maximum. It is enough that the growth and decay of phytoplankton occurs more slowly in the pycnocline than in the mixed layer.

3.3.6 The poleward migration of the spring bloom

The use of undulating towed sensors, often referred to as 'batfish', has made possible higher horizontal resolution of vertical profiles along transects than was possible with traditional 'bottle stations'. Strass and Woods (1988) reported on a series of such transects running 2000 km north–south in the mid-Atlantic from the Azores (38° N) past ocean weather ship C (52.5° N, 35.5° W) into the cyclonic sub-Arctic gyre at 54° N. The batfish cycled between 10 and 200 m every 1.4 km. Representative transects were obtained in late April 1985, June–July 1986 and August–September 1984. The records of near-surface chlorophyll fluorescence (Fig. 3.15) showed patchy bloom conditions to about 49° N in late April, a bloom centred on 49° N in June–July, with oligotrophic conditions well established south of 42° N, and in August–September oligotrophic conditions over the whole transect except for bloom conditions north of the polar front at 52° N.

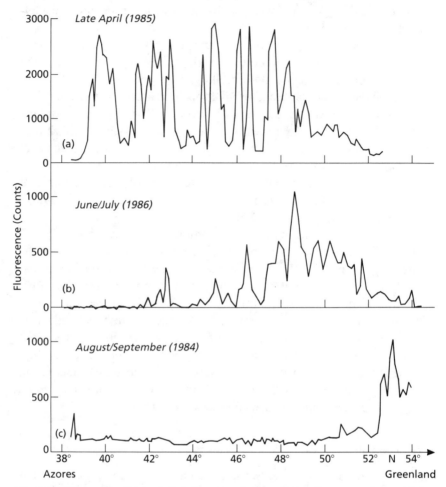

Fig. 3.15 Near-surface chlorophyll fluorescence at three times of year on a transect lying between the Azores and Greenland, showing poleward migration of chlorophyll maximum. From Wolf and Woods (1988).

Examination of the vertical structure showed that the horizontal migration of the near-surface chlorophyll maximum did not keep pace with the northward movement of the region of mixed layer shallowing, but followed the slow propagation of the 12 °C isotherm outcrop. This suggests that in addition to the need for stable stratification before a bloom can occur, there may be a temperature limitation on bloom processes.

Strass and Woods (1988) also found that during summer the deep chlorophyll maximum descended through the seasonal thermocline at approximately 10 m month^{-1}. They interpreted this as indicating that phytoplankton was using new nutrients from greater and greater depths. Since events occurred at progressively later dates as one moved north, the August–September transect showed the chlorophyll maximum sloping upwards from south to north (Fig. 3.16).

The sites where the highest concentrations of chlorophyll were found, between 1150 and 1300 km for example, were sites where the isopycnals were displaced upwards and closely spaced, indicating frontal upwelling probably associated with a cyclonic gyre or eddy. In the next section we shall review the evidence for the occurrence of events inducing localized upwelling.

3.3.7 Transient upwelling events

In addition to the biological enrichment that occurs in open waters as a result of upwelling at divergences at relatively fixed locations, it is found from satellite observations that there are transient pulses of increased surface chlorophyll that indicate meso-scale (10^3–10^5 m) upwelling events widely distributed in otherwise oligotrophic waters (Gower *et al.* 1980). Two main mechanisms have been proposed to explain them:

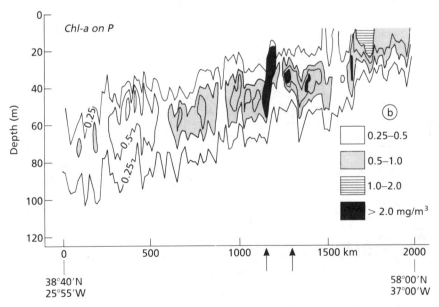

Fig. 3.16 Vertical distribution of chlorophyll on a transect between the Azores and Greenland in late summer. Note that the zone of maximum chlorophyll gets progressively more shallow as one moves north. From Wolf and Woods (1988).

internal waves, and fronts. Fasham and Pugh (1976) proposed that patches of high chlorophyll which they encountered were best explained by the presence of internal waves, and Holligan *et al.* (1985) showed that at a station with a water depth of 4000 m there was upwelling of nutrient-rich water into the surface layer caused by large soliton-like internal waves that had been generated more than a day earlier at the shelf edge (see section 6.4).

Woods (1988) has pointed to meso-scale patches of high chlorophyll concentrations with horizontal dimensions of about 10 km, that are a frequent component of satellite images, and has proposed that they are caused primarily by the jets associated with meso-scale fronts. Shelf-edge processes are considered in more detail in Chapter 6, and open-ocean fronts in Chapter 8.

3.3.8 The Antarctic divergence

The waters of the Southern Ocean exhibit a phenomenon not found in temperate or arctic waters elsewhere — a region of strong upwelling known as the 'Antarctic divergence'. Close to the Antarctic continent the prevailing winds are from the east (Fig. 3.17), giving rise to a westward flowing coastal current known as the 'east wind drift'. Since the Coriolis force is to the left in the southern hemisphere, this water

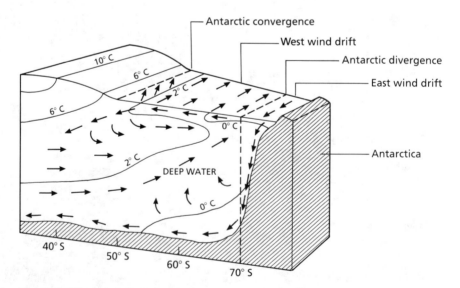

Fig. 3.17 Diagram of the Southern Ocean to show the Antarctic divergence and the Antarctic convergence. Antarctica is on the right, north is to the left.

tends to flow towards the coast where it tends to pile up, and therefore causes downwelling. Five to ten degrees of latitude north of Antarctic-aone enters the 'west wind drift', or 'Antarctic circumpolar current', flowing in the opposite direction under the influence of strong westerly winds. Again the Coriolis force flows to the left of the wind, towards the equator. Hence, the surface waters diverge between the east and west wind drifts, and the Antarctic divergence constitutes a site of upwelling that circles the globe at a latitude of approximately 70° S. As Fig. 3.17 shows, there is a major flow of Antarctic surface water northward as far as the 'Antarctic convergence' or 'polar front', which lies at a latitude of approximately 50° S and constitutes the northern boundary of the Southern Ocean.

The upwelled water at the Antarctic divergence is rich in nutrients, with nitrates around 25×10^{-6} M. As Holm-Hansen (1985) has pointed out, one would have expected that in the course of the northward drift to the Antarctic convergence, which takes about 200 days, the phytoplankton would strip most of these nutrients from the water. In practice it is found that the nitrate concentrations tend to remain relatively high all the way from the Antarctic divergence to the Antarctic convergence. Silicates, on the other hand, drop sharply over the same distance. If surface water is isolated in a container and exposed to ambient light and temperature the phytoplankton grow and strip the nitrates in 15–20 days. Holm-Hansen (1985) concluded that there must be a considerable amount of upward mixing of nitrate into the photic zone throughout the Southern Ocean. He also observed that about 75% of the production in the Southern Ocean is based on recycled ammonia, and the work of Biggs (1982) indicates that most of the ammonia for this is produced by microbial organisms <200 μm in diameter. Holm-Hansen suggested that the drop in silicate concentrations is caused by uptake by diatoms and silicoflagellates which sink out of the euphotic zone, leaving behind phytoplankton with a low silicate requirement.

An earlier view of the Southern Ocean was that the relatively high levels of dissolved nitrate in surface waters, and the known abundance of consumers such as krill, seals and whales, all indicated a high level of productivity during the short Antarctic summer. Holm-Hansen *et al.* (1977) and El-Sayed (1978) have pointed out that many of the primary productivity measurements tending to support this view came from inshore areas. On the basis of the most recent measurements in the open ocean they concluded that daily primary production averages about 0.134 g C m^{-2} d^{-1} during the growing season of about 120 days and that average annual production is only about 16 g C m^{-2}y^{-1}.

3.4 SECONDARY PRODUCTION AND THE MIXED LAYER

3.4.1 Oligotrophic waters

So far in this chapter we have paid scant attention to the role of the zooplankton. The truth is that, lacking readily available automated equipment to obtain continuous records of the abundance of zooplankton, we have far less information about interactions between zooplankton and physical processes than we have about phytoplankton. The relationship between vertical physical structure and zooplankton production is complicated by the ability of zooplankters to make extensive vertical migrations. At one time of year most of the population may be at great depth, surviving a period of food shortage by living a lethargic life in cold water. At another time they may be concentrated in the mixed layer, making use of abundant phytoplankton food and higher temperatures to grow rapidly and reproduce.

There have been many laboratory studies of the feeding rate of zooplankton, but turbulence of the water has seldom been a factor that was controlled. Recently, evidence has been brought forward suggesting that small-scale turbulence has a powerful effect on the rate of contact between a zooplankter and its food. If this is found to be true, classical studies of feeding rates will have to be re-evaluated.

Longhurst (1981) has provided a review of the vertical distribution of zooplankton. As before, it is most convenient to start with the relatively stable oligotrophic tropical systems. A very distinct pattern of vertical distribution is found in stable water columns with very shallow mixed layers, such as those found in low latitudes in the eastern parts of oceans, or in mid-latitudes towards the end of summer. The abundance of zooplankton is much greater in the mixed layer than below it, and within the mixed layer the greatest abundance is found just above the thermocline. Near the bottom of the thermocline there is typically a sharp decrease in the abundance of zooplankton, in a region which the specialists refer to as the planktocline.

It is reasonable to expect that the abundance of zooplankton may be related to the abundance or productivity of the phytoplankton. Longhurst (1976) showed for the eastern tropical Pacific that the maximum density of zooplankton coincided with the depth at which phytoplankton *productivity* was maximal, rather than with the depth at which phytoplankton *biomass* was maximal (the subsurface chlorophyll maximum). This was true both by day and by night, even though a proportion of the zooplankton made diurnal vertical migrations. From this he concluded that zooplankton grazing is a major determinant of the shape of chlorophyll profiles, and should be taken into account in modelling the dynamics of primary production.

On the other hand, Ortner *et al.* (1980) analysed data from various offshore sites in the western north Atlantic and found that the zone of maximum zooplankton biomass was coincident with the deep chlorophyll maximum. They argued that since the deep chlorophyll maxima in most cases coincided with the thermocline it was impossible to say whether the zooplankton were reacting to food abundance or to physical cues. They pointed out that it was difficult to envisage them as responding to the rate of primary production *per se*. The debate continued in papers by Longhurst and Herman (1981) and Ortner *et al.* (1981) without any clear resolution.

More recently it has been suggested that the larger zooplankton graze on the healthy phytoplankton and become most abundant at the phytoplankton production maximum during bloom conditions, while the micro-zooplankton, especially protozoa, thrive on the senescent phase of a bloom and find their most abundant food supply at the biomass maximum. Hence, both Longhurst and Ortner could be right, depending on biological conditions at the time of the observations. The evidence is as follows.

King *et al.* (1987) studied the vertical distribution of zooplankton in the Gulf of Maine, making separate assessments for macro-zooplankton (over 153 μm) and micro-zooplankton (35–153 μm). They indexed nitrogen metabolism by measuring glutamate dehydrogenase (GDH). Subsequently Le Fèvre and Frontier (1988) re-examined the data and pointed out that in September the macro-zooplankton maximum and the GDH maximum occurred at the phytoplankton production maximum, while in June the GDH maximum coincided with the micro-zooplankton maximum, which was located at the chlorophyll (phytoplankton biomass) maximum. There are other data from coastal waters which support this point of view (see section 6.4.1).

3.4.2 Zooplankton and seasonal changes in vertical structure

Longhurst and Williams (1979) showed that at the ocean weather ship *India* located south of Iceland (59° N, 19° W) the situation at the end of winter (late March) was that most of the zooplankton biomass lay at depths greater than 350 m. By day, 85% of the biomass was down there, and at night, when many species made upward migrations, 56% of the biomass was still below 350 m. During April, when the spring bloom was initiated, there was a major upward shift of the zooplankton biomass, brought about chiefly by the upward shift of the *Calanus finmarchicus* population, so that by May *Calanus finmarchicus* had 94.5% of its population in the upper 100 m. At the same time, *Calanus* began rapid reproduction so that between late March and early May the proportion of copepodites I–IV changed from 1 to 65 % of the individuals present. Longhurst and Williams (1979) pointed out that any realistic model of the development of a phytoplankton bloom in temperate waters must allow for changes in grazing pressure due to two factors, the upward translocation of biomass and the growth of the zooplankton population *in situ*. Before the extent of the grazing pressure can be calculated it is necessary to know the feeding rate of the individual zooplankters.

3.4.3 Calculation of zooplankton feeding rates

There has been a lack of agreement between measurements of zooplankton feeding rates in the laboratory and calculations made on the basis of budgets constructed for field situations. In general, the latter suggest much higher figures. The evidence has been reviewed by Cushing (1975). One of the few laboratory data sets to obtain high rates of feeding approaching those deduced from field observations is that obtained by Corner *et al.* (1972). These workers carried out the experiments in glass columns that were rotated in the vertical plane and completed 1 revolution every 2 min. The purpose was to ensure that the phytoplankton cells remained in suspension throughout the experiment, but it is worth noting that their method also ensured a relatively high level of turbulence in the containers. They found that the maximum daily rations of the zooplankton were equivalent to 47.5% of the body nitrogen.

Cushing (1968) developed a theory of grazing in which the herbivore is assumed to search a volume which is the product of the area swept by the sensors and the distance swum. As the animal stops to eat, the volume searched becomes reduced. Others have modified the concept by introducing irregular swimming tracks and varying speeds of swimming, and by introducing threshold densities of food below which the animals did not feed. However, many of the assumptions underlying this work have recently been challenged by Rothschild (1988).

His basic thesis is that the rate of contact between a planktonic organism and its food is strongly affected by small-scale turbulence in the water. Feeding studies which do not take this into account are in need of revision. Beginning with the formulation of Gerritson and Strickler (1977), the relative velocity of a prey particle to a predator particle is:

$$(u^2 + 3v^2)/3v \quad \text{for } v > u,$$

and

$$(v^2 + 3u^2)/3u \quad \text{for } u > v, \tag{3.13}$$

Table 3.2 Georges Bank scenario. Densities of organisms approximate those reported for May at the terminus of the spring bloom.

	Number cm^{-3}	Diameter (μm)	Velocity (cm h^{-1})
Bacteria	10^6	1	0.1
Phytoplankton	100	20	0.1
Nannoplankton*	2000	10	36[†]
Pseudocalanus	0.002	1000	720[‡]

*Mostly at surface and heterotrophs. [†]100 μm s^{-1}. [‡]0.2 cm s^{-1}.

where u and v are the velocities of the prey and predator respectively. Contact rate is calculated by multiplying the relative velocity by the cylindrical volume swept by the predator and the number of prey contained in that volume, $N\pi r$, where N is the density of prey and r is the radius of the cross-sectional area searched by the predator. Rothschild (1988) modifies this by introducing a term w for the rms velocity difference of two points in a homogeneous isotropic turbulent medium and states that the rms relative velocity of the predator and prey is:

$$\frac{u^2 + 3v + w^2}{3(v^2 + w^2)^{1/2}},$$ (3.14)

where the predator is faster than the prey. He developed an example using data (Table 3.2) for Georges Bank in May, from Backus and Bourne (1987). Using wind velocities of 10 and 20 knots he made some assumptions about levels of turbulence in the mixed layer and calculated the rms turbulent velocities w shown in Table 3.3. He then calculated the contact rates shown in Table 3.4 which shows that turbulence has an important effect. It is greatest for the smallest organisms because their rates of locomotion are small compared with the velocity of the turbulent water. The rates of contact of bacteria with phytoplankton are enhanced 3–4 orders of magnitude by high turbulence, and even the contacts of macro-zooplankton with phytoplankton are almost doubled.

Table 3.3 A range of turbulent rms velocities, w, given wind velocities and particle separation distances. The wind velocities are only given as examples as tides or other sources might also generate turbulent motions. From Rothschild (1988).

Particle separation distance (m)	Wind velocity (cm s^{-1})	
	10 knot wind	20 knot wind
10^{-4}–10^{-3}	0.005	0.01
10^{-2}–10^{-1}	0.1	0.2

Table 3.4 Relative velocity and number of contacts among different kinds of organisms. Zero, low (0.005 cm s^{-2}) and high turbulence (0.2 cm/ sec) rms velocity, w is incorporated in the computation. From Rothschild (1988).

	Turbulence	Bacteria		Phytoplankton		Nannoplankton	
		Relative velocity	Contacts per hour	Relative velocity	Contacts per hour	Relative velocity	Contacts per hour
Pseudocalanus	zero	720	22×10^6	720	2262	721	4.5×10^4
	low	720	22×10^6	720	2262	721	4.5×10^4
	high	1187	37×10^6	1187	3729	1188	7.5×10^4
Nannoplankton	zero	36	113	36	–	–	–
	low	43	135	43	–	–	–
	high	961	3019	961	0.03	–	–
Phytoplankton	zero	0.1	1	–	–	–	–
	low	24	301	–	–	–	–
	high	960	12064	–	–	–	–

These ideas have not yet been applied in any detailed model of the effect of zooplankton grazing on the distribution and fate of phytoplankton in the mixed layer. It is clear that the next generation of models should take into account the recent findings on two topics: the vertical distribution of zooplankton in relation to seasonal changes in the mixed layer, and the effect of turbulence on the activities of grazers.

3.5 CONCLUSIONS

In this chapter we have focussed attention on vertical processes on a scale of metres and tens of metres, generated by wind stress at the sea surface, convective circulation resulting from surface cooling and turbulence at the pycnocline generated by internal waves, etc.

We have seen that phytoplankton cells are at times carried from the surface to great depth on a time scale of 1–10 h, and are thus exposed to a fluctuating light regime. It appears that there are some biochemical processes in the cells that adapt rapidly to the changing light, while others change too slowly for there to be any noticeable response. In any case, the metabolic response of cells exposed to fluctuating light is not very different from the response of those exposed to the same amount of light delivered at a constant rate.

More than half of the world's ocean is in tropical or subtropical latitudes and a typical tropical profile of the water column shows a warm, stable, mixed layer separated from the cooler underlying water by a sharp pycnocline. Nutrient concentrations are very low above the pycnocline but relatively high below it. The peak of phytoplankton biomass, the chlorophyll maximum, is found at the nutricline. It appears that nutrients are slowly transported through the pycnocline, and these are rapidly taken up by the phytoplankton cells in the chlorophyll maximum. Light intensity is relatively low at the chlorophyll maximum and increases exponentially as one moves towards the surface. As a result, the greatest rate of primary production is usually found a few metres above the chlorophyll maximum.

A useful distinction can be made between the primary production that results from the utilization of upwelled nitrate, called new production, and that resulting from the utilization of ammonia excreted by heterotrophs in the mixed layer, called regenerated production. The ratio of new production to total production is called the f ratio. In steady state conditions the nitrogen removed from the mixed layer by sinking, or by predators which leave the area, must be balanced by the vertical transport of new nitrogen.

The chief exception to the 'typical tropical structure' (TTS) of the open tropical ocean is found in the equatorial upwelling zone, where major ocean currents cause upwelling and divergence of large volumes of nutrient-rich water from below the thermocline. Associated with this circulation pattern are several major cyclonic gyres which cause a dome-like structure in the isopycnals and an enhanced rate of vertical transport of nutrients. Other deviations from the TTS occur irregularly but not uncommonly and are associated with seasonal and longer-term climatic changes in tropical latitudes.

In temperate and sub-Arctic latitudes there is typically a burst of primary production known as the spring bloom. The mechanism underlying this is shallowing of the mixed layer after a period of vertical transport of nutrients. It was first recognized and described by Gran (1931), and was put on a quantitative physical basis by Sverdrup (1953). It has been shown that a wave of high chlorophyll biomass migrates poleward in the Atlantic in spring, approximately coincident with the 12 °C surface isotherm. Investigation of the vertical structure of the ocean while this migration is in progress shows that shallowing of the mixed layer to less than a critical depth is a precondition for the spring bloom. We shall find that the alternation of a period of strong vertical transport of nutrients with a period of stratification is a recipe for high primary production in a wide variety of marine habitats.

4

Vertical Structure in Coastal Waters: Freshwater Run-off and Tidal Mixing

4.1 INTRODUCTION

Coastal waters (which we take as extending from the edge of the continental shelf to the high water mark) are subjected to the same seasonal cycles of warming and cooling as the open ocean, and in temperate climates the mixed layer may alternate between being shallow and deep in the same way as in open water. However, the process is greatly complicated by factors peculiar to the coastal zone. The first of these is the shallowness, which leads to a situation in which a relatively shallow mixed layer may extend to the bottom. Since dead biological material, detritus, tends to accumulate and decompose on the bottom, the nutrients released by it may be carried to the surface waters and rapidly used in pho

111

tosynthesis. The second important factor is the presence of tidal currents which create turbulence in the water. If the depth is not too great in relation to the strength of a tidal current, tidally-induced mixing may extend all the way to the surface. The third peculiar feature is the barrier to advection posed by the coastline itself. For example, if surface water is driven by wind action away from the coast, the only way for it to be replaced is by upwelling from below. Since there is a good chance that the upwelled water has been enriched in nutrients, an upwelling area is likely to be a site of enhanced biological production. As we saw in the previous chapter, stratification of the water column is fundamental to biological production. When stratification first sets in, some phytoplankton are trapped in a well-lighted mixed layer and production is enhanced. In an area where stratification has been present for a relatively long time, the nutrients in the mixed layer may become depleted and their renewal from below is inhibited by the pycnocline, so that primary production tends to be depressed. Hence, stratification may act in a positive or a negative way on primary production, but its effects are always important.

In the open ocean, stratification is almost always induced by temperature differences between the layers. In coastal waters a very important additional factor must be considered, namely the flow of fresh water from the land. Having a salinity close to zero, it is much lighter than sea water and by lying on top of the sea water creates a stratification that can be independent of temperature differences between the layers. Furthermore this surface layer, being less dense, rides higher at the sea surface, creating a slope along which water flows. Since these flows depend on the buoyancy they are known as buoyancy-driven currents. One way of trying to understand the complex relationships existing between physical and biological processes in coastal waters is to view freshwater run-off as a mechanism tending towards greater stratification while wind-driven and tidal currents are mechanisms tending to cause turbulence in the water column and to break down stratification. In the open ocean, horizontal gradients are small and vertical processes control the distributions of heat, salt, nutrients, etc. In estuaries and on continental shelves, however, horizontal movement of water and large horizontal gradients tend to determine the property distributions. The Coriolis force and geostrophic balance are important features of horizontal flows so we begin by reviewing them.

4.2 THE CORIOLIS EFFECT

The Coriolis effect, as mentioned in the last chapter, is an apparent deflection to the right in the northern hemisphere and to the left in the southern hemisphere of objects travelling over the surface of the earth. The deflection is caused by the rotation of the earth as illustrated in Fig. 4.01. Consider first the person at O in Fig. 4.01(a) who throws a ball at the sun as it rises in the east (E). (Let us ignore the friction of the air and the gravity that would bring the ball down to earth.) After an hour (Fig. 4.01(b)) the sun has risen above the horizon along a path slanting up to the right in the northern hemisphere and up to the left in the southern hemisphere. Is the ball which was thrown towards E still heading towards E or is it heading towards the sun?

Because there is no friction between the earth and the ball, the ball moves in a straight line relative to the sun and stars while the earth rotates underneath. The ball continues to head straight towards the sun while the horizon moves with the earth as shown in Fig. 4.01(b). Each point on the eastern horizon moves down and to the left of the ball and the point E on earth is in a different direction relative to the fixed stars than when the ball was thrown.

The rate of the deflection may be calculated as follows. The point E, like all other points on the earth, moves around the axis of the earth once a day in a circle which lies perpendicular to the axis. The motion along this circle in degrees or radians per second is indicated in Fig. 4.01(b) by the arrow labelled Ω. This motion may be split into two components one parallel and one perpendicular to the horizon. The parallel component is $\Omega \sin \phi$, where ϕ is the latitude of the observer's position, while the component perpendicular to the horizon is $\Omega \cos \phi$. The component along the horizon is the one of interest to us because it is the cause of the apparent deflection of the ball. The ball moves in a straight line while the horizon rotates towards the left at an angular rate of $\Omega \sin \phi$.

Earthlings, however, find it much more convenient to ignore the fact that the earth rotates and assume that it remains motionless while the sun and stars move across the sky. The fact that objects, like the ball, are deflected from their original course is accounted for by a force which was invented by mathematicians to represent the effects which are due to the rotation of the earth. This imaginary force, the Coriolis force, results in the situation illustrated in Fig. 4.01(c). Here the earth remains motionless while the Coriolis force pushes the ball to the right, following the rising sun, at the angular speed of $\Omega \sin \phi$. The derivation of the Coriolis force from the angular rate of the deflection is given in Box 4.01.

The 'thought' experiment suggested in Fig. 4.01 is of course impossible to perform because gravity brings the ball back to earth before the sun has moved an appreciable amount, but the deflection is a very

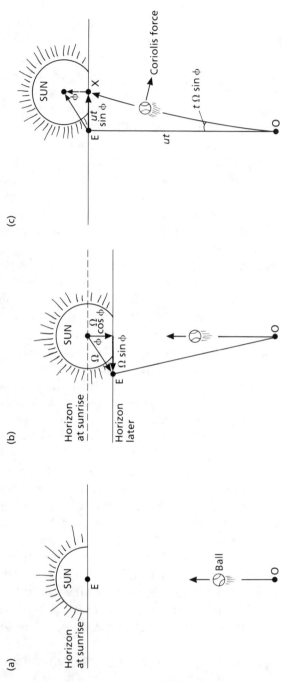

Fig. 4.01 (a) The person at O has just thrown a ball towards the sun which is rising in the east (E). (b) An hour later the earth has rotated on its axis and moved the horizon to the left and down relative to the sun and stars. The ball continues towards the sun because it is not connected to the earth. (c) If the horizon is assumed to remain motionless, the rising of the sun appears to be due to the sun's own motion rather than the rotation of the earth and the fact that the ball turns towards the sun is due to the 'fictitious' Coriolis force which pulls the ball to the right in the northern hemisphere but to the left in the southern hemisphere.

Box 4.01 The Coriolis parameter

The angular rate of the deflection cannot be used directly in the equations of motion which require that the effect be described in terms of a force or acceleration per unit mass. The acceleration which must exist to account for the observed deflection can be derived following Pond and Pickard (1983) with help from Fig. 4.01(c). Assume the ball is originally thrown from O towards E but is deflected to the right to land at X. The distance from O to E is the speed of the ball, u, times the time of flight, t. The angle EOX is the angular rate of the deflection, Ω sin ϕ, times the time of flight, t. The deflection from E to X is then the product of the angle, $t\Omega$ sin ϕ and the distance ut, i.e.

$$\text{EX} = d = ut^2 \Omega \sin \phi. \tag{4.01}$$

The acceleration that exists to produce this displacement starting from rest can now be calculated from the formula relating displacement, d, and acceleration, a, which is usually discussed in elementary physics courses, namely $d = \frac{1}{2} at^2$. Substituting $d = ut^2 \Omega$ sin ϕ in this equation gives the acceleration or force per unit mass as $2u\Omega$ sin ϕ. This is usually written fu where $f = 2\Omega$ sin ϕ and is called the Coriolis parameter.

The expression 2Ω sin ϕ shows that the Coriolis effect is a maximum at the north pole where $\phi = 90°$, and zero at the equator where $\phi = 0$. In the southern hemisphere ϕ, the latitude, is negative and the Coriolis force is in the opposite direction; objects moving over the earth's surface are deflected to the left. The direction of the earth's rotation is not different in the two hemispheres but the direction of gravity does change and what looks like rotation to the right in the north is rotation to the left in the south.

important factor in the flight of artillery shells which are separated from the earth for significant periods of time. Another common way of demonstrating the earth's rotation is with Foucault's pendulum, Fig. 4.02. When carefully set up as is often done in science centres there is no friction between the long swinging pendulum and the rotating earth. Since there is no force to twist the plane of the pendulum's swing to follow the rotation of the earth it stays fixed relative to the sun and stars as the earth rotates underneath. In the north the plane of the pendulum rotates clockwise relative to the earth at the rate Ω sin ϕ, which is the same as the component of the sun's motion parallel to the horizon noted above.

The time for a complete revolution of the pendulum relative to the earth is called the pendulum day and is 24/sin ϕ h long. At the pole the

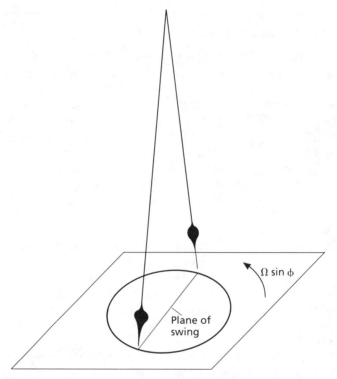

Fig. 4.02 The plane of the pendulum swing stays fixed relative to the stars, etc., but appears to rotate to the right as the room rotates to the left with the earth at $\Omega \sin \phi$ rad h^{-1}.

pendulum day is 24 h, at 45° it is 34 h and at the equator it is infinitely long, when $\sin \phi = 0$. As with the thrown balls, etc., it is hard to see, at first glance, how the pendulum day must be longer than the day. One must, of course, be careful to separate the two components of the rotation which are each less than the total rate of rotation and remember that the period of each component is the inverse. The periods of the components are not components themselves and do not add up to the period of the whole. This is a good illustration of the fundamental difference between vectors (rotation) and scalars (period).

4.3 THE GEOSTROPHIC BALANCE

The Coriolis force is an important factor in ocean currents because friction in the water is so small that water can move, like the ball in Fig. 4.01(a), over the earth's surface and remain isolated from the spinning earth. The Coriolis force is not the force that makes the water move but is the apparent force which deflects the flow when the water is moving relative to the earth. Suppose, for instance, that all the surface water in a region in the northern hemisphere starts to move south because there is a pressure gradient from north to south. The Coriolis force deflects the flow around to the right of its southward path. Over the course of about one pendulum day the direction of flow comes right around to the right by 90° and is towards the west. At this point the water is moving west while the north to south pressure gradient which started the water on its flow is still in the same direction and the Coriolis force is towards the north or 90° to the right of the flow direction. The Coriolis force and the force of the pressure gradient are now in opposite directions and balance each other. This is the geostrophic balance.

The geostrophic balance is an everyday occurrence in the atmosphere and is evident in the daily weather charts which display the distribution of isobars. The flow of air around the low and high pressure systems is parallel to the isobars not across them because the air is close to being in geostrophic balance. The mathematical expressions defining the geostrophic balance are discussed in Box 4.02.

Box 4.02 Equations of the geostrophic balance

Mathematically, geostrophic motion is expressed by two equations which define the balance of forces for the east–west (x axis) and north–south (y axis) directions. In the x direction the equation is

$$fv = \frac{1}{\rho}\frac{dP}{dx},$$ (4.02)

where the Coriolis force (fv) along the x axis is balanced by the x component of the pressure gradient on the right side of the equation. The equivalent equation for the y direction is,

$$fu = -\frac{1}{\rho}\frac{dP}{dy}.$$ (4.03)

Here the y component of the pressure gradient is balanced by the Coriolis force arising from the flow in the x direction.

These equations express only a balance of forces. There are no sources or sinks of energy included in them and because the flow is perpendicular to the forces, the forces do no work and the motion goes steadily on forever. This may seem a little unrealistic as friction must eventually convert the motion to heat but that is a very slow process leaving the geostrophic approximation as very accurate, especially in the deep sea where friction is small.

4.4 ESTUARIES

'An estuary', according to the definition of Pritchard (1967), 'is a semi-enclosed coastal body of water which has a free connection with the open sea and within which sea water is measurably diluted with fresh water derived from land drainage.' Estuaries are economically important features of the ocean because of their high biological productivity, their proximity to large cities with their wastes, and their increasing use as sites for aquaculture. In the following three sections we examine the processes which contribute to the typical circulation patterns and describe the different types of estuaries.

4.4.1 Estuarine circulation

From the definition it is evident that an estuary is just a bay, part of a bay, or a narrow inlet in which freshwater flow from the land has reduced the salinity of the ocean water. The flow of fresh water causes a characteristic circulation pattern to be set up in the estuary in which fresher and therefore lighter water flows out of the estuary in the surface layer and a deeper flow brings water from the open sea into the estuary (Fig. 4.03). This so called 'estuarine circulation' is observed to be the dominant circulation pattern in estuaries where run-off is moderate and where mixing by tidal currents is weak.

The main force that drives the estuarine circulation is a horizontal pressure gradient created by the density difference between the newly-added fresh water and the resident salt water. The force lies in the surface layer and is directed down the estuary away from the source of fresh water. The resulting flow, being dependent on the buoyancy of the fresh water, is usually referred to as a buoyancy-driven flow.

The origin of the pressure gradient is illustrated in Fig. 4.04 in which

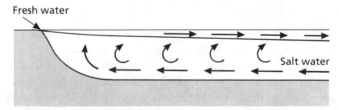

Fig. 4.03 The partially-mixed estuary. Fresh water entering at the head of the estuary on the left creates a near-surface flow of light water out of the estuary and a compensating deep flow of saltier water up towards the head of the estuary. The upper layer gets thicker as it moves away from the source of fresh water because salt water is entrained from below.

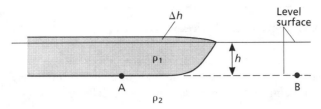

Fig. 4.04 A pool of light water, ρ_1, lies on top and beside water of greater density ρ_2. If the pressures at A and B are the same the height of the sea surface above A must be higher than above B.

a layer of fresh water of density (ρ_1) overlies water of higher density (ρ_2). We assume there is no horizontal pressure gradient at the bottom of the low density layer or at any deeper level. Thus the pressure at 'A' at the bottom of the upper layer is the same as the pressure at 'B' which is on the same level as A but beyond the freshwater layer. If the pressures are to be the same, the weight of the water above the two points must be the same and therefore the column of the lighter water above A must be higher than the column of denser water above point B. The difference in the height of the water surface in the two regions (Δh) can be calculated by equating the pressure at A with the pressure at B. Pressure is calculated by multiplying the density of the water times the height times the gravitational acceleration, i.e. $P = \rho gh$. Above A the density is ρ_1 and the height is $h + \Delta h$, so the pressure is $\rho_1 g(h + \Delta h)$, and above B the density is ρ_2 but the height is only h, giving a pressure of $\rho_2 gh$. After equating these, a little manipulation leads to the formula

$$\Delta h = h(\rho_2 - \rho_1)/\rho_1. \tag{4.04}$$

As an example let the freshwater layer with a density $\rho_1 = 1000 \text{ kg m}^{-3}$ have a thickness $h = 2 \text{ m}$ over the deeper layer of density $\rho_2 = 1025 \text{ kg m}^{-3}$. Equation 4.04 states that the water above point A, in Fig. 4.04, is about 5 cm higher than above point B.

Since the pressures at A and B are the same there is no force to move the water horizontally between the two points. At all points above A, however, the pressure is greater than at the corresponding level above B. This higher pressure pushes the water in the upper layer to the right in the figure over the top of the denser water.

As the lighter water flows over the heavier layer towards the open sea, turbulence in the water causes the two layers to mix together which reduces the differences existing between the two layers. The upper layer becomes thicker, increases in density and slows down. The importance of this process to the circulation in the estuary is explained in Box 4.03.

Box 4.03 The balance of forces in an estuary

The driving force of the circulation is the difference in pressure which is created down the estuary by the fresh water in the surface layer. This force (per unit mass) is expressed mathematically as

$$\frac{1}{\rho}\frac{dP}{dx}. \tag{4.05}$$

The balancing force is generated by the mixing between the upper and lower layers. When fast-moving upper water is exchanged with slow-moving lower water a force is generated which slows down the upper layer and speeds up the lower layer. A mathematical expression for the force starts with the equation

$$K_e\frac{d\bar{u}}{dz}, \tag{4.06}$$

which is the diffusion equation for the vertical flux of momentum and is analogous to Eq. 2.11. K_e is the eddy diffusivity and $d\bar{u}/dz$ is the vertical gradient in the mean velocity. The formula calculates the vertical flux of momentum through a horizontal layer of thickness dz. As long as the same amount of momentum is leaving the layer as is entering it no force is being exerted on the layer. A force arises when the layer is gaining or losing momentum which happens when the vertical flux is not constant with depth. This occurs when either the eddy diffusivity or the velocity gradient vary with depth. Usually K_e is considered constant and the convergence or divergence of momentum is attributed to variations in the vertical current shear. Since the shear is just the velocity gradient, the gradient of the shear is the gradient of the gradient of the velocity which is commonly called the second derivative of the velocity. Thus if K_e is constant the force is

$$K_e\frac{d^2\bar{u}}{dz^2}, \tag{4.07}$$

and the balance between the pressure gradient down the estuary and the drag of the underlying water is expressed by

$$\frac{1}{\rho}\frac{dP}{dx} = K_e\frac{d^2\bar{u}}{dz^2}. \tag{4.08}$$

The equation does not include the Coriolis force because we assume that the estuary is too narrow for this force to be important. The flow is considered to be two dimensional or the same along any section cut lengthwise along the estuary. One problem with the equation is that it is not possible to go to an estuary and directly measure the magnitude of either the pressure gradient or the drag force. It is also well known that the eddy diffusivity is not a constant but a function of time and of position. The equation is, however, very useful because it can be used, in association with the other required equations, to construct a mathematical model of the estuarine flow (Rattray and Hansen 1962; Dyer 1973). Then the flow and the property distributions predicted by the model can be compared to the observed distributions. By adjusting the model to produce distributions which are similar to those observed it is possible to determine which are the most important factors controlling the circulation.

One feature of estuarine circulation that is observed in the models is the deep flow of saline water into the estuary. This flow is a direct consequence of the turbulent exchange between the upper fresh layer and the layer immediately below. As the water in the upper layer proceeds out of the estuary it becomes slower and thicker because of the vertical diffusion of momentum. At the same time diffusion causes the fresher upper water to be exchanged with the deeper saltier water and the surface layer becomes saltier as it progresses out of the estuary. Salt then is transported out of the estuary by the outward flow in the upper layer. In time this would cause the water in the estuary to become completely fresh which does not happen because the deep flow develops and brings salt into the estuary.

4.4.2 Sources of turbulence: the Richardson number

The turbulent motions that mix the fresh upper layer of an estuary with the saltier layer below may be generated by a number of different processes. Breaking waves at the surface and rough flow over the bottom and sides of the estuary could generate significant amounts of turbulence, but the largest contributions are thought to be provided by the turbulence generated by the breaking of waves on the interface between the two layers (Farmer and Freeland 1983). These internal waves are generated on the density interface by wind waves at the surface, pressure fluctuations in the atmosphere, flow over irregularities in the bottom bathymetry, etc.

The likelihood that these internal waves will become unstable and break up into turbulence is usually estimated with the Richardson number. This is a ratio of two opposing forces; one which tends to keep the density interface level and the other which tends to increase the size of the waves. The first of these is generated by the density change across the interface which creates a buoyancy restoring force (section 3.2.4) that tends to limit the amplitude of the waves. The energy to increase the size of the waves is generated from the difference in velocity between the two layers. The strength of the buoyancy force is usually represented by the square of the Brunt–Väisälä frequency (Eq. 3.06) while the energy available in the shear flow is represented by the square of the velocity gradient. The Richardson number is the ratio of these two:

$$\text{Ri} = \frac{g}{\rho}\frac{d\rho}{dz}\Big/\left(\frac{du}{dz}\right)^2.$$

(4.09)

If this number is high the buoyancy forces associated with the density stratification are strong, and waves on the interface do not grow and break but tend to be damped out. In fact if the ratio is greater than 0.25 waves of all wavelengths are stable (Turner 1973). If the density difference decreases or the shear increases such that $\text{Ri} < 0.25$ waves will grow in amplitude by taking energy out of the velocity shear and eventually become unstable. This is the Kelvin–Helmholtz instability and causes the waves to roll up and disintegrate into a patch of turbulence. Theoretical and experimental results, also reviewed by Turner (1973), show that as the Richardson number decreases the most unstable waves have a wavelength that is about eight times the thickness of the interfacial layer between the two layers of unequal density. Some interesting *in situ* photographs of this process were obtained in the Mediterranean by Woods (1968).

4.4.3 Different types of estuaries

The well-behaved estuarine circulation outlined above is found in many estuaries but there are also many processes that can interfere to alter the basic pattern. In some estuaries the depth slopes gently from where the river reaches sea level and the river flow is strong enough to drag the sea water layer further out than it would be if there was no freshwater flow. In these cases the upper and lower layers appear as in Fig. 4.05. Because of the wedge-like shape of the lower layer these are called salt wedge

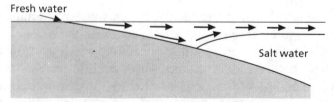

Fig. 4.05 The salt wedge estuary. At the mouths of shallow rivers, high run-off may cause the salty water to be pushed outward in a wedge shape.

estuaries. The position where the lower layer meets the bottom changes with the strength of the flow in the upper layer, and in the case of the Mississippi this point can vary by more than 200 km throughout the year. Calculations to determine the shape of the salt wedge are reviewed by Dyer (1973).

Changing salinity at the boundary of the salt wedge leads to flocculation and sinking of fine particles carried in suspension. These particles get carried landwards in the deeper waters and concentrate at the tip of the salt wedge, thus producing a zone of high turbidity, the 'turbidity maximum', which also tends to move up and down the estuary according to the strength of the freshwater flow. In this way, sinking particles are distributed over a wide area and lead to the formation of intertidal mud flats so characteristic of estuaries.

In some estuaries friction between the bottom and the tidal currents produces enough turbulence to completely eradicate the estuarine circulation. These tidally-mixed estuaries (Fig. 4.06) tend to be shallow and in areas of high tidal currents. The rate at which turbulent energy is converted from the tidal flow into turbulent flow can be estimated with the equation given by Pingree *et al.* (1978a) (for further details see Box 4.04):

$$\varepsilon = 2.5 \times 10^{-7} \times \frac{u^3}{h}. \tag{4.10}$$

Here u is the absolute speed of the tidal flow in m s^{-1} averaged over one tidal cycle and h is the depth of the water in metres. The value 2.5×10^{-7} is an average drag coefficient, and can be expected to differ according to the nature of the sea floor. The production of turbulent energy ε is in units of W kg^{-1}. Substituting 2 m s^{-1} for u and 10 m for h the energy conversion works out to 2.2×10^{-7} W kg^{-1} which, as was pointed out in Chapter 2, is a relatively high level of turbulence in the ocean. Direct measurements of turbulence in the Strait of Gibraltar, (Kinder and Bryden 1988), indicate dissipation rates as high as 10^{-4} W kg^{-1}. Thus, suitable conditions exist to create very high levels of turbulent energy which will tend to destroy the vertical stratification created

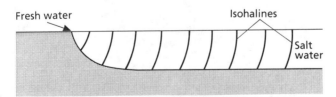

Fig. 4.06 The tidally-mixed estuary. Highly turbulent estuaries tend to be well-mixed vertically. Isohalines, or lines of equal salinity, are nearly vertical, with highest values near the sea, lowest values near the river.

by the fresh water, so that buoyancy-driven flow is eliminated. In this situation the turbulent flow that replaces the estuarine circulation is then the main agency for the transport of properties up and down the estuary. The movement of salt up the estuary, for example, is not by advection as in the estuarine flow but by turbulent diffusion.

A fourth type of estuary has a deep inner basin separated from the continental shelf by a shallow sill near the mouth. The waters in the deep basin are isolated from the waters of the continental shelf and are flushed only at irregular intervals when there is an influx over the sill of water dense enough to replace them (Gade and Edwards 1980). In the upper part of the water column, above the sill depth, fjords often exhibit the classic estuarine circulation. Areas noted for an abundance of fjords include the coasts of western Canada, Norway and Chile.

The longitudinal section of the estuary in Fig. 4.07(a) shows the bathymetry of a fjord with a sill at 'B' separating the waters of the continental shelf from the deep basin at 'A'. The density of the water at the points A and B are plotted against time in Fig. 4.07(b). In the deep basin the density decreases slowly due to vertical turbulent diffusion, which is the only mechanism tending to exchange water between the light upper layer and the denser deep water. The density of the water

(a)

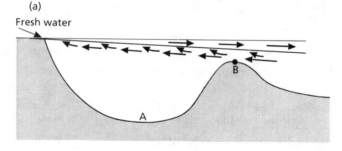

(b)

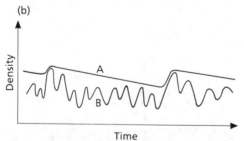

Fig. 4.07 (a) Longitudinal section of a fjord showing the sill at B and the deep interior basin. The estuarine circulation is confined to the waters above the sill. (b) The density in the deep basin at A decreases slowly with time because of vertical diffusion. The deep water is replaced and increases in density when the water coming over the sill at B is denser than the water in the deep basin.

coming into the fjord over the sill at B, on the other hand, fluctuates in time following changes in tidal currents, run-off and wind stress. When these factors cause the density at B to be greater than the density at A the inflowing water sinks down and replaces the water in the deep basin.

The influence of the tidal currents on the density of the water coming over the sill at B depends on its ability to create turbulence, which tends to mix the light outflowing water with the denser inflowing water. As the amplitude of the tide decreases between the spring and neap tides (see Chapter 7) the generation of turbulence and the degree of mixing decrease. The stratification at the sill, at this time, is more likely to be maintained and the density of the inflowing water is less likely to be diminished by turbulent mixing. Renewal of the deep water is therefore most likely during neap tides. This process is discussed by de Young and Pond (1988) in relation to a fjord on the west coast of Canada.

The density outside the sill can also be affected by wind stress at the surface which can produce an increase or decrease in the depth of the isopycnals. According to the studies of Muench and Heggie (1978), this is the main mechanism causing deep water renewal in late summer in the deep silled and low run-off fjords of Alaska. Other aspects of water exchange in estuaries induced by passing storms are addressed by Lewis and Platt (1982) and Goodrich (1988).

The rate of run-off is also an important factor limiting the inflow of dense water into fjords. A high run-off can create a thick, rapidly flowing upper layer that completely dominates the processes at the sill and which blocks any inward flow of dense water from outside. In fjords in cold climates with high run-off, deep water renewal tends to occur in the winter when the run-off is a minimum.

Let us now examine some real-life examples of the various mechanisms described above. We shall begin with the effect of the freshwater run-off, then consider the effect of tidal mixing.

4.5 THE EFFECT OF FRESHWATER RUN-OFF ON BIOLOGICAL PRODUCTION IN ESTUARIES

4.5.1 Primary production in the St Lawrence Estuary

A detailed study of the interaction of physics and biology at a station in the St Lawrence Estuary in Canada was made by Levasseur *et al.* (1984). The data were collected from June 1979 to July 1980. From December until April the water column was only weakly stratified, characterized by temperatures around 0 °C and salinities of 27–31‰. With the melting of the winter snow the freshwater run-off peaked in May and June. This, together with the spring warming, caused stratification which persisted until the following winter. However, the degree of stratification varied. The more stratified periods coincided with surface waters of lower salinity and higher temperature, and vice versa. Correlational analysis suggests that these changes were related to the monthly spring-neap tidal cycle. Turbidity was always high at the sampling station, the depth of 1% incident light intensity being always between 5 and 10 m.

Phytoplankton biomass was low from October to May and high from June to September. Naked flagellates were present in the water column all the year round but diatoms were present only from June to September and revealed a succession of dominant forms, with *Thalassiosira* and *Chaetoceros* dominant in July and *Leptocylindricus* and *Nitzschia* dominant at the beginning and end of September respectively. The role of physical factors in controlling diatom succession was investigated by discriminant analysis and it was found that the three diatom groups were mainly discriminated along the axis for which surface salinity and temperature are diagnostic. Inspection of the data revealed that the diatom bloom was not initiated until freshwater run-off and spring warming had made the mixed layer shallow enough that the mean light intensity exceeded the critical light intensity. There was a period in August when strong tidal mixing generated internal tides and broke down the stratification. At this time there was a drastic decrease in diatom numbers. Hence it is clear that from 1979 to 1980 the formation of a shallow mixed layer by the influence of freshwater run-off and surface warming, and in opposition to the effect of tidal mixing, was the primary influence controlling diatom abundance and productivity in the St Lawrence Estuary.

There was, however, a considerable delay between the first formation of the shallow mixed layer at the beginning of May and the growth of the diatom bloom in July. If our ideas developed earlier, about stratification being the prime condition for a spring bloom, are correct, why the delay? The authors suggested that since there were no diatoms in the surface waters during winter, the growth of the bloom depended on horizontal or vertical advection of cells. The most likely source would be resting spores on the bottom, and the strong stratification resulting from the spring run-off would have the effect of decreasing vertical eddy transport and thus create a barrier to the seeding of the surface layer. In the estuary, unlike in most coastal waters, diatoms are not readily available in surface waters at the end of winter.

When they considered the factors responsible for the succession of diatom species, they ruled out nutrient concentrations because the concentrations of nitrate plus nitrite, phosphate and silicate were almost always above 1.0, 0.2 and 10.0 μg-at l^{-1} respectively and were not considered to be limiting. This could be explained by the intermittent breakdown in stratification by tidal mixing, which replenished the supply of nutrients in the surface waters. On the other hand partial correlation analysis suggested that temperature had the most influence on diatom abundance during the period of summer stratification. The authors therefore concluded that there exists at the study site in the estuary a hierarchy in the relative importance of physical factors controlling phytoplankton succession (Fig. 4.08). The most important factor is the frequency of destabilization of the water column which determines the level of nutrients in the mixed layer. The second is the mean light intensity in the mixed layer, which is a function of freshwater run-off and solar radiation. The third is the mean temperature in the mixed layer. As this changes, the phytoplankton community responds by a succession of species. The cycle of stabilization/destabilization may be compared with the Gran effect mentioned in the last chapter. This is an example in which a period of mixing followed by a period of stratification leads to high phytoplankton productivity.

In a concurrent study, which took into account spatial heterogeneity in the estuary, Therriault and Levasseur (1985) showed that during the period of high freshwater run-off the freshwater plume affected the whole lower estuary and had adverse effects on phytoplankton productivity, partly, as we have discussed, by creating a barrier to seeding by

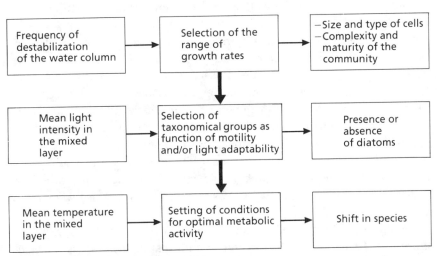

Fig. 4.08 Diagram indicating hierarchical control of phytoplankton succession by physical factors. Large arrows, direction of hierarchy; small arrows, theoretical effect of each factor on the community; last column, observed phytoplankton response in natural conditions. After Levasseur *et al.* (1984).

diatoms through vertical transport and partly by washout, i.e. rapid downstream transport of the flagellates growing in the surface mixed layer. Further downstream still, in the Gulf of St Lawrence, primary production was also under the influence of the freshwater run-off, but in this case it was enhanced because the problem of washout was less severe, while nutrient entrainment was strong and beneficial (see section 4.7.2). During the period of low run-off in summer, a spectacular enhancement of primary production occurred in the plumes of two relatively small rivers which poured zero-salinity water into one side of the estuary at a place where the main water body had a salinity of more than 26%. The stability conferred by this plume ensured that this region had the highest primary production in the estuary in summer time.

4.5.2 Primary production in Chesapeake Bay

In a recent review of the influence of physical processes on the biology of the Chesapeake Bay estuary, Brandt *et al.* (1986) demonstrated very clearly how the spring run-off at the head of the bay leads to a very strong pycnocline with seaward-flowing surface waters and landward-flowing deeper waters. The combined effect of salinity and temperature is integrated into the plots of sigma-t in Fig. 4.09. (Sigma-t or σ_t is the density of the water minus $1000 \, kg.m^{-3}$.) They show clearly the movement of the pycnocline from a region near the head of the bay in February to a layer at 8–12 m depth extending about 160 km along the bay in May. In an open-ocean situation this would pose a formidable barrier to the vertical movement of nutrients, but the strong tidal currents in combination with net mean flows in opposite directions in the two layers, results in strong internal wave activity, as revealed by acoustic records of the plankton (Fig. 4.10). The internal waves cause nutrients to be mixed into the surface waters (see section 4.4.2 on Kelvin–Helmholtz instability). The landward-flowing countercurrent is instrumental in causing both phytoplankton and zooplankton to move in a cyclical manner within the estuary. For example, Brandt *et al.* (1986) have documented the movement of high concentrations of the dinoflagellate *Prorocentrum* from near the mouth of the Chesapeake bay to the shallow waters 250 km landward, travelling with the bottom waters. At the head of the bay it was mixed up into surface waters and caused a 'red tide'.

Webb and D'Elia (1980), working in the York River Estuary, a tributary of the Chesapeake Bay system, found that on neap tides the estuary was stratified and at this time the water below the pycnocline became oxygen depleted while the nitrogen and phosphorus content rose.

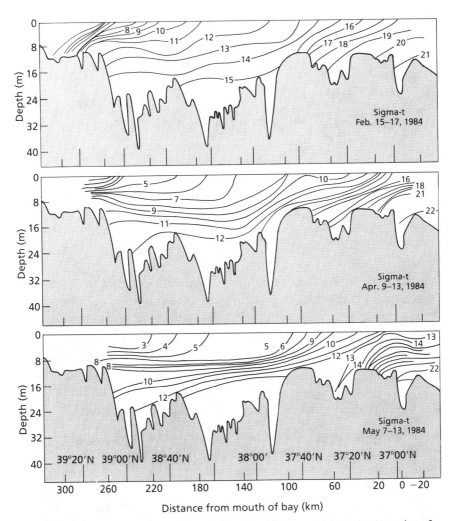

Fig. 4.09 Evolution of the pycnocline stratification in Chesapeake Bay in the spring of 1984. In February it is mainly between 200 and 300 km from the mouth. In May it extends almost the whole length of the bay. After Brandt *et al*, (1986).

A few days after the peak of the spring tides the estuary became fully mixed by tidal currents, replenishing the oxygen at depth and upwelling large quantities of nitrogen and phosphorus. There was an inverse linear relationship between nitrogen and oxygen in the water, showing that benthic organisms were adding nitrogen in proportion to the rate that they were consuming oxygen. The relationship of oxygen to phosphorus was not linear, reflecting the fact that phosphorus binds to sediment particles but is released at low oxygen concentrations. The study showed

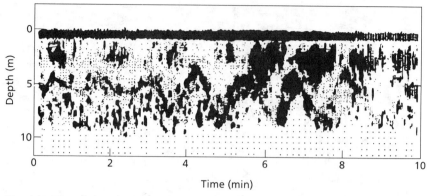

Time (min)

Fig. 4.10 Time series of acoustic backscatter intensity versus depth near Chesapeake Bay Bridge, 31 May 1984. Note evidence of vertical movement by internal waves. From Brandt *et al.* (1986).

very clearly how the balance between the stabilizing effect of freshwater run-off and the mixing effect of tidal currents can be so delicate that it shifts between neap and spring tides. This situation, alternating between upwelling of nutrients and stratification, provides the conditions for very high primary production. Once again, the parallel with the Gran effect is very striking.

4.5.3 Freshwater run-off and secondary production in estuaries

A variety of animals have adapted to the characteristic circulation of stratified and partially-mixed estuaries, with a seaward flow of low salinity water and a compensatory landward flow of bottom water. By making vertical migrations at the appropriate times they can travel upstream for part of the time and downstream for another part, thus maintaining themselves in the estuary year-round or entering and leaving the estuary on a seasonal basis. Since the net seaward and landward flows have the ebb and flow of tidal currents superimposed on them, organisms also have the possibility of making saltatory movements, e.g. moving landward on the rising tide and sitting still in the bottom boundary layer when the tide is ebbing. In this way they will progress faster upstream than if they were in the water column continuously. Bousefield (1955) invoked these mechanisms to account for the distribution of barnacles in the estuary of the Miramichi, in eastern Canada, and Grindley (1964) proposed that copepods in Southampton Water, UK, remained in the landward-flowing water by ceasing vertical migration when they encountered the lowered salinities of the seaward-flowing surface waters. Wood and Hargis (1971) demonstrated for oyster larvae in the James River Estuary (part of the Chesapeake Bay system), that the increasing salinity associated with increasing upstream flow near the bottom on each tidal cycle was used as the cue by the larvae to rise into the water column and get carried upstream. An example of seaward transport of larvae is provided by Provenzano *et al.* (1983) who found that at the mouth of Chesapeake Bay the maximum abundance of larvae of the blue crab, *Callinectes sapidus*, occurred in surface waters at night, while the salinity was falling on an ebbing tide. From this and other evidence they concluded that the crabs undergo their development offshore and reinvade the estuary as megalopa larvae or juvenile stages.

Turning to fish, Graham (1972) presented evidence to suggest that larval herring were retained within the Sheepscot estuary of Maine by travelling upstream with the landward net tidal flow near the bottom. On reaching the limit of landward penetration, they rose into surface waters and travelled with the net flow to near the mouth of the estuary before descending again into the landward net flow. A comparative study of the young stages of three fish species living in Cape Fear River estuary, North Carolina, was made by Weinstein *et al.* (1980). They found that Atlantic croaker, *Micropogonias* was concentrated near the bottom at all times and became concentrated near the head of the estuary, while spot, *Leiostomum*, and flounders, *Paralichthys*, made nocturnal excursions into surface waters, which served to carry them into the salt marsh creeks during flood tides. Flounders were thought to also use the inflowing bottom currents to make saltatory movements, as described earlier.

A particularly detailed study of the transport of fish larvae in a two-layered estuary was made by Fortier and Leggett (1982, 1983), working in the upper estuary of the St Lawrence. The dominant pelagic species are capelin (*Mallotus villosus*) and Atlantic herring (*Clupea herengus herengus*). The capelin spawn on tidal flats and beaches of the upper estuary and after emergence are flushed out towards the most productive zones of the lower estuary. The herring spawn demersally at sites downstream of the sampling location and are carried landwards in the bottom water, then retained in the upper reaches of the estuary. Fortier and Leggett (1982) showed that the initial stage of dispersal is a passive one, with the capelin larvae concentrated in the upper 20 m of the water column, while the herring larvae < 10 mm long were concentrated in the deep layer (40–60 m). Herring larvae > 10 mm made diurnal vertical migrations through the pycnocline and ceased to move systematically upstream, since their average vertical position was close to the depth of zero longitudinal velocity.

In an intensive set of hourly samples for 129 h it was found that with increasing length capelin larvae congregated closer to the surface, so that there was a length-dependent acceleration of the seaward drift. Post-larval herring made vertical migrations which brought them into the surface layer during the flood tide, thus minimizing seaward drift during their time in the upper layer. There was evidence that the lower layer was undergoing a cyclonic circulation in the upper estuary, and that the herring post-larvae were retained within it. Fortier and Leggett (1984) found that after yolk-sac absorption a coherence developed between the patterns of variation in abundance of herring and abundance of micro-plankton. The herring followed the semi-diurnal vertical displacements of the early stages of copepods over distances of a few tens of metres, but it was not possible to tell whether the total pattern of coherence was a result of herring actively searching for higher food concentrations or from differential mortality by starvation of those fish that found themselves in regions of low food concentrations.

Summarizing this section, we see that the secondary producers in an estuary are often adapted to the two-layered estuarine circulation that is found wherever the influence of freshwater run-off predominates over the influence of tidal mixing. A common form of adaptation is for the behaviour of young stages to be related to the circulation pattern in a way that optimizes their opportunities for feeding and survival. The details are complex and not fully understood. One lesson is clear, a major change in the circulation pattern of an estuary brought about by damming the freshwater flow, a tidal dam or other engineering projects may well have far-reaching effects on the primary and secondary productivity of the system. Examples will be given in section 4.8. We may note in passing that organisms are capable of adapting their life histories to circulation patterns on many scales, even up to the scale of the Gulf Stream, where eels and squid, for example, ride with the current and return by another route.

4.6 THE BIOLOGICAL EFFECTS OF TIDAL MIXING

4.6.1 Physics of tidal fronts

The idea that the turbulence generated by high tidal currents keeps the water over some shallow regions mixed all year while the quieter regions in the deeper waters become stratified in summer was first advanced more than 60 years ago. Bigelow (1927) suggested the process to explain the fact that the water over Georges Bank in the Gulf of Maine remained mixed all year while the water over the deeper regions became stratified in summer. Subsequently, Dietrich (1950) (quoted by Le Fèvre 1986) suggested that the summertime front across the western end of the English Channel exists because of an increase in the turbulent mixing in the shallower regions of the Channel. (Fronts are regions of strong gradients of temperature and other properties in the sea. The various types of front are discussed in Chapter 6.) Later Simpson and Hunter (1974) constructed an energy argument that has proved very successful in predicting the locations of these tidally-mixed fronts and before we try to understand the processes that may increase the biological activity at these fronts we first take a look in some detail at the theory which predicts the locations of such fronts.

Simpson and Hunter (1974) postulated that a front would be found where the intensity of turbulent mixing was just enough to continuously overcome the barrier to mixing presented by the stratification. They estimated the amount of energy required to mix a stratified column by calculating the difference in the potential energy of the water column before and after mixing. The details of this calculation are presented in Box 4.04. The main result of the analysis is summarized by Eq. 4.21 which can be used to locate positions of tidal fronts. One such calculation by Loder and Greenberg (1986) for the whole of the Gulf of Maine is shown in Fig. 4.12. These authors first determined the tidal velocities u and v with a numerical model, then estimated values of h/D_t. The contours of the diagram are $\log_{10}(h/D_t)$ and the transition between mixed and stratified water occurs at the contour 1.9. Values less than this indicate well-mixed conditions and values greater than 1.9 will be stratified.

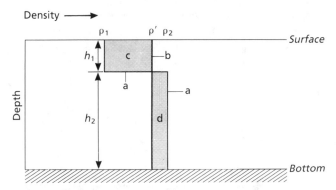

Fig. 4.11 Vertical profiles of density before and after mixing. Before mixing (a) the upper layer of thickness h_1 and density ρ_1 overlies a layer of thickness h_2 and density ρ_2. After mixing (b) the water is homogeneous at a density ρ'. The regions labelled (c) and (d) are assumed to have the same area.

Box 4.04 The Simpson and Hunter calculation

Simpson and Hunter (1974) first calculated the potential energy on the stratified and mixed sides of the front. The potential energy (PE) of an object (Box 3.02) is its mass (m) times the gravitational acceleration (g) times the height (h) of the mass's centre above some reference level, and usually written $PE = mgh$. Profile (a) in Fig. 4.11 represents the density of the stratified column before mixing while profile (b) is the profile after mixing. Before mixing the potential energy of the lower layer, whose thickness is h_2, is $mgh_2/2$, since $h_2/2$ is the height of the layer's centre of mass above the bottom reference level. The mass of the column of water is given by the density of the water times the volume which, for a column 1 metre square, is given by the height $h_2 \times 1 \times 1$. Thus the potential energy of the lower layer per square metre becomes

$$PE = \frac{1}{2}\rho_2 g h_2^2. \tag{4.11}$$

The potential energy of the upper layer is the density times the volume ($\rho_1 h_1$), times g, times the height of the centre of mass ($h_2 + h_1/2$) and therefore the total potential energy of the stratified 1 metre square column is,

$$PE_s = \frac{1}{2}\rho_2 g h_2^2 + \rho_1 g h_1 (h_2 + h_1/2). \tag{4.12}$$

After the water column is mixed (Fig. 4.11) the potential energy of the whole column is

$$PE_m = \frac{1}{2}\rho' g (h_2 + h_1)^2, \tag{4.13}$$

where ρ' is the density of the mixed water. If we assume that the equation of state for water is linear and that salt plays no role in the stratification the height of the water column will not change because of the mixing and the rectangles labelled (c) and (d) in Fig. 4.11 will be equal in area. This conservation of mass is expressed by

$$\rho'(h_1 + h_2) = \rho_1 h_1 + \rho_2 h_2. \tag{4.14}$$

From this, ρ' can be expressed in terms of ρ_1 and ρ_2 which can be substituted in Eq. 4.13 to get

$$PE_m = \frac{1}{2} g (\rho_1 h_1 + \rho_2 h_2)(h_2 + h_1). \tag{4.15}$$

Therefore we can write the difference in the potential energy between the mixed and stratified state as

$$PE_m - PE_s = \frac{1}{2} g[(\rho_1 h_1 + \rho_2 h_2)(h_2 + h_1) - \rho_2 h_2^2 - 2\rho_1 h_1 (h_2 + h_1/2)], \tag{4.16}$$

which becomes after simplifying

$$PE_m - PE_s = \frac{1}{2}gh_1h_2(\rho_2 - \rho_1). \tag{4.17}$$

If density (ρ) is a linear function of temperature (T), that is, if $\rho = \alpha T$, where α is the thermal expansion coefficient of water $(\simeq 10^{-4}\,°C^{-1})$ which we met in Eq. 3.03, then the change in density $(\rho_2 - \rho_1) = \Delta\rho = \alpha\Delta T$. Also by using Eq. 3.02 $(\Delta T = \Delta Q/mc)$ relating the temperature change to the amount of heat (ΔQ) added to the water, the mass of the water (m) and the specific heat (c) we can write $\Delta\rho$ as $\alpha\Delta Q/mc$ and then Eq. 4.17 becomes

$$\Delta PE = \frac{1}{2}gh_1h_2\alpha\Delta Q/mc, \tag{4.18}$$

and replacing the mass m by $\rho_1 h_1$ leads to the expression derived by Simpson and Hunter 1974), i.e.

$$\Delta PE = gh\alpha\Delta Q/2c, \tag{4.19}$$

where h, the total depth $(h_1 + h_2)$, has replaced h_2 which it approximately equals if we assume that h_1 is small relative to h_2.

Equation 4.19 then represents the difference in the potential energy between the stratified and mixed water columns and is equal to the amount of mixing energy which must be supplied to overcome the stratification created by the addition of the heat ΔQ into the layer of thickness h_1.

The turbulent energy which is required to keep the water mixed is generated by the friction between the tidal currents and the bottom. Following Loder and Greenberg (1986) we write the rate at which energy is extracted from the tidal currents as

$$D_t = \rho C_D <(u^2 + v^2)^{3/2}>, \tag{4.20}$$

where u and v are the horizontal components of the tidal velocity, ρ is the density of the water and C_D is a drag coefficient $\simeq 0.0027$. The angled brackets indicate that the velocity values are averaged over a tidal cycle. Not all this energy, however, will be used to mix the water. By calibrating the theory in the field, Loder and Greenberg (1986) find that only a fraction $\varepsilon_t \simeq 2.6 \times 10^{-3}$ is used to change the stratification. Thus the energy required to keep the water mixed will be provided when the energy defined by Eq. 4.19 is equal to $\varepsilon_t D_t$, that is when

$$\varepsilon_t D_t = gh\alpha\Delta Q/2c,$$

which rearranges to,

$$h/D_t = 2c\varepsilon_t/\alpha g\Delta Q. \tag{4.21}$$

By substituting $g = 10\,\text{m s}^{-2}$, $c = 4.2\,\text{kJ kg}^{-1}\,°C^{-1}$, $\alpha = 1.6 \times 10^{-4}\,°C^{-1}$, $\varepsilon_t = 2.6 \times 10^{-3}$ and $\Delta Q = 170\,\text{W m}^{-2}$, $h/D_t \simeq 80$ for the region where the stratified water meets the mixed water.

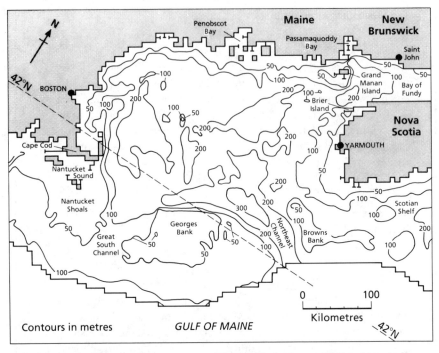

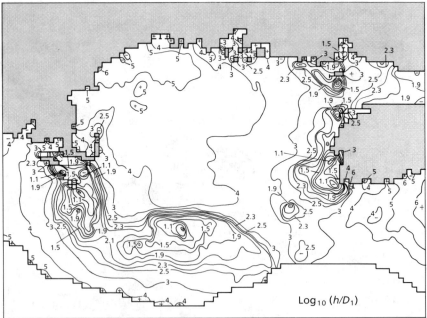

Fig. 4.12 Upper; the bathymetry and coastline of the Gulf of Maine used in the numerical model of Loder and Greenberg (1986). Reproduced with permission from Pergamon Press. Lower; the contours of $\log_{10}(h/D_t)$. Tidal fronts are expected at contours of 1.9. For explanation see text.

Pingree *et al.* (1978a) approached tidal fronts from the point of view of the turbulent-energy dissipation rate due to tidal flow (Eq. 4.10). They calculated an index $E = \log_{10}\varepsilon$ and showed (Fig. 4.13) that in areas where $E > -1.0$ the waters were tidally mixed throughout the year, and where $E < -2.0$ the waters were stratified in summer. This technique is the inverse of the Loder and Greenberg (1986) method which explains why lower values define stratified conditions rather than higher values as in Fig. 4.12.

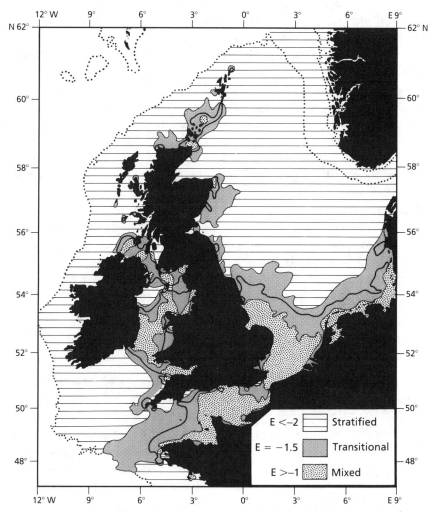

Fig. 4.13 Distribution of $E = \log_{10}(\varepsilon)$, where ε is the tidal energy dissipation per unit mass (erg g^{-1} s^{-1}). The continuous line $E = -1.5$ represents the predicted positions of frontal boundaries. Shading on either side indicates the transition zone between well-mixed and well-stratified waters. Waters that are stratified in summer are represented by light horizontal lines. Reproduced with permission from Pingree *et al.* (1978b), Pergamon Press.

Between the stratified and the well-mixed regions Pingree *et al.* (1978a) define a transition zone in which the degree of tidal mixing varies according to the stage of the lunar tidal cycle (Chapter 7). They showed lines at $E = -1.5$ which they predicted would be the average positions of the fronts between stratified and tidally-mixed waters. These predictions were, in general, confirmed by field observations. Extensive areas of tidal mixing occur in the southern North Sea, the English Channel and the southern Irish Sea. The Celtic Sea is a stratified region. Pingree *et al.* (1976) studied the development of the spring bloom in this region and found that the thermocline established first in the region of weakest tidal currents, then gradually spread to areas of stronger tidal currents. The spring bloom of phytoplankton coincided with thermocline formation. The tendency of tidally-generated turbulence to oppose stratification caused by surface heating is very clear.

Because the mixing energy of the tidal flow D_t is proportional to the cube of the tidal velocity, the criterion expressed by Eq. 4.21 is often expressed by the ratio h/U^3, where U is understood to be $(u^2 + v^2)^{1/2}$, and the fronts are referred to as '*h* over *U* cubed' fronts.

The theory of Simpson and Hunter (1974) has been refined by Simpson and Bowers (1981) and Loder and Greenberg (1986) to include the effects of wind mixing and of variable levels of mixing efficiency. Bowers and Simpson (1987) compared the various theoretical approaches to the observed positions of tidal fronts on the European-shelf seas and concluded that in shallow water the models incorporating wind mixing effects are most accurate. In deeper water the wind effects decline in importance and no theory is significantly superior to the others at predicting the front positions.

4.6.2 Tidal mixing and phytoplankton production

In light of the postulated dependence of the spring bloom on thermocline formation, what is the pattern of phytoplankton production in tidally-mixed areas where no summer stratification occurs? The lack of stratification would be expected to decrease phytoplankton productivity. On the other hand, dead organic matter is continually decomposing on and in the sediments, providing a source of nutrients that can be mixed back into the water column to stimulate phytoplankton production. Georges Bank, in the Gulf of Maine, is one such tidally-mixed area that has been intensively studied (Bumpus 1976; Backus and Bourne, 1987). Chlorophyll-a concentrations are homogeneously distributed in the water column (O'Reilly *et al.* 1981) and the depth of the euphotic layer in summer averages about 50% of the depth of the water column. Under these conditions primary production continues throughout the year, with no clearly-marked seasonal peak. O'Reilly and Busch (1984) reported fluctuating *production* levels with some indication of a peak in October. On the other hand, Riley (1941) and others had found a phytoplankton *biomass* peak in April. This is explained by the low level of zooplankton grazing at this time of year. As the season progresses, phytoplankton production remains at a high level, but most of the biomass is removed on a daily basis by the grazers.

Current estimates of total primary production (^{14}C uptake) on Georges Bank are $450\,g\ C\ m^{-2}\ y^{-1}$ on the shallow part ($< 60\,m$) and $320\,g\ C\ m^{-2}\ y^{-1}$ on the deeper part (Sissenwine *et al.* 1984). This is higher than many other tidally-mixed coastal areas in temperate waters, and the explanation offered is that a clockwise gyre (see Chapter 6) retains water on the bank, nutrient-rich water from the slopes is advected onto the bank and the continuous mixing of the waters makes nutrients regenerated from the sediments available to the phytoplankton.

Fransz and Gieskes (1984) reviewed data for both biomass and productivity in various parts of the North Sea. Their biomass data confirm the earlier results, that offshore sites within the tidally-mixed area have well-marked spring and autumn peaks while coastal sites have relatively high biomass throughout the spring and summer. Their field data (Fig. 4.14) show that coastal waters have a summer productivity almost twice as high as the offshore areas but a rapid falling off in autumn. This appears to be attributable to a larger flux of nutrients from the sediments and rivers into the tidally-mixed water column (Mommaerts *et al.* 1984) but a poorer penetration of light in the autumn on account of the sediment load. The estimated annual productivity (^{14}C uptake) for the North Sea is considerably lower than for Georges Bank, $250\,g\ C\ m^{-2}\ y^{-1}$ offshore and $200\,g\ C\ m^{-2}\ y^{-1}$ in coastal waters.

In Naragansett Bay (Rhode Island, USA) the waters are tidally mixed at all times of the year (Kremer and Nixon 1978), and this results in a marked winter–early-spring bloom of phytoplankton followed by a series of summer blooms. Regeneration of nutrients from the benthos accounts for about half the needs of the phytoplankton, the remainder coming from allochthonous inputs and from regeneration within the water column.

The conclusion from this section is that tidally-induced mixing in relatively shallow coastal waters prevents stratification of the water column, but the potentially adverse effects on phytoplankton productivity are more than compensated for by the increased nutrient flux to the water column from the sediments, so that annual primary productivity

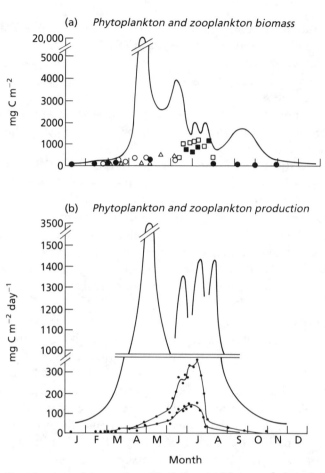

Fig. 4.14 Coastal waters of the southern North Sea. (a) Biomass of phytoplankton (line) and zooplankton (points). (b) Estimated production of phytoplankton (heavy lines) and zooplankton (fine lines). The estimates for zooplankton are of two types: 'potential production' and 'lowest estimate'. For details see Fransz and Gieskes(1984).

in tidally-mixed areas tends to be above the average for coastal waters. If the phytoplankton is being transported by turbulence throughout the water column, and is exposed to abundant nutrients, the onset of the bloom in spring depends mainly on the seasonal increase in light, and not at all on the onset of stratification. The net result is that the annual production cycle appears to start earlier in tidally-mixed areas.

4.6.3 Consequences of tidal mixing for zooplankton

We saw in section 3.4.2 that in the open ocean initiation of the spring growth in the zooplankton population is often triggered by the upward migration of a large biomass of adult and late-stage copepods that have spent the winter at depths greater than 350 m. Reproduction begins as soon as this population reaches surface waters and grazing pressure on the phytoplankton develops relatively early in the season. In part this is made possible by the rapid warming of the surface waters after stratification. In tidally-mixed waters, warming is delayed by the lack of stratification, and there is no population from deep water that can ascend to the euphotic zone and begin to feed on the phytoplankton. The net result is that tidally-mixed waters tend to have a relatively slow growth of the zooplankton population, which often does not peak until early or mid-summer. It appears that much of the biomass of the spring bloom escapes grazing by the copepods and sinks to the bottom. For example, Fransz and Gieskes (1984) determined biomass and productivity of both phytoplankton and copepods in the coastal region of the North Sea adjacent to Belgium (Fig. 4.14) and concluded that only in June and July was there a match between the productivity of the phytoplankton and the food consumption of the copepods. It should be noted that Gieskes was concerned only with zooplankton $> 50 \mu m$ and that others have noted high abundances of micro-zooplankton such as ciliates, especially in the late summer and autumn in many inshore locations.

4.7 RIVER AND ESTUARINE PLUMES ON THE CONTINENTAL SHELVES

4.7.1 Physical mechanisms

In this section we are concerned with the effect of water from either a river or an estuary pouring out onto a continental shelf, or into a semi-enclosed sea. In some situations where river flow heavily predominates over any tidal effects, the surface outflow onto the continental shelf is mainly of fresh water from the river itself, and this is called a river plume. In other situations there is strong penetration of salt water into the river valley to form an estuary, and the outflow onto the continental shelf is of river water mixed with considerable quantities of salt water. This constitutes an estuarine plume. Examples of river plumes that have been studied in an integrated biological/physical way, are the Mississippi at each of its three passes in its delta, the Amazon River, the Fraser River and the Connecticut River. Estuarine plumes are much more common, those of the Chesapeake, the St Lawrence and the Hudson River being some of the more intensively studied in North America.

When the light upper water from an estuary or a river flows out into the open ocean it leaves behind a narrow region where the flow is predominantly two dimensional and enters a less-confined world where the Coriolis force can change the direction of the flow. In the illustration in Fig. 4.15 the water moves down the estuary at a velocity u. Because this motion is relative to the earth there is a Coriolis effect directing water to the right of the flow. The water, however, is not able to change direction because of the side boundaries of the estuary but the water does move to the right causing a slight tilt to the sea surface and a pressure gradient in the direction opposite to the Coriolis force. The flow is therefore in geostrophic balance and the slope of the sea surface can be estimated in the way shown in Box 4.05 (see also Gill 1982).

When the water flows out of the estuary this surface slope is missing, so cannot provide the pressure force to balance the Coriolis force. Being unopposed, the Coriolis force causes the flow to turn to the right but it comes under the influence of the shoreline again and sets up a situation similar to the one found within the estuary. As the Coriolis force pushes the water to the right, the blocking of the coast causes an opposing pressure gradient in the form of a slight slope in the sea level to be generated against the coast, and the plume of fresher water continues on its way as a coastal current in geostrophic balance parallel to the coast. A laboratory model of this process is described by McClimans (1986).

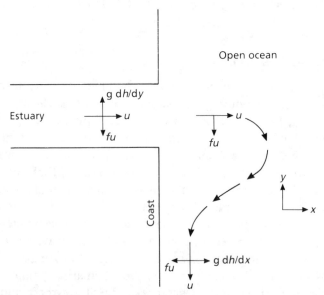

Fig. 4.15 Water moving down the estuary at velocity u is in geostrophic balance. The Coriolis force balances the pressure gradient. In the open ocean the flow is not in geostrophic balance and in the northern hemisphere the Coriolis force causes the flow to turn to the right and flow along the coast.

Box 4.05 The surface slope across an estuary

We first assume a coordinate system in which the x axis points down the estuary. Since there is no flow parallel to the y axis $v = 0$ and the balance of forces in the water is given completely by Eq. 4.03, i.e.

$$fu = -\frac{1}{\rho}\frac{dP}{dy}.$$

The pressure gradient can be written in terms of the sea surface slope by using the equation that relates pressure and water depth, i.e. the hydrostatic relation

$$P = \rho gh, \tag{4.22}$$

where P is the pressure, ρ the density, g the acceleration due to gravity (10 m s^{-2}) and h is the height of the water column. Taking the first derivative of this and inserting $dP = \rho gdh$ in Eq. 4.03 we get

$$fu = -g\frac{dh}{dy}, \tag{4.23}$$

which states that the sea surface slope in the y direction (dh/dy) is equal to fu/g. In a narrow channel where the flow may be considered constant over the whole width, the difference in the height of the sea surface from one side to the other will be $dh = fuW/g$, where W is the width of the channel. If u is 0.5 m s^{-1}, $f \simeq 10^{-4} \text{ s}^{-1}$ and $W = 200 \text{ m}$ the change in height across the estuary is $\simeq 1 \text{ mm}$.

The Coriolis turning of the water coming out of rivers and estuaries is a common phenomenon and is often observed from aircraft and in satellite photographs. The fresher water in the current is usually distinguishable from the resident water of the ocean because it has more suspended material in it. A northward flowing coastal current driven by the run-off of rivers in British Columbia is described by LeBlond *et al.* (1986).

When the water leaving a river or estuary flows into a region where there is already a current the movement of the fresh water plume is not as simple as in the case described above. The position of the plume in these cases will be the result of the relative strengths of the background flow in the ocean and the new flow entering from the river or estuary. A strong coastal flow will overwhelm a weak outflow and the plume may turn in an unexpected direction. In most cases the flows are tide dependent and the plume changes position over the tidal cycle. For example, the Connecticut River has a well-developed river plume at its mouth

(Bowman and Iverson 1978). Under light wind conditions a layer of fresh water only about 2 m deep spreads out over the coastal waters for a distance of about 10 km from the river mouth. The tendency for the plume to be turned to the right by the Coriolis force is obscured by the local tidal currents, which deflect the plume alternately to the right and to the left. The seaward flank of the plume is marked by a sharp front (see Chapter 6).

Estuarine plumes are normally of much larger volume than river plumes because the freshwater run-off has entrained considerable quantities of sea water within the estuary. For example, the Chesapeake Bay estuary discharges large volumes of water at a salinity of about 16‰ and the plume is deflected south by the Coriolis force to form a strong coastal current. In many parts of the world estuaries occur sufficiently close together around the main ocean basins that their plumes may overlap and reinforce one another to form a continuous anticlockwise coastal current in the northern hemisphere or a clockwise current in the southern. In a workshop reported by Skreslet (1986) accounts were given of such currents in Scotland, Norway, Iceland, Greenland, eastern Canada and the eastern USA and Drinkwater (1986) suggested that they were part of a nearly continuous flow round the perimeter of the North Atlantic. A modelling study has predicted the existence of a similar anticlockwise flow round the perimeter of the North Sea (Müller-Navara and Mittelstaedt 1985).

Others have pointed out that the plumes are sometimes deflected by wind stress. For example, the Hudson River plume, off New York, took up a position approximately at right angles to the coast after 3 days of southwest winds (Bowman and Iverson 1978). The same river plume also weakened under conditions of low river flow, to the point where it was temporarily obliterated by local tidal currents. More investigations are needed before the strength and variability of buoyancy-driven coastal currents can be assessed.

4.7.2 Biological effects of river and estuarine plumes

The biological effects of fresh water discharge may be considered under three headings: (a) direct effects of the materials carried by the river on biological production in the plume; (b) entrainment and consequent upwelling of nutrient-rich water, which is likely to enhance primary and secondary production; and (c) enhancement of the stability of the water column, which may be expected to enhance productivity at the time of a spring bloom, but which may inhibit vertical mixing and hence reduce primary productivity at other times of year.

(a) Effects of material carried by rivers

The nutrients carried into coastal waters by river plumes have a marked effect on productivity in a region surrounding the river mouth. The time of peak run-off of the river is normally a time of diatom bloom, and commonly the diatoms are deposited on the sea floor around the river mouth. Riley (1937) found an area of enhanced chlorophyll and phosphorus extending about 50 km to the east and south of the Mississippi delta, and about 125 km westward in the coastal current. Revelante and Gilmartin (1976) found high nutrient uptake by phytoplankton near the mouth of the River Po in northern Italy. Off the mouth of the Amazon peak diatom biomass occurs about half way across the continental shelf, when the surface salinity is about 5‰. It appears that the diatoms produced in the river plume are carried landward in the bottom water and deposited on mud banks (Milliman and Boyle 1975). This localized effect is to be constrasted with other effects, such as lowered salinities and enhanced plankton production, attributed to the presence of Amazon water as far away as Barbados (Kidd and Sander 1979).

The foregoing examples are of distinct river plumes. In estuarine plumes the story is rather different because the nutrient load carried by the river is frequently utilized within the estuary itself. On the other hand, many major estuaries around the world are the sites of dense human settlement, so that the estuaries receive a heavy load of nutrients from land drainage or sewage. Frequently the estuary does not have the capacity to assimilate this material and the estuarine plume is strongly enriched with nutrients. The plume of the Hudson Estuary, off New York is a good example (Malone 1982, 1984). The sewage input into the estuary itself is 1.6×10^5 kg N d^{-1}, which leads to a nitrogen concentration in the water of about $60\,\mu$g-at l^{-1}. Phytoplankton production within the estuary is severely light-limited on account of the high turbidity, so only a small proportion of the nutrients are utilized within the estuary and most pour out into the estuarine plume, where the amount of chlorophyll in the water column runs at 40–60 mg m^{-3} for most of the year, and may have a monthly mean close to 200 mg m^{-3} during the spring bloom period. In the spring, grazing mortality is low and most of the production sinks to the bottom. Later in the year the zooplankton biomass increases and in summer is thought to consume about 30% of the phytoplankton productivity. Malone (1982) demonstrated that the area of high phytoplankton productivity associated with the sewage enrichment is concentrated within 20 km of the mouth of the estuary. Beyond that, chlorophyll-a concentration decreases as salinity increases. A budget for nitrogen supply and utilization indicates that the area required for the phytoplankton to assimilate all of the dissolved organic nitrogen from the sewage run-off varies from $670 \, km^2$ during summer, when phytoplankton is most active, to $1350 \, km^2$ during winter. In effect, all of the nutrients derived from sewage and land run-off are utilized within that region of the New York bight close to the estuary, known as the Apex.

(b) Effects of entrainment

From the circulation pattern of a partially mixed estuary (section 4.4.3) we can see that nutrients released by decomposition of organic matter on the bottom will be carried to the surface waters when salt water is entrained by the fresh water, and will enhance primary production. In an attempt to determine the importance of this effect relative to other upwelling mechanisms, Sutcliffe (1972) produced a nitrogen budget for St Margaret's Bay, Nova Scotia. From data on primary production he concluded that there was an average total uptake by the plants of $59 \, g \, N \, m^{-2} \, y^{-1}$. From consideration of the volume of freshwater run-off and its average nutrient content he calculated that a total of $7 \, g \, N \, m^{-2} \, y^{-1}$ was supplied from this source. He estimated that $12 \, g \, N \, m^{-2} \, y^{-1}$ was supplied by zooplankton excretion and $14 \, g \, N \, m^{-2} \, y^{-1}$ from bacterial decomposition of organic matter, leaving $26 \, g \, N \, m^{-2} \, y^{-1}$ to be supplied from nitrogen upwelled from the lower layer of the two-layered estuarine circulation. He argued that this new nitrogen could be upwelled either by entrainment in the river plumes, or by wind-driven circulation. As a test of whether entrainment processes were important, he proposed to test for correlations between annual freshwater run-off and annual biological production. If the correlations were good, he would conclude that variation in river run-off significantly influenced biological production through the entrainment mechanism. If not, he would assume that wind-induced upwelling was the major mechanism supplying the required nitrogen.

Unfortunately, information on year-to-year variation in freshwater run-off into St Margaret's Bay was not available. Undeterred, Sutcliffe turned to the St Lawrence River, for which excellent data on run-off were available over a long period. He showed (Sutcliffe 1973) that there was an excellent correlation between the discharge of the St Lawrence River in April and the landing of lobsters in the Gulf of St Lawrence 9 years later. Since lobsters take about 9 years to grow to marketable size, he concluded that in a year of high river run-off there was good survival of lobster larvae, and that this was reflected in good catches of that year-class 9 years later. From this it was concluded that the upwelling of nutrients associated with entrainment in the estuarine plume is a major factor influencing biological production in any given year.

The correlation has held good for all of the intervening 15 years (Drinkwater 1987), probably one of the best correlations of its kind anywhere. However, Sutcliffe thought he was making a simple choice between entrainment and wind-induced upwelling, but in recent years a third possible mechanism has been invoked. As we saw in earlier parts of the chapter, freshwater run-off can lead to earlier stratification of the water column and to greater resistance to vertical mixing. As a result, the surface layer is likely to become warmer in summer, and this may easily affect the productivity of the plankton and the survival of lobster larvae. In the Gulf of St Lawrence it seems that entrainment may be the key factor influencing plankton productivity in one area, while stabilization of the water column is more important in another (Bugden *et al.* 1982; Sinclair *et al.* 1986).

In places where estuaries discharge onto the continental shelf it has repeatedly been shown that deeper waters are drawn towards the mouth of the estuary from a considerable distance. Norcross and Stanley (1967) found that bottom drifters released 70 km seaward of the mouth of Chesapeake Bay (Fig. 4.16) were consistently drawn towards the mouth of the bay, and Pape and Garvine (1982) found a similar circulation off Delaware Bay. It is therefore reasonable to assume that nutrients regenerated into this bottom water are carried towards the coast and are eventually entrained into the estuarine plume, thus increasing biological productivity. Direct evidence of the magnitude of this effect is lacking mainly because of the difficulty of defining the boundaries of a plume which changes its direction under the influence of wind and tides (section 4.7.1).

(c) Effects of fresh water on stability of water column

We have mentioned in passing that addition of fresh water to salt water causes earlier stratification and greater resistance to breakdown of thermal stratification by tidal or wind-induced mixing. This leads naturally to the idea that in temperate latitudes the spring bloom may start earlier close to the coast where freshwater influence is most strong.

In the St Lawrence Estuary the presence of a low salinity surface layer inhibited mixing in the water column and favoured phytoplankton production in the spring (section 4.5.1). The effect was particularly dramatic where a plume of fresh water was superimposed in the brackish water in the lower estuary. In this plume were found some of the highest

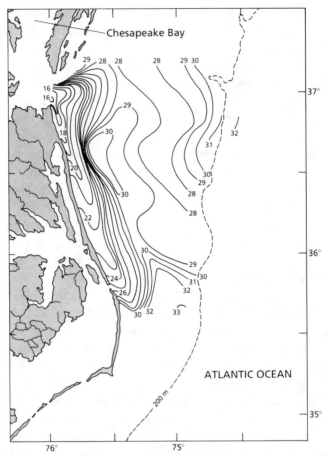

Fig. 4.16 The Chesapeake Bay plume, as revealed by salinity contours. The mouth of the bay is near the upper left corner, and the plume is directed south, close to shore. After Garvine (1986).

concentrations of chlorophyll in the whole estuary. It is possible that similar effects occur where river or estuary plumes flow out onto continental shelves.

Thordardottir (1986) explained interannual variations in the timing of the onset of the spring bloom of phytoplankton off the coast of Iceland by interactions between freshwater run-off and the wind regime. It is interesting to note that when the primary production is averaged for the years 1958–1982 there are several distinct seasonal patterns. At the station furthest from shore there is little primary production in March and April, a single major bloom in May and relatively low production for the rest of the season (Fig. 4.17). Close to shore, production begins in March and remains at a relatively constant level through the summer. At intermediate distance from shore there is a

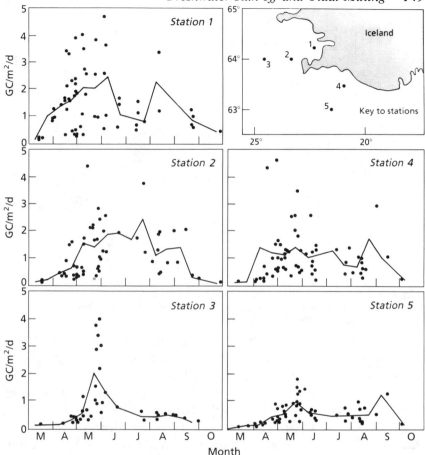

Fig. 4.17 Data on integrated photic zone primary production, pooled for the years 1958–1982 at five stations off the southwest coast of Iceland. Note the tendency for an earlier start to the spring bloom, and higher summer production, at stations close to shore. After Thordardottir (1986).

pattern with more or less equal blooms in spring and autumn. When the geographical distribution of the rate of primary production was plotted (Fig. 4.18) using data averaged over the years 1958–1980, it was clear that the phytoplankton bloom normally occurred first in a bay on the southwest corner of Iceland known as Flaxafloi. In discussing the factors determining the time of the bloom along different parts of the coast of Iceland, Thordardottir (1986) was able to rule out light and nutrients, both of which were non-limiting from March onwards. As we might expect, the timing of the bloom was clearly related to the formation of a shallow mixed layer. Detailed analysis of the temperature and salinity profiles at various stations showed that the stratification associated with the beginning of bloom conditions was attributable to reduced salinity

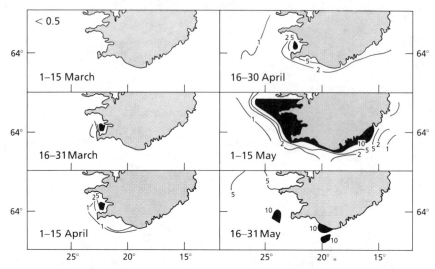

Fig. 4.18 Seasonal progression of the spring bloom off the southwest of Iceland. Contours show primary production (mg C m^{-3} h^{-1}) at a depth of 10 m. Pooled data 1958–1980. After Thordardottir (1986).

caused by freshwater run-off, rather than to surface warming. Breakdown of stratification was associated with strong winds. Hence the explanation of the pattern of primary production in time and space shown in Fig. 4.18 is that the bloom normally begins first in Faxafloi which receives surface run-off and is sheltered from winds on three sides. It then begins close to shore along the south coast, where stratification is facilitated by freshwater run-off, but is delayed somewhat by wind mixing. The bloom begins last of all in the open ocean where there is negligible influence of freshwater run-off, and stratification depends on surface warming.

The general picture for Iceland is that there is a mean residual surface current that flows clockwise round the island, with a speed of 0.06–0.11 m s^{-1} in the area south and west of Iceland. The rate of flow shows considerable variation according to wind conditions and the freshwater run-off. As we saw earlier, the detailed mechanism driving such currents is still under investigation, but it is clear that freshwater run-off is an important factor. It follows that in other parts of the world where coastal currents associated with freshwater outflows are recognized, it is probable that the additional stability conferred by the lowered salinity is instrumental in bringing about an earlier start to the primary production cycle, at least in temperate and sub-Arctic environments. A systematic investigation of the phenomenon has not yet been made, but the data from the Norwegian Coastal Current in spring, provided by Peinert (1986) indicate that the same phenomenon is to be

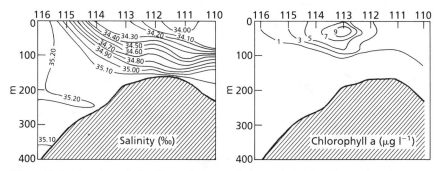

Fig. 4.19 Section through the Norwegian coastal current in spring. For explanation see text. After Peinert (1986).

found there (Fig. 4.19). At the time of the cruise, offshore waters (to the left of the figure) were not yet stratified. The chlorophyll maximum was present at station 113, and according to the author, the bloom in the lowest salinity waters (station 111) had already passed.

We now see that there are two mechanisms that lead to geographic migration of the spring bloom in temperate latitudes. The movement of the spring bloom from south to north in the northern hemisphere has been shown (section 3.3.6) to be associated with surface warming as indicated by the outcrop of the 12 °C isotherm, and it now appears that the presence of river and estuarine plumes in coastal waters leads to a migration of the spring bloom from the coast out to deeper water as the season progresses.

4.7.3 Effects of river and estuarine plumes on secondary production

In section 4.7.2(b) brief reference was made to Sutcliffe's (1973) discovery that there was a good correlation between river run-off into the Gulf of St Lawrence and the landing of lobsters 9 years later. This implies that the St Lawrence estuarine plume has a profound effect on the survival of lobster larvae, yet in the intervening years little progress has been made in unravelling the mechanism of the interaction. In Iceland, there is also a commercial interest in the effect of freshwater run-off, because the main spawning grounds of the rich Icelandic fish stocks are off the south and southwest coasts, in those areas most affected by fresh water. The early stages of copepods are an important component of the diet of first-feeding cod larvae and it has been found that the years in which zooplankton densities were highest were the years in which phytoplankton production started early under favourable conditions provided by freshwater-induced stratification (Thordardottir 1986).

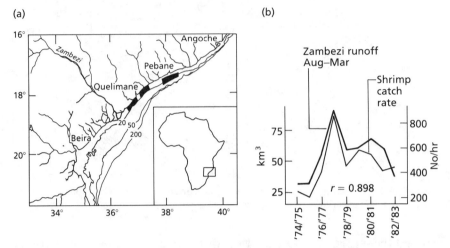

Fig. 4.20 (a) Map of part of the Zambezi River and adjacent coastline showing productive shrimp grounds (shaded) and depth contours, (b) Plots of Zambezi run-off (km^3) and catch-rate of shrimps. Pooled data, 1974–1983. After da Silva (1986).

Another interesting situation of commercial importance is the Zambezi River, which discharges into the waters of the east coast of Africa in the southern hemisphere and has an estuarine plume which turns northwards along the coast. The region of this plume is marked by high concentrations of organic matter in the sediments and by very productive fisheries for the shrimp *Penaeus indicus*. Da Silva (1986) has demonstrated that there is a very high correlation between the catch-rate for the shrimps (taken as an index of abundance) and the run-off of the Zambezi from August to March (Fig. 4.20). Once again, the mechanism is not clear. The shrimps use the two-layered flow of the estuarine plume to make shoreward and offshore migrations during their development, but it is not clear which part of the life cycle benefits from high run-off.

In general it seems that our understanding of the relationship of the connection between river run-off and secondary production is very much at the correlational and descriptive phase, and that much more work is needed to understand the mechanisms at work.

4.8 EFFECTS OF MAN-MADE MODIFICATIONS TO RIVER RUN-OFF

The ecological effects of man-made changes in freshwater outflow to coastal waters have recently been reviewed by Drinkwater (1988). In many parts of the world major rivers have been dammed to provide hydroelectric power generation. Since the economic ideal is to have a year-round constant supply of electricity, the tendency is to store water at times of high flow and release it during times of low natural river flow. This has the effect of modifying the seasonal pattern of river run-off to which the organisms have become adapted over long periods of time, often with major disruption of biological events in coastal waters. In warm dry climates storage of water in reservoirs is associated with considerable loss by evaporation, so that the total annual run-off is considerably diminished.

4.8.1 The Black Sea

A well-documented example is that of the Dniester and Dnieper Rivers discharging into the Black Sea (Tolmazin 1985). Major hydroelectric projects were begun in the 1950s and more-or-less completed by the early 1970s. Figure 4.21 shows the seasonal pattern of the run-off of the Dnieper before and after the construction of dams. The loss of a major spring peak, and its replacement by a series of much smaller peaks is very evident. Accompanying this was a major loss of water quality, since much of the water from the reservoirs was used in agriculture, industry and by municipalities, returning to the system loaded with various contaminants.

The lowering of the amplitude of the peak discharge of the Dnieper River resulted in the low-lying marshes in the Dnieper estuary not being

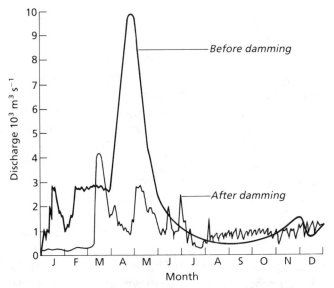

Fig. 4.21 Run-off of the Dnieper River before and after the construction of a series of hydroelectric dams. Note the major reduction in the spring peak of run-off. After Tolmazin (1985).

covered during the period of peak fish migration. The average salinity in the Dnieper estuary increased, with an upstream extension of the near-bottom high salinity layer. These environmental disruptions led to the disappearance of many species of zooplankton and fish from the estuary. Commercial landings of fish from the Dnieper estuary decreased by a factor of five, and in the Dniester River they dropped to close to zero.

Nutrients from agricultural run-off and sewage led to greatly increased phytoplankton production both in the estuaries and in their plumes. The phytoplankton biomass was largely ungrazed and sank to the bottom. With the increased stratification caused by increased summer freshwater discharge, oxygen concentrations in the lower layers were reduced to zero and mass mortalities of invertebrates and fish resulted. Oxygen depletion had occurred near the estuarine plumes even before the construction of hydroelectric dams, but the anoxic events became more frequent and widespread, resulting in a sharp decrease in the catches of turbot, flounder and crab.

A lesson to be learned from this case is that adverse biological effects are not necessarily caused by factors which decrease biological production. The total primary productivity in the estuaries and their plumes increased, but the changed distribution of salinity and oxygen killed the invertebrates and fish that had adapted over long periods to the more natural conditions. The primary production was consumed mainly by bacteria, and the commercial fish landings were severely reduced.

4.8.2 San Francisco Bay

Another example quoted by Drinkwater (1988) is the extensive modification of the Sacramento–San Joaquin river system which flows into northern San Francisco Bay. The unmodified discharge of these rivers was about $34 \, km^3$ per year. Of this, about 40% is now removed for local consumption, and 24% diverted to central and southern California, leaving only 36% entering the estuary (Nichols *et al.* 1986). Correlated with these changes there has been a marked reduction in primary production in northern San Francisco Bay, which is accentuated during summers of abnormally low flow. Two mechanisms for this drop in production have been suggested. Cloern *et al.* (1983) pointed out that stratified and partially-mixed estuaries normally have a zone of high turbidity around the region of maximum penetration of the deep saline water, and suggested that reduced freshwater flow allowed the zone of high turbidity to penetrate further upstream than before so that primary production was reduced on account of lower light penetration. Nichols (1985) pointed out that during years of particularly low freshwater flow benthic filter feeding animals associated with the more saline water penetrated further upstream into the estuary and may have reduced the phytoplankton biomass by their grazing. Whatever the mechanism, reduced river flow into northern San Francisco Bay is correlated with reduced primary production, and this in turn with reduced secondary production. The adult population of striped bass has decreased by 75% since the mid-1960s and the chinook salmon population by 70% since the early 1950s (Stevens *et al.* 1985; Kjelson *et al.* 1982). In the summer of 1977, when freshwater run-off was exceptionally low, the phytoplankton biomass dropped to 20% of normal levels, zooplankton and shrimp abundance decreased, and the striped bass population declined severely. The possible mechanisms involved include: (i) a decline in primary production as a direct result of reduced entrainment by the fresh water plume; (ii) degradation of fish spawning areas caused by the upstream migration of the turbidity maximum; and (iii) accumulation of higher concentrations of pollutants, resulting from reduced dilution by the freshwater run-off. At present it is not possible to tell which of these is the most important factor. The correlation of reduced run-off with reduced fish production is confirmed by examination of the historical records, which show, for the years before diversion (1915–1944), a strong correlation between mean annual run-off and commercial catches of salmon, striped bass and shad. These results are an interesting parallel with those of Sutcliffe (1973) for the Gulf of St Lawrence.

As Drinkwater (1988) points out, these examples are not the only ones available. Though less thoroughly investigated, there are clear examples of alterations in biological productivity resulting from changing river run-off at the mouth of the Nile (Egypt), in the Sea of Azov and the Aral Sea (USSR), from the Santee River, North Carolina, and from the Zambezi in southern Africa. We are left in no doubt whatsoever that freshwater run-off is a major factor influencing biological productivity in coastal waters, and that human perturbations usually have the effect of reducing biological productivity.

4.9 THE EFFECT OF WIND ON VERTICAL STRUCTURE

As has been mentioned in earlier sections, wind blowing across the sea surface sets up a stress which causes water to begin moving in the same direction as the wind. The Coriolis effect soon modifies the direction of flow, causing the current to veer to the right in the northern hemisphere, to the left in the southern. When the situation has persisted for some time the net movement of the water is at right angles to the direction of the wind, and this is known as Ekman transport. When the direction of

Ekman transport is away from the coast, surface waters move offshore and their place is taken by deeper water that upwells close to shore (see Chapter 5). In situations where that deeper water is nutrient-enriched, this upwelling serves as a stimulant to primary production.

There are parts of the world where upwelling is a continuous process for many months of the year. These distinctive upwelling systems will be treated in Chapter 5. However, in the context of the present chapter it is instructive to see how coastal waters that are normally stratified in summer as a result of freshwater run-off and seasonal warming can have their vertical structure strongly modified by intermittent wind-driven circulation. In Nova Scotia, Canada, the circulation of water in numerous coastal inlets was studied between 1934 and 1953 and summarized by Platt *et al.* (1972). It was clear that Ekman transport caused major changes in the vertical structure of the water column in every year of study. The following is a typical example: winds blew from the south and southwest, roughly parallel with the coast of Nova Scotia, for about 10 days in early September 1967. The surface temperature in St Margaret's Bay dropped from 15.6 to 6.9 °C in less than 7 days, while the surface salinity increased from 29.36 to 31.26% and the thermocline was destroyed. The nitrate content of the cold water was more than double that of the warm surface water that it had replaced. An increase in primary production occurred at this time. After 11 September, winds blew mainly from the east and northeast, the surface temperature changed back to 14.6 °C, and the thermocline was re-established. From the records back to 1924, Platt found that similar incidents had occurred in late summer and autumn every year, and the changes in temperature structure were coherent in 10 different inlets along the coast. Heath (1973) constructed a simple two-layered physical model of the system and also found good correlations between the observed flows and atmospheric phenomena. He suggested that the upwelling should be of general occurrence in bays on other coasts subjected to changes in wind and air pressure associated with the passage of atmospheric disturbances.

In this example, which is only one of many that could be given, the effect of the wind is to cause upwelling of nutrient-rich water and stimulate primary production. In section 4.7.2(c) we saw that off the coast of Iceland, wind-driven mixing delays the onset of the spring bloom by preventing the formation of a shallow mixed layer under the influence of freshwater run-off. In the previous chapter we saw that shallow mixed layers formed under the influence of solar warming can be broken down by wind mixing. In coastal waters, then, we find four main factors influencing the vertical structure of the water column: (i) warming and cooling which result from seasonal changes in solar radiation; (ii) freshwater run-off, which tends to produce a low-density surface layer and intensify vertical stratification; (iii) tidal currents which tend to generate turbulent mixing from below, and may prevent the formation of a thermocline or halocline; and finally (iv) wind-driven currents which may lead to coastal upwelling, or which may generate enough turbulence in the surface layer to break down or prevent stratification. No general statements can be made about the way in which these factors interact. At some times of year the formation of a shallow stratified layer enhances primary production, while at others it depresses it. Situations in which shallow stratified layers are formed for a time but intermittently broken down are frequently associated with high levels of primary production.

4.10 THE EMERGENCE OF GENERAL PATTERNS

Cushing (1975) drew attention to the work of Colebrook and Robinson (1965) who had summarized the results from continuous plankton recorders towed behind ships-of-opportunity moving in and out of UK ports over many years. They had found four main seasonal patterns of phytoplankton biomass (Fig. 4.22). These were: (i) the *oceanic* cycle, with a pronounced spring bloom in May and a low-amplitude peak in September–October; (ii) the *shelf* cycle, similar to the oceanic cycle but starting earlier in the year; (iii) a *bank* cycle, characteristic of the central North Sea, which has an early spring bloom and an autumn bloom of approximately equal magnitude; and finally (iv) a *coastal* cycle, characteristic of coastal waters of the southern North Sea, having a very early spring bloom and variable but relatively high biomass throughout the summer.

Since Colebrook and Robinson (1965) drew attention to these patterns, examples have been found in many parts of the world. The *oceanic* type is found in almost all offshore waters in temperate climates, and the initiation of the spring bloom is now understood in terms of the shallowing of the mixed layer under the influence of spring warming, as permitted by the relaxation of wind-induced mixing. The earlier start of the spring bloom in the *shelf* cycle is, as we saw in relation to Iceland, partly a function of freshwater run-off and partially dependent on the amount of shelter from prevailing winds afforded by the local topography. The maintenance of a relatively high level of production throughout the summer, as in the *coastal* cycle, is made possible in part by

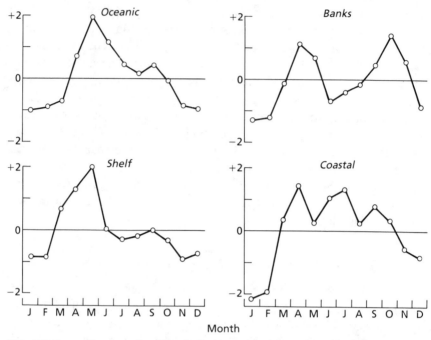

Fig. 4.22 Seasonal patterns in chlorophyll abundance around the British Isles. Data were normalized to give a zero mean and unit variance, for purposes of comparison. Vertical axis: standard deviations. Modified from Colebrook and Robinson (1965).

tidally-induced mixing, which constantly replaces the nutrients in surface waters, and may also be sustained by the pattern of circulation characteristic of stratified estuaries. The *bank* cycle, with its spring and autumn peaks of biomass, separated by a summer period where biomass is low, appears to be characteristic of areas, such as Georges Bank, where the water column is tidally mixed, but where the summer biomass of phytoplankton is held to low levels by an abundant zooplankton. In places where there is a very large spring bloom that is not consumed by the zooplankton, there appears to be a massive downward transport of phytoplankton, which may be used by the benthos or even carried to areas where it accumulates and presumably undergoes bacterial decomposition.

PART B
PROCESSES ON A SCALE OF 1–1000 KILOMETRES

5

Vertical Structure in Coastal Waters: Coastal Upwelling Regions

5.1 INTRODUCTION

A theme running through the two preceding chapters on vertical structure is that the key to high biological productivity is the upwelling of 'new' nutrients from deep waters into the euphotic zone and the retention of phytoplankton in well-lighted waters by stratification of the water column. We have seen that optimum conditions for phytoplankton production can be produced by patterns of alternating convective mixing and thermal stratification, by periods of strong tidal mixing alternating with stratification caused by freshwater run-off, or by some combination of these. Wind-induced mixing modifies these patterns by breaking down the stratification, but commonly the stratification is re-established soon after the wind abates.

In this chapter we consider the special places where wind-induced upwelling is the dominant mechanism for bringing new nutrients to the surface. These upwelling areas have been well studied, partly because they are associated with economically important fish stocks. Cushing (1971) in a review entitled 'Upwelling and the Production of Fish' estimated the total production of fish and squid in upwelling areas in the late 1960s. He calculated that over 26 million tonnes, mainly sardines, were produced in the Benguela current system off the southwest coast of Africa and the Canary current system off northwest Africa, over 12 million tonnes in the Peruvian anchoveta fishery, over 5 million tonnes of anchovy and hake in the California current system, and so on. It is impossible that all of this production could be harvested, but for comparison we may note that this weight of fish is equivalent to more than half of the world's annual commercial fish landings. If reckoned in monetary value, the percentage would be much lower because these species have traditionally been used primarily for the production of fish meal. What is the explanation of this enormous biological production? The subject has been extensively reviewed by Boje and Tomczak (1978), Richards (1981) and Barber and Smith (1981).

The basic concept of wind-induced upwelling was mentioned in section 4.9. Wind blowing across the surface of the sea causes water to begin to move in the same direction. The Coriolis force, resulting from the earth's motion, causes the current to deviate to the right in the northern hemisphere, to the left in the southern. After some time, the net movement of surface water is at right angles to the direction of the wind, and this is Ekman transport. When Ekman transport is away from the coast, surface waters moving away from the shore are replaced by deeper water that upwells close to shore. This water is normally nutrient-rich and primary production is stimulated.

As we shall discuss in more detail in Chapter 8, the major ocean basins each have a western boundary current that tends to be fast and deep, and an eastern boundary current that is broad, shallow and less well defined. Winds of the appropriate strength and direction to cause upwelling are more prevalent on the eastern sides of ocean basins. Nutrient-rich water below the pycnocline is closer to the surface in these areas, and is therefore more readily upwelled, giving a boost to phytoplankton production. Wind-driven upwelling does occur on the western sides of ocean basins, but tends to be of short duration and, because the pycnocline tends to be deeper, often brings up water from above the pycnocline that is not nutrient-rich. Such upwelling is biologically less important.

World-wide, there are five major coastal currents associated with upwelling areas (Fig. 5.01): the California current (off Oregon and Cali-

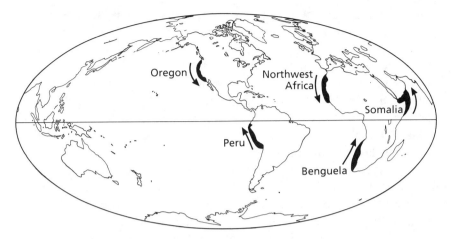

Fig. 5.01 Major coastal upwelling regions of the world, adapted from Thompson (1977). Arrows indicate prevailing winds.

fornia), the Peru current, the Canary current (off northwest Africa), the Benguela current and finally a rather anomalous example, the Somali Current in the western Indian Ocean. Here the southwest monsoon sets up Ekman transport away from the coast of Somali and the Arab states: this is the only major upwelling area occurring on the western side of an ocean basin. The first four of these are plainly visible on our satellite image (see Plate 1).

As we explore the processes occurring in these upwelling areas we shall find that there is great variability. For example, off Peru the upwelling continues more or less year-round but there are years when biological production almost ceases. Off Oregon and Portugal there are large seasonal changes in the strength of upwelling, clearly related to seasonal differences in wind strength and direction. Even within the 6-month season of upwelling off Oregon, there are four or five periods of strong upwelling separated by periods of little or no upwelling. Each of these gives rise to a burst of productivity equivalent to a spring bloom (Barber and Smith 1981). On still shorter time scales, the diurnal build-up of an onshore seabreeze, and the semi-diurnal patterns of tidal current give rise to corresponding fluctuations in biological productivity.

5.2 THE PHYSICS OF COASTAL UPWELLING

5.2.1 The Ekman spiral

When the wind blows over the surface of the water it clearly generates waves but it also drags the water along. This is how wind-driven currents begin. But because the transfer of momentum from the air to the water occurs under the influence of the Coriolis force, the direction of the water movement is not the same as that of the wind. This was first pointed out in a theoretical analysis by V. W. Ekman in 1905 and now most phenomena connected with the process have his name attached such as Ekman spiral, Ekman layer, Ekman drift, Ekman pumping, etc. Our main concern in this chapter is with the Ekman drift which creates the coastal upwelling zones. However, we start with the Ekman spiral which is the name given to the arrangement of the currents generated in the upper layer of the ocean by the wind.

We begin with an elementary physical explanation based on the simple situation outlined in Fig. 5.02. This shows the adjustment of the water at the surface to the drag of the wind. Each stage of the adjustment is illustrated on the left with a diagram of the horizontal forces on the water. A separate diagram of the resulting current is on the right. In stage I the wind is blowing parallel to the positive x axis and begins to drag the water along in the direction of the positive x axis with speed V. (Note that the force and velocity scales are not the same.) The friction in the water creates another force, the water drag, opposing this current, and the rotation of the earth generates the Coriolis force directed 90° to the right of the current. These three forces are clearly not balanced, in stage I, as there is no force opposing the Coriolis force and the drag of the wind along the positive x axis is greater than the drag of the water along the negative x axis.

In the second step the wind and its drag remain in the same direction along the positive x axis but the Coriolis force begins to cause the direction of the surface current to rotate around to the right of the wind. The Coriolis force, always perpendicular to the current, also rotates with the current as does the water drag which is always directly opposed to the current.

In stage III the forces have come into equilibrium. The component of the Coriolis force along the negative y axis is balanced by the component of the water drag along the positive y axis. The drag of the wind along the positive x axis is balanced by the sum of the components of the Coriolis force and the water drag along the negative x axis. The balance is attained when the surface water is moving at 45° to the right of the wind. This angle is a consequence of assuming that the Coriolis and the drag forces are the only forces acting on the water and that they are equal in magnitude. Ekman's theoretical analysis is developed more fully in Box 5.01.

The water just beneath the surface is not dragged directly by the wind but by the surface water which lies directly above and which is moving at 45° to the right of the wind. An adjustment of the forces in

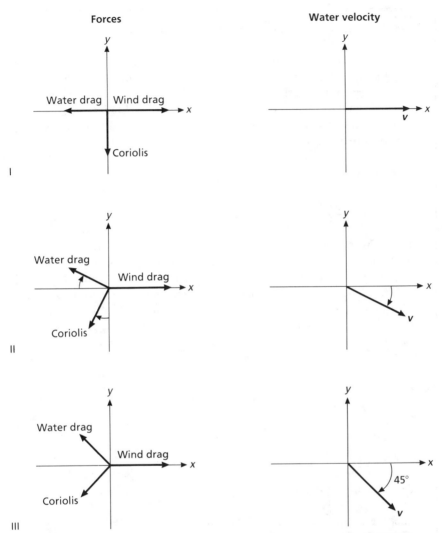

Fig. 5.02 The beginning of a wind-driven surface current in three stages showing the forces on the left and the water velocity on the right. In stage I the wind drag creates a flow of water which gives rise to the water drag and the Coriolis force. In stage II the Coriolis force causes the current in the water to rotate around to the right in the northern hemisphere. The force due to the water drag and the Coriolis force rotate with the current in the water. In the final stage the current has rotated the amount required to have the force due to the wind drag balanced by the combined effects of the Coriolis force and the drag of the water.

this second layer like the adjustment in the top layer causes the flow to be slightly to the right of the surface water, that is at an angle greater than 45° to the right of the wind. This process continues down through

Box 5.01 Mathematical derivation of the Ekman spiral

Mathematically the Ekman spiral shown in Fig. 5.02 is derived (Pond and Pickard 1983) by first assuming that the only forces involved in the process are the Coriolis force and the frictional drag force. Forces arising from horizontal pressure gradients are assumed to be negligible. The balances between Coriolis and drag forces, along the x and y axes are given by the equations

$$fv = -K_v \frac{d^2u}{dz^2},$$
(5.01)

$$fu = K_v \frac{d^2v}{dz^2},$$
(5.02)

where f is the Coriolis parameter, u and v are the x and y components of velocity, z is the depth and K_v is the vertical eddy diffusion of momentum which was introduced in section 2.2.7. The terms on the right of these two equations for the frictional drag are the same ones discussed in the previous chapter in relation to estuarine circulation (Eq. 4.08). Ekman solved these equations for u and v and found that the velocity is described by the spiral illustrated in Fig. 5.02. The equations for the u and v components of velocity are

$$u = V_0 \cos\left(\frac{\pi}{4} + \frac{\pi z}{D_E}\right) \exp\left(-\frac{\pi z}{D_E}\right),$$
(5.03)

$$v = V_0 \sin\left(\frac{\pi}{4} + \frac{\pi z}{D_E}\right) \exp\left(-\frac{\pi z}{D_E}\right),$$
(5.04)

where V_0 is the water velocity at the surface, D_E is the thickness of the Ekman layer and z is the depth which is positive downward. At the depth D_E the magnitude of the velocity in the spiral is $V_0 e^{-\pi}$ or $\simeq 4\%$ of the surface value. The direction of the current at D_E is directly opposite to the surface current V_0. Pond and Pickard (1983) show that the depth of the Ekman layer can be related to the wind speed by the equation

$$D_E \simeq 4.3 \, W/(\sin \phi)^{1/2},$$
(5.05)

where W is the wind speed in m s^{-1} and ϕ is the latitude. Thus for a given wind speed the depth of the Ekman layer increases from the pole to the equator where it is infinitely deep. One example in Pond and Pickard (1983) gives an Ekman depth of 100 m at 10° latitude for a 10 m s^{-1} wind. The same wind conditions give a depth of 50 m at 45° latitude.

the water column until all the momentum that is transferred from the wind to the water is converted to motion in the water. The speed of the water in each layer gets progressively less with depth and turned more around to the right in the form of a spiral — the Ekman spiral. A horizontal projection of the currents at 11 equally-spaced levels between the surface and the bottom of the Ekman layer is shown in Fig. 5.03. This bottom limit of the Ekman layer (D_E) is also discussed in Box 5.01.

The form of the velocity spiral shown in Fig. 5.03 requires that the friction in the water, K_v, be a constant throughout the depth of the Ekman layer. This is a poor assumption in most regions of the upper ocean because of the vertical stratification which strongly affects K_v. It is also difficult to extract from velocity observations the part that is due only to the wind. For these reasons the spiral is a difficult phenomenon to observe in the ocean. However, Stacey *et al.* (1986), Richman *et al.* (1987) and Price *et al.* (1987) have been able, through the analyses of long current meter records obtained in the upper layer of the ocean, to demonstrate a spiralling of the flow with increasing depth which is in close agreement with the theory.

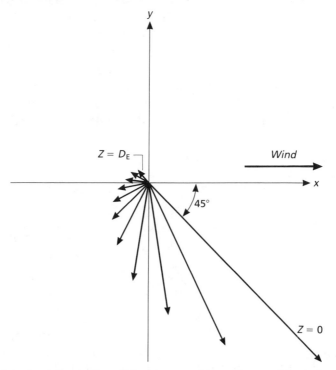

Fig. 5.03 A horizontal projection of the currents at 11 equally-spaced levels from the surface to the bottom of the Ekman layer (D_E). The currents are generated by a wind blowing parallel to the positive x axis. For explanation, see text.

5.2.2. Ekman drift and coastal upwelling

One important consequence of the velocity spiral is that the net movement of the wind-driven flow, after averaging over the Ekman layer, is 90° to the right of the wind. This result, which is one of the foundations of the wind-driven circulation in the oceans and of coastal upwelling, is derived mathematically by Pond and Pickard (1983). The result may also be derived, in a less rigorous manner, from the simple schematic diagram given in Fig. 5.04.

Here we assume that at the bottom of the Ekman layer the frictional forces which transfer the momentum of the wind down into the water become very small. This follows because the change in the velocity gradient (i.e. the terms on the right of Eqs 5.01 and 5.02), becomes vanishingly small at this depth. This means that all the wind energy transferred to the water is confined to the Ekman layer. Consequently the whole layer can be treated as a slab which absorbs all the momentum transferred from the wind and which moves without friction over the ocean. This simplification of averaging over the layer of frictional influence (the Ekman layer) allows us to ignore the details of the velocity distribution. We need only consider the force due to the drag of the wind and the Coriolis force, as illustrated in Fig. 5.04.

The wind drag shown in the figure as acting on the whole of the Ekman layer tries to pull the water in the direction of the wind. The Coriolis force which is the only other force involved must balance the wind drag and therefore must be equal in magnitude to the wind drag and in the direction opposite to the wind. The flow of water which gives

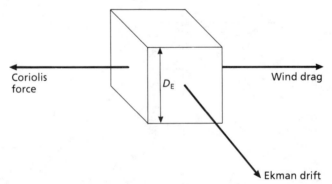

Fig. 5.04 A cube of water in the Ekman layer treated as a layer without frictional coupling to the remainder of the ocean lying below. The force due to the wind drag is assumed to act on the whole cube rather than just at the surface and is balanced by the Coriolis force which is generated by the Ekman drift moving perpendicular to both forces.

rise to the Coriolis force must be 90° to the right of both the Coriolis force and the wind. This flow is the Ekman drift. The reduction of the Ekman spiral to its net effect of a current perpendicular to the wind allows us to easily calculate the transport of water due to the wind.

The Ekman transport, commonly labelled M_E, is derived by Pond and Pickard (1983) and given by

$$M_E = -\tau/f, \qquad (5.06)$$

where τ is the wind stress at the surface of the water and f is the Coriolis parameter. A typical wind stress of 0.1 N m^{-2} and Coriolis parameter at a latitude of 45° of $\simeq 10^{-4} \text{ s}^{-1}$ gives an M_E of $1000 \text{ kg m}^{-1} \text{s}^{-1}$, that is one metric tonne of water per second flowing at 90° to the right of the wind for every metre along a line parallel to the wind.

The relationships between the wind, Ekman drift and coastal upwelling are illustrated in the perspective drawing in Fig. 5.05 which

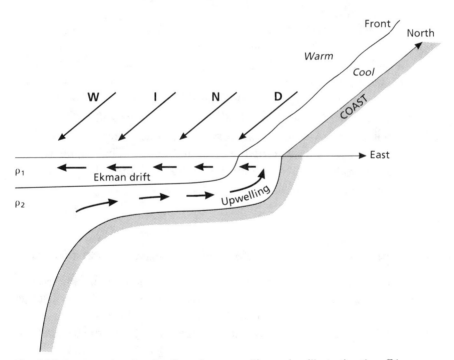

Fig. 5.05 A perspective drawing through an upwelling region illustrating the offshore Ekman drift in the upper layer being replaced near the coast by upward moving water from the lower layer. The upwelling water, usually cool, is separated from the offshore warm water by a surface front parallel to the coast. The wind blows from north to south.

suggests waters off California with an equatorward wind parallel to the coast. The Ekman drift, to the right of the wind, is directed offshore and decreases to zero at the coast. The water which replaces the offshore drift cannot be supplied by horizontal flow because of the coastal boundary. Instead it is upwelled from the deeper layers. The illustration also indicates the commonly observed front in the near surface layers which separates the cooler upwelled water next to the coast from the warmer offshore water.

The upwelling creates a current parallel to the coast in the same direction as the wind. This is generated by the horizontal density gradient that arises because the upwelling water close to shore is denser than the water offshore. This gradient, shown in Fig. 5.06(a) by the rising interface between the two layers of different density, results in a horizontal pressure gradient directed towards the land which in turn creates a geostrophic current parallel to the shore, Fig. 5.06(b). The horizontal pressure tries to push the water back to the shore but the Coriolis force turns the flow into a current parallel to the coast and towards the equator.

In the deeper levels the pressure gradient developed by the upwelling tends to be towards the offshore direction rather than towards the shore and a poleward current develops. Such equatorward and poleward density-driven flows are predicted by mathematical models (Gill 1982) and are commonly observed in the upwelling areas of the world as will be illustrated in later sections. The widths of the upwelling zones and the associated poleward and equatorward currents are related to a natural horizontal length scale of the stratified ocean. This is the Rossby deformation scale and is so important in oceanography that we make a slight detour to examine it in detail.

5.2.3 The width of coastal upwelling and the Rossby deformation scale

The vertical section through a coastal upwelling area in Fig. 5.06(a) indicates two horizontal lengths of interest. The first is D, the distance of the front from shore, while R_i is the width of the region where the interface between the upper and lower layer rises up to the sea surface. The distance D is discussed by Csanady (1981) who demonstrates mathematically that after the interface reaches the sea surface the front starts to move offshore. In his example of an upper layer 20 m thick, the interface reaches the sea surface with a 7 m s^{-1} wind blowing parallel to the coast for $\simeq$28 h. If the wind is greater than 7 m s^{-1} over the 28 h,

(a) *Vertical section*

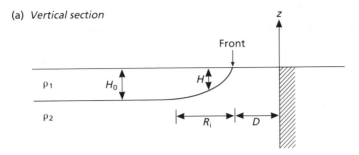

(b) *Plan view*

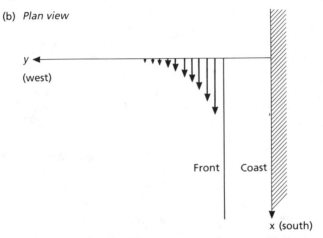

Fig. 5.06 (a) A section across an upwelling region in which the upper layer of density ρ_1 and depth H overlies a deep layer of density ρ_2. The interface between the layers rises from the undisturbed depth H_0 to the sea surface in the distance R_i. At the sea surface the interface lies a distance D from the shore. (b) A plan view of the currents parallel to the coast which are generated by the distribution of density in (a).

the interface first moves to the surface then moves offshore in proportion to the increase above $7\ \mathrm{m\ s}^{-1}$.

The width of the region where the interface rises to the sea surface, shown as the length R_i in Fig. 5.06(a), is approximately equal to

$$(g'H)^{1/2}/f, \tag{5.07}$$

where g' is the reduced gravity (see Box 5.02) and H is the depth of the upper layer. R_i is called the Rossby internal deformation scale after C.G. Rossby who first discussed it. Because it is one of the fundamental length scales in oceanography we demonstrate mathematically how it arises in Box 5.02.

Box 5.02 Derivation of the internal Rossby deformation scale

In Eq. 5.07 f is the Coriolis parameter, H is the depth of the upper layer and g' is called the reduced gravity and is equal to the gravitational acceleration (g) times the fractional increase in density between the layers. Thus

$$g' = \frac{\Delta\rho}{\rho} g .\tag{5.08}$$

The reduced gravity is therefore the gravitational acceleration acting on a parcel of water which has been displaced vertically from its equilibrium position. If there is no vertical change in density there is no buoyancy force as given by Eq. 3.05 and particles in the water 'feel' no gravitational restoring force when they are moved vertically.

The pressure gradient that drives the current in Fig. 5.06(b) is derived from the horizontal gradient in density where the upper layer rises up to the surface. The shape of the interface between the layers is set by the internal adjustment of the layers and depends on two laws. The first is expressed by the balance of forces in the geostrophic balance and the second is the conservation of angular momentum which says that if a fat object which is rotating becomes thinner (without losing mass) it will rotate faster. Everyday examples of this rule can be found in the twirling skater who changes her rate of spin by moving legs and arms closer and further away from the centre of mass. The water draining a sink speeding up as it approaches the central drain is another example of this conservation rule.

In oceanography conservation of angular momentum is discussed in terms of vorticity which is a mathematically convenient term equal to twice the rate of rotation. Physically the two are identical except for the factor 2. In relation to the situation shown in Figs 5.06(a) and (b) the conservation of vorticity is given roughly by

$$\frac{f + \zeta}{H} = \frac{f}{H_0} = \text{constant} .\tag{5.09}$$

Here again f is the Coriolis parameter representing the vorticity ($2 \times$ rotation) of the earth at the latitude ϕ and which is constant for this discussion. H is the depth of the upper layer with H_0 representing the thickness of the upper layer in the region where there is no current and ζ is the vorticity of the water relative to the earth, i.e.

$$\zeta = \frac{dv}{dx} - \frac{du}{dy} .\tag{5.10}$$

Since the v component of velocity is zero in Fig. 5.06(b), we can replace ζ by $- \, du/dy$, in Eq. 5.09, and the conservation of vorticity is now expressed by

$$f - du/dy = \frac{H}{H_0} f, \tag{5.11}$$

which states that the thickness of the upper layer (H) gets thinner as the horizontal velocity gradient (du/dy) increases. If there is no horizontal velocity gradient the depth of the upper layer is always equal to H_0.

The equation for the other law, the geostrophic balance, is (Eq. 4.23)

$$fu = -g' \frac{dH}{dy}, \tag{5.12}$$

where g' is the reduced gravity. Now we assume that the thickness of the upper layer decreases exponentially across the current, that is let

$$H = H_0(1 - e^{-y/L}), \tag{5.13}$$

and we wish to determine L, the horizontal scale of the deformation of the upper layer, in terms of the other variables. This is easily done by substituting the value for H given in Eq. 5.13 into Eq. 5.11, which gives

$$fu = \frac{g'H_0}{L} e^{-y/L}. \tag{5.14}$$

From this an expression for du/dy may be derived which when equated to the value for du/dy derived from Eq. 5.11 leads to the value of

$$L = (g'H_0)^{1/2}/f. \tag{5.15}$$

L is then the natural horizontal scale of currents in the ocean and coastal upwelling regions when a geostrophic balance and the conservation of vorticity are the dominant physical laws. The scale is usually represented by the symbol R_i and called the Rossby internal deformation scale. Towards the equator as f goes to zero the Rossby deformation scale becomes very large which as we shall see in Chapter 9 is an important feature of equatorial dynamics and El Niño. At the mid-latitudes $f \approx 10^{-4}$. If $g' \approx 0.02$ a 100 m upper layer will be deformed over ≈ 20 km. In other words the current will be about 20 km wide. The length L is also often called the internal Rossby radius. In this context it refers more to the natural horizontal scale of baroclinic eddies, i.e. those eddies with a vertical velocity gradient.

5.2.4 Variations in upwelling

The coastal upwelling described so far is a process that is assumed to be constant along a straight coast on the western side of a continent under the influence of a constant equatorward wind parallel to the shore. There are, however, many factors that can produce variations in this picture. Changes in the strength of the wind component parallel to the shore, the vertical structure of the water, variations in the bottom bathymetry as well as instabilities in the currents can all create variations in the upwelling process.

The easiest way to view some of the variations in the upwelling is with infrared satellite images which are able to show the horizontal temperature gradient in the surface water generated by the upwelling. Such images usually show the boundary between the upwelling water and the resident water to be very irregular with eddies and plumes of near-shore water moving offshore. The processes that lead to these time-dependent irregularities, some of which have been reviewed by Brink (1983), are difficult to observe in the ocean. Narimousa and Maxworthy (1985, 1987), however, have studied coastal upwelling in a rotating tank in the laboratory and have observed similar irregularities. The authors believe they are created by waves which become unstable through complex interactions in the flow. Petrie *et al.* (1987) attribute the plumes moving offshore of an upwelling band next to Nova Scotia to baroclinic instability which is a common instability of vertically sheared flows in stratified fluids. It is the process whereby eddies are created from the energy in the mean flow. In the atmosphere these are the passing lows or cyclones of the mid-latitudes that make the weather change from day to day.

Changes in the bathymetry can change the strength of the upwelling. A submarine ridge extending out from the coast, for example, produces conditions more favourable to upwelling than exist in the neighbouring regions. Upwelling usually begins first at such ridges and remains stronger at the ridge even when it has developed in the other locations. Numerical models by Peffley and O'Brien (1976) and rotating tank models by Narimousa and Maxworthy (1985, 1987) show that a rise in the bottom is effective in creating preferred conditions for upwelling, whereas a point of land sticking out into the ocean unaccompanied by a rise in the bottom does not significantly alter the rate of upwelling. Mountains close to the coast can also create a preferred location for upwelling by causing the wind to increase in speed as it is deflected by the mountains (Gill 1982; Roden 1961).

All coastal upwelling sites are subject to seasonal and shorter term variations in the wind strength. The well-studied upwelling west of the continental United States, for example, occurs mainly in the summer but disappears in the winter with the decline of the upwelling-favourable winds (Huyer 1983). The upwelling west of southern Africa in the Benguela current is perpetual at some latitudes but varies with the seasons at the northern and southern limits as the wind gyres that cause the upwelling move north and south (Shannon 1985). Variations in the wind due to storms and the daily heating and cooling are also shown by Shannon (1985) to have significant impacts on the upwelling off southwest Africa.

A change in the depth of the mixed layer in the upwelling zone can result in very large changes in the effects of upwelling. Off Peru, for example, when the upwelling brings nutrient-rich water up into the euphotic zone and stimulates phytoplankton production, the mixed layer is about 20 m deep. In times of El Niño the mixed layer is $\simeq 100$ m deep because of the tilting of the pycnocline across the Pacific (see section 9.2.1), and the upwelled water comes from above the pycnocline where the nutrients have already become depleted. In this case the upwelling can be just as strong as when the mixed layer is shallow but the effects on the biology are drastically different as there is no nutrient rich water rising up to the euphotic zone.

5.3 THE CANARY CURRENT SYSTEM

Of the various upwelling systems that have been intensively studied, that off the Spanish Sahara of northwest Africa is perhaps the most straightforward to understand, as a basis for comparison with other systems. The results of a cruise in 1974 known as JOINT I have been summarized by Huyer (1976). Progress on a multinational study of this region under the umbrella of the Cooperative Investigation of the Northern Part of the Eastern Central Atlantic (CINECA) was reviewed in Hempel (1982).

5.3.1 Upwelling and primary production

Figure 5.07 shows that during the JOINT I cruise the alongshore currents at all depths at a mooring in a mid-shelf position about 30 km from shore were well correlated with the wind strength. At a time of weak wind a gentle equatorward flow averaging about $10 \, \text{cm s}^{-1}$ extended almost 100 km offshore and to a depth of about 200 m on the continental slope, while from 200 to 400 m on the continental slope there was a poleward counter-current with a velocity up to $5 \, \text{cm s}^{-1}$ (Fig. 5.08a). At a time of strong wind an equatorward coastal jet developed with current speeds greater than $30 \, \text{cm s}^{-1}$ while the poleward counter-current had slightly increased velocities. During the strong

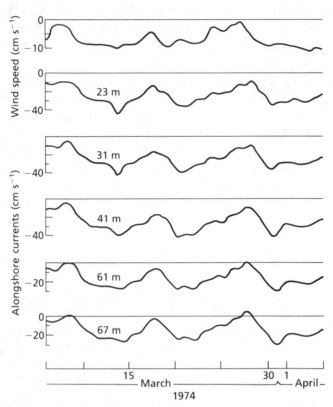

Fig. 5.07 Wind speed and alongshore currents at various depths at a station in the middle of the continental shelf off northwest Africa.

winds a two-layeredcirculation developed over the shelf. There was off-shore transport at rates >15 cm s^{-1} in the upper half of the water column and onshore transport at speeds up to 20 cm s^{-1} in the lower half (Fig. 5.08b).

When isopycnals next to a western coast slope upwards we infer that upwelling is in progress. Figure 5.08(c) shows that σ_t contours sloped gently upwards towards the coast during the time of weak winds, but turned sharply upwards during the strong winds. Barton et al. (1977) contoured the surface temperatures from 26 February to 9 April, showing that there were five upwelling events during the JOINT I cruise. The cold water first appeared close to shore, but moved out across the shelf as the upwelling event persisted (Fig. 5.09). The distribution of nutrients (Codispoti and Friederich 1978) shows that the nitrate-rich waters followed the same pattern, being brought to the surface first in the inshore

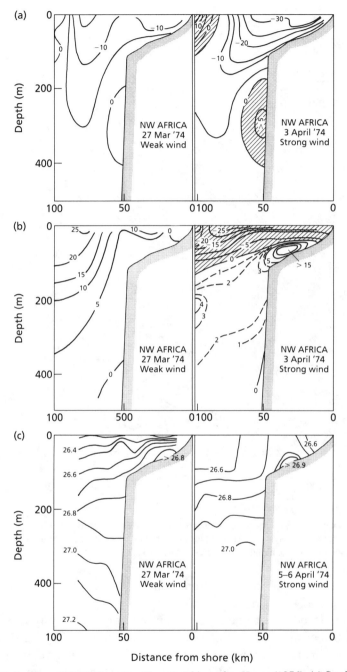

Distance from shore (km)

Fig. 5.08 Results of JOINT I cruise off north Africa, after Huyer (1976). (a) Sections showing the distribution of alongshore flow at times of weak wind and strong wind. Contour intervals are 5 cm s^{-1}. Positive values (shaded) are poleward and negative values equatorward. (b) Sections showing the distribution of onshore and offshore components of the flow during a weak wind and a strong wind. Offshore components (shaded) are negative. (c) Sections showing distribution of σ_t during weak winds and strong winds.

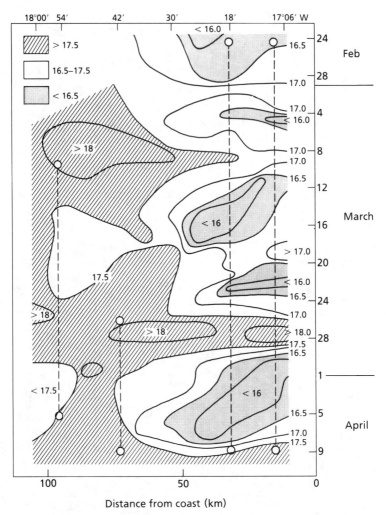

Fig. 5.09 Contours of surface temperatures from 26 February to 9 April, during the JOINT I cruise of northwest Africa. Note that pulses of cold water (shaded) correspond with periods of strong alongshore winds. From Barton *et al.* (1977).

area, then carrying the maximum concentrations out to the mid-shelf area (Fig. 5.10).

It would be nice to be able to continue the story by reporting that upwellings of nitrate-rich waters were quickly followed by bursts of primary production. Huntsman and Barber (1977) reported on primary production during the JOINT I cruise, and the averages for the entire cruise show a consistent pattern with highest levels of nutrients in the 10 km zone closest to shore, and with the rate of primary production

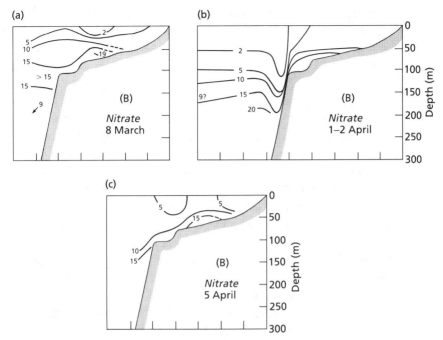

Fig. 5.10 Patterns of nitrate distribution (μg-at l^{-1}): (a) during weak winds, (b) soon after the onset of strong winds, and (c) after persistent strong winds. Reprinted with permission from Codispoti and Friederich (1978), Pergamon Press.

increasing steadily as they moved offshore then levelling off to 2–3 g C m^{-2} d^{-1} across the entire shelf. Pulses of high productivity during early April and early May coincided with periods of high and steady wind stress and one might have thought that the wind stress generated upwelling of nutrient-rich water, which in turn led to increased primary production. This is too simplistic. For one thing, there is a time delay between the onset of the wind and the arrival of nutrient-rich water out on the shelf. Secondly, there has to be a quantity of nutrient-limited phytoplankton cells ready to respond to the availability of nutrients.

Huntsman and Barber (1977) put forward the hypothesis that the observed high productivity results from the alternation of upwelling events and relatively calm periods. The upwelling brings nutrients into the surface waters, but during the calm periods stratification develops and the phytoplankton grows and multiplies while held in the shallow mixed layer. In other words, there is a miniature 'spring bloom' during each calm period.

To test this hypothesis they made numerous measurements of the assimilation number, which is the amount of carbon fixed per unit time, per unit of chlorophyll-a. It is well known that when phytoplankton is growing and dividing rapidly under the stimulation of nutrient enrichment, the assimilation number is high. They found that the assimilation numbers were high during periods of calm and lower during periods of upwelling, thus supporting their hypothesis. Jones and Halpern (1981) published further data supporting this point of view.

Minas *et al.* (1982) took a broad geographical view of the northwest African upwelling region. They showed that the region is divided into two major zones by a front which separates North Atlantic central water (NACW) from South Atlantic central water (SACW), at about the latitude of Cap Blanc. (JOINT I was conducted in the Cap Blanc region). The SACW, to the south of Cap Blanc is richer in nutrients than the NACW, but the poleward subsurface counter-current of the upwelling region carries SACW well into the Cap Blanc region. Upwelling occurs year-round at Cap Blanc and northwards, but south of Cap Blanc it occurs mainly in winter and spring, because in summer the Azores high-pressure cell which drives the equatorward winds, Fig. 5.01, has moved further north. Hence primary production is maximal in the Cap Blanc region, where upwelling is from nutrient-rich SACW and occurs year-round. To the north it is lower because the nutrient content of NACW is less, while to the south it is lower because the upwelling season is limited. In the Cap Blanc region, and probably in several other areas, the annual production is thought to be about $730 \, g \, C \, m^{-2} \, y^{-1}$, or $2 \, g \, C \, m^{-2} \, d^{-1}$ on average.

5.3.2 Upwelling and zooplankton

Up to this point we have been able to follow physical and biological processes on the 'event' scale, or meso-scale. Upwelling and primary production follow the onset of a strong wind within a few days. The next step in the food chain, the growth of zooplankton populations, is decoupled from these events because typical zooplankton organisms, such as copepods, require weeks rather than days to complete a life cycle. Trumble *et al.* (1981) have reviewed the available data from JOINT I on seasonal changes in zooplankton biomass. Adult copepods from deep water are upwelled in spring and begin to reproduce. Their offspring thrive on the abundant phytoplankton but tend to be carried offshore during the periods of strong upwelling. The intensity of upwelling falls off in autumn so the zooplankton are able to stay over the shelf and the populations reach their peak density. The annual mean value has been estimated at about $60 \, g \, m^{-2}$ (wet weight), with low value about 40 and high values about $120 \, g \, m^{-2}$. It must be remembered that we are talking about a dynamic situation in which the abundance of phytoplankton is a function of the grazing pressure as well as other environmental variables, and the abundance of the zooplankton is determined partly by the grazing pressure of the fish stocks. The consensus seems to be that since phytoplankton populations are able to respond rather rapidly to the favourable conditions provided by upwelling areas while the zooplankton respond only slowly, the phytoplankton production normally exceeds the consumption by the zooplankton. In other words, the zooplankton are seldom food limited.

5.3.3 Upwelling and fish

The yearly cycle for fish species in the Cap Blanc region has also been summarized by Trumble *et al.* (1981). Seventy-five per cent of the fish catch comprised only four types. Most abundant were clupeids (*Sardina pilchardus*, the sardine, and *Sardinella aurita*). Next most abundant were jack mackerel, *Trachurus* spp. and *Carynx rhonchus*. Redfish (Sparidae) and other species followed. *S. pilchardus* tends to dominate in the cooler water in the northern part of the upwelling, while *S. aurita* occupies the warmer, more southern waters. Their ranges tend to be adjusted with the changing seasons, with northward migrations taking place as summer approaches. In a study of the fish eggs and larvae present in the Cap Blanc region in spring, Palomera and Rubies (1982) found 65 different taxa, but 94% of all larvae examined were the sardine *Sardina pilchardus*. This heavy dominance of clupeids is characteristic of upwelling systems worldwide.

There is lack of agreement in the literature on what food is taken by the sardines and related clupeids, yet resolution of this question is needed for any calculation of a biological budget for an upwelling ecosystem. Longhurst (1971) concluded from a literature review that the sardine-like fishes in upwelling areas ate mainly phytoplankton. Mathisen *et al.* (1978), reporting on JOINT I, stated that 'Sardines eat mostly small zooplankton while horse mackerel prefer the larger ones'. Nehring and Holzlöhner (1982) stated that 'stomach investigations have shown that the African sardine feeds mainly on phytoplankton'. Nieland (1982) reported detailed studies of the contents of the oesophagus and cardiac stomachs of *S. aurita* and *S. eba* at the southern end of the upwelling region and came to the conclusion that the fish filtered phytoplankton, small zooplankton and detritus from the waters in proportions that differed with time and place. They also preyed on fish larvae several centimetres in length.

A critical review of the question was published by Cushing (1978). He pointed out that most sardines and sardinellas feed by filtering the water and have a finer mesh of gill-rakers than do the herring-like fishes. One consequence is that if they pursue the smaller zooplankton they are bound to collect the larger phytoplankton as a by-product. The question then is: are they adapted to make efficient use of these diatoms? Cushing (1978) pointed out that there is one truly phytoplankton-feeding clupeoid, *Cetengraulis mysticetus*, and this has a gizzard-like organ in which it crushes the diatoms. The menhaden (*Brevoortia tyrannus*), which feeds on benthic diatoms also has a gizzard-like organ for crushing them, but in addition has a very long gut, four to five times the body length, which is needed for the digestion of algae. Comparing these phytophagous fish with the clupeoids, we find that the guts of the carnivorous clupeoids are about half the length of the body while the guts of sardines and anchovies are about equal in length to the body. Since the sardines and anchovies lack a gizzard and have only a moderately elongated gut, Cushing (1978) suggested that they are not predominantly phytophagous, but feed on small zooplankton and make some use of the diatoms which they collect inadvertently. He conceded that it was possible that during the period of first feeding the larvae may depend completely on algae.

As mentioned earlier, Mathisen *et al.* (1978) asssumed that sardines fed mostly on small zooplankton while horse mackerel fed on larger zooplankton. Taking the biomasses of fish estimated acoustically during JOINT I, and some reasonable estimates of daily food requirements, these authors demonstrated that the estimated supply of food in the zooplankton was always in excess of the requirements of the fish stocks.

We have now followed the process through from the wind stress to the upwelling of nutrients, to the bursts of phytoplankton production during the periods of slack winds, to the build-up of zooplankton populations and their consumption by fish. The end result, from the point of view of human welfare, is the landing of large quantities of fish. Ansa-Emmin (1982) summarized the fish landings for the northwest Africa upwelling area. The total reported catches of all species in 1974 totalled 2.68 million tonnes. Almost 1 million tonnes were clupeidae, 0.67 million tonnes being sardines. Over 0.5 million tonnes were carangidae, the horse mackerels, and 0.2 million tonnes were squid. The other major catches were true mackerel (*Scomber* spp.), hakes (*Merluccius* spp.) and sparids. The catch was taken by a dozen industrialized countries, the USSR alone taking 287,000 tonnes of sardines and 55,000 tonnes of sardinellas, 360,000 tonnes of horse mackerel and nearly 200,000 tonnes of mackerel. Clearly, upwelling areas of the ocean can be very productive indeed.

5.3.4 Regeneration of nutrients

No account of an ecosystem would be complete without reference to the process which closes the loop, nutrient recycling. While upwelling of 'new' nutrients is the distinctive feature of coastal upwelling systems, we should not lose sight of the fact that a proportion of the primary production that takes place makes use of nitrogen, usually in the form of ammonia, regenerated by benthic invertebrates, bacteria, zooplankton, and fish. Two mechanisms are thought to be at work: one in the water column, the other on the bottom. There are two main possibilities for phytoplankton in the surface waters over the shelf. They may be consumed by zooplankton or they may sink to greater depths, even to the bottom. If they are consumed by zooplankton their nitrogen may appear as faecal pellets (which sink) or as excreted ammonia. If the phytoplankters sink to the bottom they may release ammonia during decomposition, or they may be consumed by benthic animals which in turn excrete ammonia. One way or another, a high proportion of the phytoplankton nitrogen ends up being released in the shoreward-moving lower layer of the water column. As a result, there is an accumulation of ammonia close to shore. This is exactly what has been found during the chemical investigations of JOINT I. Measurements of concentrations do not permit estimations of flux rates but Rowe *et al.* (1977) measured the flux of nutrients out of bottom sediments off Cap Blanc during JOINT I. The average total flux of nitrogen was $410\,\mu$g-at N m^{-2}h^{-1}, which at the time of measurement would account for 30–40% of the nitrogen required in the water column for photosynthesis. When we remember that there would be additional regeneration of nitrogen in the water column, it is clear that upwelling of new nitrogen is by no means the only method of stimulating primary production in the area, and regenerated nitrogen may account for more than half of the primary production. The two-layered cross-shelf flow acts as a kind of nutrient trap in much the same way as the two-layered estuarine flow discussed in Chapter 4. When primary production is stimulated by upwelling, much of the nitrogen taken up by the phytoplankton ends up being regenerated in the shoreward-flowing deeper layer and returns to the coastal waters to further stimulate primary production. Barber and Smith (1981) (quoting Whitledge, unpubl.), estimated that on the shelf off Cap

Blanc the regenerated N comprises 72% of the total N, with 33% coming from zooplankton excretion, 24% from benthos and sediments, and 15% from pelagic fish.

5.4 COMPARISON WITH THE PERU UPWELLING SYSTEM

The Peruvian upwelling system at $15°$ S was studied intensively in 1976–1977 in multidisciplinary cruises known as JOINT II (Fig. 5.11). There are three important differences between the upwelling site off Peru and that off northwest Africa. The first is that the shelf off Peru is narrower (20 km versus 50 km) and drops off more steeply (200 m at the shelf break versus 110 m). The second is that the deep water off Peru has higher nutrient concentrations (upwelled water contains 20–25 μg-at l^{-1} of nitrate compared with 5–10 μg-at l^{-1} off northwest Africa). Finally, the wind stress is less, and is more constant, averaging 0.79 ± 0.4 dynes cm^{-2} compared with 1.55 ± 1.0 off northwest Africa. The net result is that when the Ekman transport is active, offshore

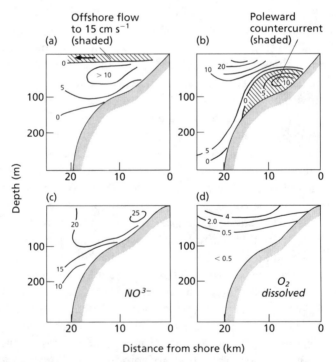

Fig. 5.11 Conditions off Peru, April–May 1977. From Brink *et al.* (1980) and Codispoti *et al.* (1982). (a) Onshore–offshore currents; (b) alongshore currents; (c) nitrate concentration; and (d) dissolved oxygen concentration.

transport occurs mainly in the top 20 m, and shoreward transport is in an intermediate layer over the shelf, at depths of about 30–80 m, instead of in the whole of the water column down to the bottom, as it is off northwest Africa. Off Peru, the water close to the bottom is relatively still, and there is marked accumulation of organic matter to give a reducing sediment.

Two further differences are consequences of those listed above. One is that since the wind stress is less, the wind-induced mixing does not penetrate so deeply. During periods of strong upwelling, phytoplankton is still retained within the euphotic zone, which means that primary production is maintained at a relatively constant level, whether the wind stress is high or low. Secondly, the poleward counter-current, flowing beneath the equatorward coastal jet, is situated at intermediate depth over the continental shelf off Peru, whereas off northwest Africa it is located on the shelf slope (Fig. 5.08a). Figure 5.11 summarizes the features of the Peru upwelling system discussed above.

A detailed analysis of the primary production cycle in a segment of the Peru upwelling system at 15° S was published by MacIsaac *et al.* (1985). Because the upwelling is relatively constant it is possible to trace a distinct plume of cold water moving out from the coast and to recognize a number of zones along the axis of the plume. Zone I is the area of intense upwelling within about 7 km of the coast, where nutrients are abundant but phytoplankton biomass is relatively low. The phytoplankton cells are growing and taking up nutrients at rates considerably less than those of which they are inherently capable. In zone II the water column is stabilized by solar warming and the phytoplankton cells are found to increase their rates of nutrient uptake, photosynthesis and synthesis of macro-molecules, a process known as 'shift-up'. Zone III is characterized by the rapid depletion of nutrients by the 'shifted-up' phytoplankton, so that there is a rapid accumulation of biomass and all processes occur at maximal rates. In zone IV nutrient depletion occurs, so that the cells experience nutrient limitation. In this environment 'shift-down' of physiological rates occurs. Using drogues to track water masses, MacIsaac *et al.* (1985) estimated that phytoplankton cells moved from zone I to zone IV in 8–10 days, during which time they travelled 30–60 km away from the coast. The chlorophyll maximum occurred about 18 km offshore.

5.4.1 Interannual variability in the Peru upwelling system

As Fig. 5.11 shows, the source of upwelled water off Peru in March 1977 was at a depth of 30–60 m. The water was at a temperature of

15.5–16.5° C, and contained 20–25 μM nitrate. This state of affairs was only part of a larger structure extending across the whole of the South Pacific (see Chapter 9). Trade winds blowing across the Pacific Ocean from east to west drive the south equatorial current and cause the mixed layer to be only 10–20 m deep in the eastern Pacific but up to 80 m deep in the western Pacific. In other words, the thermocline is tilted upwards at its eastern end. Then the winds blowing from south to north along the coast of Peru cause cool, nutrient-rich water to upwell from below the thermocline. This is the normal state of affairs.

It is now known that in anomalous years the trade winds weaken or reverse, and the thermocline off the coast of Peru sinks to a depth of about 100 m (see section 9.2.1). Ekman transport along this coast continues, but the water that is upwelled is now much warmer and not rich in nutrients. As a result, there is a sharp reduction in the biomass and productivity of the phytoplankton. The phenomenon, originally recognized from the warm water anomaly which typically begins about Christmas time, was referred to as El Niño which means the Christchild. More recently it has come to be referred to as an El Niño–southern oscillation (ENSO) event. The physics of the southern oscillation will be dealt with in Chapter 9, but we should note that the Peru upwelling system seems to be uniquely vulnerable to such drastic changes, so that its interannual variability in productivity is very great. Sea surface temperature anomalies were noted in 1965, 1969, 1972 and 1976 (Barber et al. 1985) but in 1982–1983 an ENSO event occurred with a severity that is considered very rare, occurring with a periodicity of 100 years or more (Rasmusson and Wallace 1983). Barber et al. (1985) showed that at the height of the 1982–1983 anomaly, in May 1983, the upwelling waters were at 29 °C instead of the usual 16–18 °C, and mean primary productivity was only $10 \, \text{mg} \, C \, m^{-3} \, d^{-1}$. Two months later conditions had returned to normal and mean primary productivity was $219 \, \text{mg} \, C \, m^{-3} \, d^{-1}$.

5.4.2 Total primary production in the Peru upwelling system

Many attempts have been made to calculate the total primary productivity of the Peru upwelling system, and to compare this with the total fish production. The most recent attempt is that of Chavez and Barber (1987). They give the references to earlier calculations, which differed widely among themselves. One of the main reasons for the great discrepancies was the lack of agreement about the area of ocean influenced by the upwelling. For example, Cushing (1969) used 479,000 km^2 in his calculation while Ryther (1969) used 60,000 km^2. Chavez and Barber (1987) argued that the offshore dimension of coastal upwelling is limited by the Rossby deformation scale (section 5.2.3). From Eq. 5.15 we see that the formula for this is $(g'H_0)^{1/2}/f$. Thus the radius is a function of the Coriolis force f, (which varies with latitude), the depth H_0, and the vertical density gradient. Taking these factors into account Chavez and Barber (1987) calculated that the width of the Peru upwelling system varied from 270 km at 4° S to 60 km at 18° S. Calculating the width at degree intervals they arrived at an area of 182,000 km^2, intermediate between the two values quoted above. The mean of the large number of determinations of primary production in 1983–1984, after the recovery from the El Niño event, was 2.28 g C m^{-2} d^{-1} or 834 g C m^{-2} y^{-1}. This converted into a total production of 1.52×10^{14} g C y^{-1}. With increased availability of satellite pictures of chlorophyll distribution, it should be possible to refine the estimate of the average area affected by upwelling, and determine its variability.

5.4.3 Secondary production in the Peru upwelling system

Cushing (1971) described the fish and birds of the Peru upwelling as follows:

> Anchoveta ... live in the upper part of the thermocline and presumably migrate towards the surface at night. On the bottom live flatfish in rather shallow water, and near the bottom in deeper water live hake and rosefish. The hake probably migrate upwards at night and feed on euphausids. Between the anchoveta and the hake live the horse mackerel, possibly in the lower parts of the thermocline and below. At the surface live guano birds which are surprisingly abundant. In the Peru current there are 30 million birds, mainly pelicans, boobies and cormorants; during El Niño, the numbers are sharply reduced, because some migrate southwards out of the area when the fish are no longer accessible and some die of starvation. In 1958 during El Niño, the bird populations were sharply reduced but by 1962 the numbers of birds had fully recovered. So they must eat very large numbers of anchoveta.

It is clear that the same types of fish are found in the Peru system as in the northwest Africa upwelling system. Cushing (1971) went on to point out that the fish community is a specialized one, and the species are not caught much outside of upwelling areas. The sardines and anchovies usually spawn in the areas of most intense upwelling, close to shore, and Cushing speculated that both the juveniles and the adults make use of the two counter-currents (onshore and poleward) to maintain themselves within the upwelling system. As with the northwest African system, there is lack of agreement about whether the anchovies and sardines are predominantly herbivorous. Walsh *et al.* (1980), summarizing earlier work, concluded that for the Peruvian anchovy, *Engraulis ringens*, larvae of 0.3–0.9 cm consume 'some phytoplankton', while larvae of 0.9–3.0 cm eat mainly zooplankton. In juveniles from 4.6 to 12 cm 'increasing amounts of algae are found' and adults over 12 cm long are almost entirely phytophagous. It is therefore not surprising that the drastic reductions in primary production associated with an ENSO event should have a major effect on the biology of anchovies.

Barber *et al.* (1985) showed (Fig. 5.12) that the years of temperature anomalies (El Niño years) were associated with reduced landings of anchoveta. There are several possible explanations: the adults could have starved for lack of phytoplankton food, the fish could have migrated away from the areas where they are usually caught or the larvae could have failed to survive through lack of both phytoplankton and zooplankton. There is some evidence for each of these hypotheses. Barber *et al.* (1985) report that anchoveta seek out the upwelling areas by exhibiting a preference for water of 16–18 °C. As an El Niño builds, the cool water is found only in certain more persistent centres of upwelling

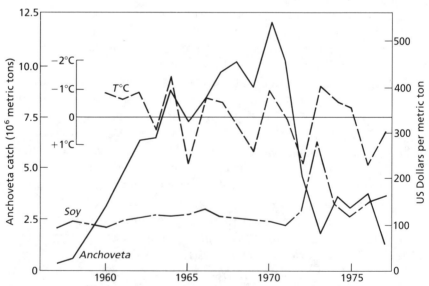

Fig. 5.12 Heavy line, anchoveta catch by Peru; (T, °C), temperature anomaly of the eastern Pacific; Soy, the cost of soya meal ($ tonne^{-1}). After Barber *et al.* (1985).

and the anchoveta concentrate there. In 1982–1983 they were found in large numbers in isolated upwelling centres surrounded by warm water on three sides and land on the fourth. When the cool pockets of water were overridden by the anomalously warm water at the peak of the event, the anchoveta died and many were found floating. There were reports of anchoveta having moved to cooler, deeper water at about 100 m, but such water has very little phytoplankton and it seems likely that the fish would not survive there long.

On the question of migration, Valdivia (1978) reported a rapid southward migration of the stocks over about 180 nautical miles in less than 15 days during 1975, at a time when the 1976 El Niño was building. Barber *et al.* (1985) and others have argued that the anchovy stocks had presumably evolved an ability to survive ENSO events, since there is evidence from sediment cores of a dominance of anchoveta off central Peru for more than 2000 years. Migration south to find cooler waters may well have been one of the strategies for survival.

The question of larval survival is a complex one. The anchovies spread their spawning over 7–8 months of the year, presumably as an adaptation to the occurrence of unfavourable conditions at particular times and places. However, there are two peak periods: the austral winter–spring spawning (July–September) and the summer spawning (February–March) (Valdivia 1978). Walsh *et al.* (1980), investigating the survival of larvae off the northern coast of Peru during winter, found that survival was higher when dinoflagellates in relatively high concentrations were available to the first-feeding larvae. Strong wind events had the effect of dispersing the dinoflagellate concentrations and adversely affected survival. When comparing the northern waters with those further south, they found that the strong wind events were more common in the south and these waters were therefore less favourable to the survival of larvae. Under El Niño conditions, more of the spawning adults were found in the south, and their reproductive success was judged to be lower on this account. Pauly and Tsukayama (1987) conducted a detailed study of historical records of the northern and central stocks of anchoveta, to see whether dispersal of dense layers of food by storms could be used as an explanation for poor recruitment of anchoveta. They concluded that they could not. The relationship between the distribution of food for first-feeding anchovy larvae and their subsequent survival has been investigated in detail by Lasker (1978) and will be discussed in connection with section 5.5 on the California current.

Barber *et al.* (1985) commented on a situation where there was a superabundance of dinoflagellate food for the larvae. At the onset of the

1976 El Niño there was a bloom of the dinoflagellate *Gymnodinium splendens* along 1000 km of the coast from March until the end of May. It was attributable to the stabilization of the water column as the warm water invaded the area. While it provided an excellent feeding environment for the early larvae, it apparently was not a suitable food for the adult fish, which had a reduced fat content, a reduced weight at a given length and reduced length at sexual maturity. The 1977 recruitment, of fish spawned in 1976, was extremely poor and the stock along the Peru coast fell to the lowest levels ever observed.

5.4.4 Exploitation of the Peru anchoveta stocks

As Fig. 5.12 shows, the ENSO events of 1965 and 1969 led to minor decreases in the landings of anchoveta, but the 1972 event was associated with a drastic decline from which, in fact, the stock has never recovered. The history of the anchoveta fishery in Peru has been reviewed by Glantz (1985). For a century, beginning in the 1840s, the major industrial activity along the Peruvian coast had been the mining of the bird-droppings, guano, from the rocky islands. Large populations of fish-eating birds are characteristic of upwelling populations worldwide, and prominent white accumulations of droppings at their roosting sites are an inevitable concomitant. As Cushing (1971) remarked, it is no accident that upwelling areas commonly have a Cabo Blanco, Cap Blanc or Cape Blanc. The guano of Peru was mined and exported for fertilizer to many parts of the world. Beginning in the 1950s, a lucrative industry to harvest the anchoveta and convert them to fish meal was developed in Peru. The landings increased rapidly to a peak of about 12 million tonnes in 1970, then dropped to 2–3 million tonnes for a few years. Since 1977 the catch has hovered around 1 million tonnes. In fisheries circles a debate has centred around the question of whether the population crash is part of a natural cycle of events that has occurred many times in the past, or whether it is primarily the result of gross overfishing.

In discussing the issue, Barber *et al.* (1985) pointed out that during the period of low anchoveta abundance there were increases in the abundance of sardine, jack mackerel and mackerel, suggesting a possible natural shift in species dominance. (We shall see in section 9.3.2 that low offshore Ekman transport appears to favour these species.) However, the sediment records do not support the view that there has been dominance by sardine and mackerel for significant periods in the past. In 1975 conditions for anchoveta reproduction were excellent, the index of recruitment was one of the highest ever recorded, and by early 1976 the anchoveta stocks were estimated at 8.3 million tonnes. However, 1976 was the year of the *Gymnodinium* bloom mentioned earlier, and this was followed by an ENSO event. Then, from 1977 to 1982 there was an anomaly in the global pressure system which resulted in a weakening of the southeastern Pacific trade wind system and a decrease in the thermocline tilt, which in turn meant poorer nutrient conditions in the Peru upwelling system. According to this view, anchoveta went into the 1982–1983 ENSO event at a low population level because of the 1976 event followed by 6 years of unfavourable environmental conditions. On the other hand, Barber *et al.* (1985) pointed out that by 1984 the levels of primary production in the upwelling system were among the highest ever observed, but the anchoveta have not responded by a large increase in stock size.

The explanation for this could be the predation on anchoveta by mackerel and horse mackerel. Muck and Sanchez (1987) reviewed historical data on stock size, stomach contents and energetics of these predators and concluded that their consumption of anchoveta could at times equal or exceed the commercial fish catch. They therefore concluded that years of high abundance of mackerel and horse mackerel could explain, at least in part, years of low abundance of anchoveta.

At this point in the discussion Barber *et al.* (1985) drew attention to the model of Steele and Henderson (1984) which has two components. The first is a population model with multiple equilibrium states, the second is an environmental variable that interacts with the population model. This environmental variable is made to vary stochastically, and the variance increases as the periodicity increases from days to decades, a pattern that conforms with observed and theoretical values. The Steele and Henderson (1984) model predicts rapid changes in abundance of pelagic fish stocks with intervening quiescent periods of about 50 years. Increased fishing pressure leads to an increased frequency of the marked changes in abundance.

While the question is not yet fully resolved, there seems to be both theoretical and empirical evidence for thinking that the low abundance of anchoveta since 1977 has been primarily determined by unfavourable environmental conditions, but that the vulnerability of the stocks to these influences was increased by the heavy fishing pressure to which they were subjected. The failure of the anchoveta to return to their former dominance even when the conditions became extremely favourable in 1984, may reflect the greater abundance of sardine, jack mackerel and mackerel in the pelagic system. There have been analogous shifts in relative abundance of pelagic fish in other upwelling systems. A well-known textbook example is the rapid decline of the sardine populations and the rise of anchovy stocks which took place in the California current in the 1940s and 1950s. This will be discussed in the next section.

Walsh (1981) assumed that the sharp decline in the anchoveta landings in Peru in 1972 was clearly the result of overfishing, and produced an interesting carbon budget for the Peru upwelling system before and after the overfishing (Fig. 5.13). The numbers are admittedly tentative, but are supported by a considerable amount of indirect evidence. He suggested that during the period 1966–1969 the primary production of about $1000 \, g \, C \, m^{-2} \, y^{-1}$ was divided equally between the anchoveta stocks and the copepods. The production of the copepods was used by

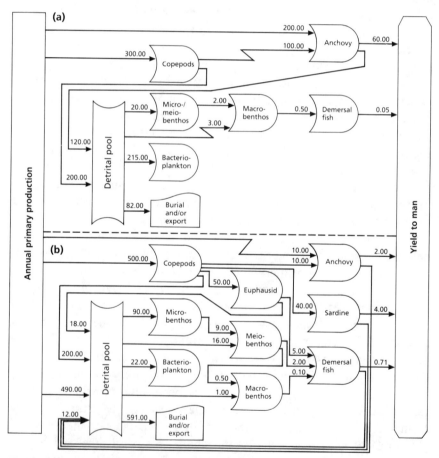

Fig. 5.13 A budget for the flux of carbon ($g \, C \, m^{-2} \, y^{-1}$) through the Peru food web: (a) before overfishing, and (b) after overfishing of anchovy. From Walsh (1981).

the anchovies while the faecal matter from both anchovies and copepods fuelled a benthic food web which supported a modest production of benthic fish. During the period 1976–1979 the copepod food consumption and production was thought to remain about the same, but the copepods were taken mostly by sardines and euphausids. Anchovy consumption of the phytoplankton production declined sharply and the surplus primary production sank to the bottom to support the benthic food web and enhanced benthic fish production. Walsh suggested that in the later period there was also a massive burial and/or export of phytoplankton detritus from the benthic subsystem. Clearly, the stage is set for a return to high levels of anchoveta biomass, an event for which numerous Peruvian fishermen and fishery scientists are waiting with keen interest.

5.5 THE CALIFORNIA CURRENT SYSTEM

As we see from Fig. 5.01 there is a northern hemisphere analogue of the Peru upwelling system. It is driven by prevailing northerly winds, and upwelling occurs along the Pacific coast of the USA from the Canadian border south to Baja California and beyond. This is the California current system. Detailed studies of upwelling events have been made, particularly off Oregon, and a comparison of northwest Africa and Oregon was made by Huyer (1976). In several respects the situation off Oregon resembles that off Peru. The poleward undercurrent appears over the shelf as well as the slope, and the shoreward flow is strongest at mid-depths over the shelf. Upwelling events are less strong and of shorter duration off Oregon than they are off northwest Africa.

Bakun (1973) examined records of wind strength over the ocean off northwest USA over a period of 20 years. He calculated the wind stress from the formula

$$\tau = \rho_a \, C_d \, v^2 \; ,$$

where ρ_a is the density of the air, taken as $0.00122 \, \mathrm{g \, cm^{-1}}$ and the drag coefficient C_d was taken as constant at 0.0026. He then computed the Ekman transport, M, from wind stress, τ, and the Coriolis parameter, f, according to the equation

$$M = \tau/f.$$

Finally, he integrated the data to give the Bakun upwelling index, a 20-year average of monthly mean Ekman transport for different parts of the coast, expressed as cubic metres per second per 100 m of coastline. The range is from $300 \, \mathrm{m^3 \, s^{-1}}$ (offshore direction) to $-212 \, \mathrm{m^{-3} \, s^{-1}}$

(onshore). Some of the results are displayed in Fig. 5.14. It can be seen that the index indicates year-round upwelling off southern California, with stronger upwelling in summer, but off Oregon and Washington in the north there is strong downwelling in winter, and upwelling is confined to the period April–September.

Figure 5.15 shows that in summer there is a negative temperature anomaly, indicative of upwelling of cold water, all the way along the coast from Oregon in the north to Baja California in the south, with the exception of a warm anomaly off San Diego. The upwelling is obviously most intense between Cape Mendicino and Monterey. Yet when Bernal and McGowan (1981) analysed a 21-year time series of biological and physical data from the California Current, in which the upwelling occurs, they found no significant correlations with upwelling indices. There have been large positive and negative anomalies in zooplankton biomass, but

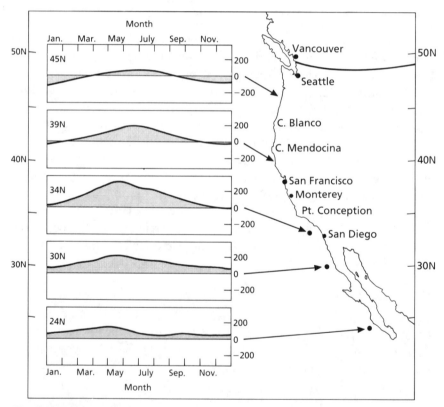

Fig. 5.14 Each plot on the left hand side of the diagram is the upwelling index for each month of the year, averaged for the 20 years 1948–67. Each is for a different latitude, as indicated on the map on the right. Note that the strongest upwelling occurs at 34° N. In the southern half of the range the index is positive all year round, but at latitudes 39–45° N the index is negative in winter. Modified from Bakun (1973).

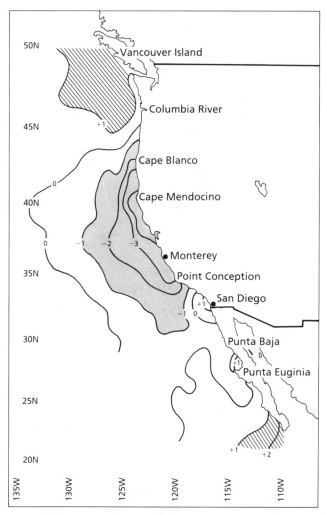

Fig. 5.15 Average coastal temperature anomaly (°C) in summer, calculated as the difference between each location and a smoothed reference temperature for offshore conditions at the same latitude. From Bakun and Parrish (1982).

these were best correlated with anomalies in the transport of cold, low-salinity, nutrient-rich water from the north. We shall show in Chapter 9 that these anomalies are correlated with changes in circulation patterns in the whole North Pacific.

Although the temperature anomaly in Fig. 5.15 shows an apparently uniform area of cold water, this is an artifact of the method of calculation. The situation at any one time, as seen by satellite, is extremely complex, with coastal upwelling systems tending to be centred on topographical features such as capes and canyons, and with plumes of upwelled water extending far out into the California current. As in other coastal upwelling systems, the strength of upwelling is strongly dependent on wind speed and direction, and changes from day to day. Figure 5.16 shows the distribution of temperature and nitrate off Pt Sur, California, on 9 June 1980, as inferred from satellite imagery supplemented by shipboard observations (Traganza *et al.* 1983). This situation is typical of an early phase of an upwelling event at this site. If the event persists for many days, interaction with the California current may give rise to a cyclonic structure about 100 km in diameter, with high biological production along the associated fronts, or it may extend into a plume up to 250 km long (Traganza *et al.* 1981, 1987). In summertime, patterns of this kind may be found in various stages of development or decline all the way from Oregon to Baja California. As winter approaches the upwelling is progressively restricted to the southern portion of the region and in spring the region of upwelling spreads north again. There seems to be no good synthesis of the biological productivity of this dynamic system. In the southern California bight, where upwelling is not particularly strong, Smith and Eppley (1982) found that the 16-year average for primary production was $0.402 \, \mathrm{g\,C\,m^{-2}\,d^{-1}}$, about $150 \, \mathrm{g\,C\,m^{-2}\,y^{-1}}$. Using the temperature anomaly at the Scripps pier as an index of upwelling, they found that the highest daily rates were associated with the maximum amount of upwelling, and vice versa. In the California current itself Hayward and Venrick (1982) found wide variation in the biomass and productivity of phytoplankton. Carbon fixation rates varied over an order of magnitude, from about 20 to nearly $200 \, \mathrm{mg\,C\,m^{-2}\,h^{-1}}$. These might be roughly converted to a range of $0.2–2.0 \, \mathrm{g\,C\,m^{-2}\,d^{-1}}$. These differences reflect the heterogeneous nature of the California current, with its admixture of advected and upwelled water. In trying to calculate the carbon flow from primary production to the pelagic fish stocks, Lasker (1988) used a value of $0.5 \, \mathrm{g\,C\,m^{-2}\,d^{-1}}$ for the southern California bight and adjacent California current, but commented: 'It also seems likely that to the north, off central and northern California, the highly energetic "jets and squirts" which move large quantities of cold, nutrient-enriched water offshore and become entrained in the southward moving California current . . . add more primary production to the habitat of the sardine than has been included in the gross overall average.'

5.5.1 Fish production in the California current system

As in other eastern boundary current systems, the most abundant fishes in the California current system are sardines, anchovies, hake, jack

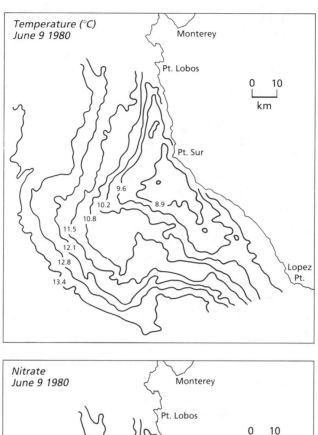

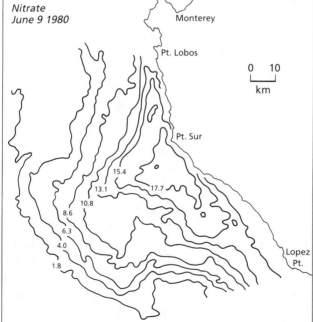

Fig. 5.16 Temperature and nitrate distributions off Pt Sur, California, from satellite and shipboard operations. Note the difference between actual field data and the averaged long-term trends in the previous figure. From Traganza *et al.* (1983).

mackerel and mackerel. Sardines, *Sardinops sagax*, were heavily exploited from 1916 to 1967. The peak landings were in 1936–1937 and exceeded 700,000 tonnes. The catch fell drastically in the 1950s and 1960s and in 1967 the California State Legislature imposed a moratorium on the sardine fishery. The next most abundant species is the northern anchovy, *Engraulis mordax*, closely related to *E. ringens* of the Peru upwelling. Its numbers began to increase as the biomass of sardines declined, and some postulated that the two were in competition so that the decline of sardine stocks released resources for the anchovies. However, Soutar and Isaacs (1969) studied the 1850-year record of fish scales in the anaerobic sediments off California and concluded that northern anchovy scales were present in large numbers throughout the series, while sardine scales appeared intermittently for periods of 20–150 years, with absences that averaged 80 years in duration. They concluded that the two species were not in competition.

The sardines of the California current system are divisible into four stocks, but of these the largest by far before overfishing was the one that spawned in the southern California bight and migrated to the upwelling areas off northern California to exploit the dense zooplankton stocks that are associated with the coastal upwelling. The anchovies also have several sub-populations (Fig. 5.17). The stock off Oregon spawns at about 44–46° N, mainly in July at the time of the northern upwelling. The central sub-population spawns principally in the southern California bight. Eggs and larvae can be found throughout the year, but the peak abundance is in the spring, while the minimum is in the autumn. The fish remain in the southern California bight throughout their lives and in recent years this has been the largest stock. There is a southern stock off Baja California, for which peak larval abundance is from January to March.

It thus appears that the largest stocks of both sardine and anchovy spawn in the southern California bight. Upwelling is relatively weak and phytoplankton production is lower than in the California current proper. Bakun and Parrish (1982) have suggested that strong offshore flow associated with Ekman transport is likely to carry eggs and larvae too far offshore, to positions from which they may never return, and that the choice of the southern California bight for spawning area reflects a need to avoid areas of strong upwelling. They also suggested that areas of strong Ekman transport are areas where there are strong winds which may destroy the fine-scale strata of food organisms needed by first-feeding larvae. This idea, attributable to Lasker (1975) will now be examined in more detail.

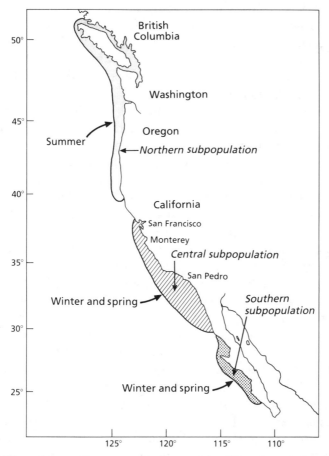

Fig. 5.17 The major spawning areas and seasons of the three subpopulations of the northern anchovy off the west coast of North America. From Smith and Lasker (1978).

5.5.2 The survival of first-feeding larvae

Anchovy eggs are most abundant in the southern California bight during February, March and April (Smith and Lasker 1978). After the absorption of the yolk sac, they must obtain sufficient food to meet their metabolic requirements within 2–3 days or they will die. Hunter (1972) studied the feeding behaviour of the larvae and found that the food capture success rate of 3-day-old, first-feeding larvae is very low but increases exponentially with age. A consequence of this is that 3-day-olds require a very high density of food organisms in their immediate environment in order to survive. He estimated that it was at a minimum about 1790 dinoflagellates per litre.

Lasker (1975) made a field investigation of the feeding conditions for anchovy larvae in the southern California bight. Eggs were taken from anchovies held in the laboratory and were incubated under controlled conditions. They were then taken to sea so that a ready supply of first-feeding larvae was available for shipboard experiment. In March and April 1974 he found at several stations that there was a marked chlorophyll maximum at a depth of 30–35 m, and that the most abundant organism was the naked dinoflagellate *Gymnodinium splendens* with a diameter of 40–50 μm, occurring at densities of 30–40 particles ml^{-1}. Larvae placed in waters from near the surface took very little food, but larvae placed in water from the chlorophyll maximum fed extensively. The experiments were repeated at stations where other kinds of phytoplankton dominated. It was concluded that spiny or chain-forming diatoms such as *Thalassiosira* or *Chaetoceros* did not stimulate the larvae to feed, nor did any phytoplankton of diameter much less than 40 μm. Finally, the larvae were stimulated to feed only when the phytoplankton was at a population density of at least 20–30 cells ml^{-1}. The latter fact was confirmed in a very striking manner. Towards the end of the study a strong wind caused a deepening of the mixed layer and obliterated the chlorophyll maximum containing the *Gymnodinium*. After that it was impossible to stimulate the anchovy larvae to feed with water taken from any depth in the water column.

Lasker (1978) then investigated the horizontal and vertical extent of areas in the southern California bight suitable for anchovy feeding. Between September and December 1974 suitable food was abundant within 5 km of the shore in a chlorophyll maximum layer. During January 1975, gyral circulation redistributed the water so that adequate concentrations of dinoflagellates could be found at some depth all throughout the bight. In February 1975 northerly winds strengthened and strong upwelling began. The dinoflagellate populations were gradually replaced by small diatoms on which the anchovy larvae would not feed. These conditions persisted through the early spring and summer months.

On the basis of Lasker's (1975, 1978) results we may conclude that physical factors have a strong influence on year-class success of northern anchovy. The first-feeding larvae rely on high concentrations of phytoplankton such as naked dinoflagellates of a size class close to 40 μm. These high concentrations occur mainly at the chlorophyll maximum which forms near the thermocline under relatively calm conditions in the southern California bight. The onset of strong winds at a critical time may mix the water column to such a depth that the chlorophyll maximum is dispersed, or the winds may set up strong Ekman transport so that the water in the southern California bight is replaced by upwelled water containing mostly small diatoms in relatively low concentrations. Either of these situations render the environment unsuitable for successful feeding, growth and survival of small anchovy larvae. To put it another way, while the presence of upwelling and the consequent high biological productivity creates optimum conditions for the growth of juvenile and adult sardines, it is the absence of strong upwelling in late winter and early spring in the southern California bight that provides the optimum conditions for survival of first-feeding larvae. The timing of strong winds and the onset of upwelling in this region may well be a major determinant of year-class strength.

To summarize our understanding of factors influencing anchovy and sardine production in the California current, we see that strong wind stress and associated upwelling that are the most characteristic features of eastern boundary currents appear in themselves to be detrimental to the success of the larvae. For good survival the larvae require a well-developed horizontal layer of high phytoplankton density, in which dinoflagellates are the dominant form. We have seen in Chapter 2 that diatoms, with their high sinking rate, tend to dominate in highly turbulent or newly upwelled water, while dinoflagellates, which are active swimmers, tend to replace diatoms in more stable water columns with lower nutrient concentrations. Hence, the optimum conditions for success of first-feeding anchovy larvae tend to occur in waters that have been free of strong upwelling and strong wind-mixing for a considerable period. Strong wind-mixing and upwelling at the wrong time can cause heavy mortality. Once past the critical early stages of development, the juveniles migrate towards the areas of intense upwelling and exploit the high biological productivity found there.

It is interesting to compare this conclusion with that reached in section 5.3.1, where we saw that upwelling provides the supply of nutrients to the euphotic zone but it is the relatively calm periods between upwelling events that permit the development of high phytoplankton biomass, in effect a little 'spring bloom'. On a longer time scale, it is the season of active upwelling that brings nutrients to the surface waters, but it is the season of minimal wind stress and upwelling that provides favourable conditions for first-feeding anchovy larvae.

A second factor to be considered in relation to larval survival is the risk that strong offshore transport will carry the larvae away from the favourable coastal environment. The two mechanisms occur on very different time scales and have quite different characteristics. The destruction of a chlorophyll maximum layer can occur within a few days and is in effect irreversible within the period of development of a cohort of larvae. On the other hand, offshore transport in a system as complex as the California current can take weeks to transport the larvae far enough from shore that their chance of completing their life history is placed in jeopardy. Furthermore, the offshore transport may be part of a gyral circulation that eventually begins to carry the larvae back towards shore, or the larvae may be able to migrate down into a shoreward-moving counter-current. In our present stage of understanding, there is no way to integrate these two mortality factors, beyond pointing out that both sardine and anchovy appear to avoid the regions of strongest upwelling in their choice of spawning location. By spawning down-stream of those regions they appear to minimize the hazards while plac-ing their larvae in a location from which, with a short migration, they can benefit from the high productivity associated with coastal upwelling.

5.6 THE BENGUELA UPWELLING SYSTEM

The last of the four major upwelling areas located in eastern boundary currents is the Benguela system, to the west of southern Africa (Fig. 5.01). The physics, chemistry and biology of this system have been thoroughly reviewed in recent years (Andrews and Hutchings 1980; Chapman and Shannon 1985; Shannon 1985; Payne *et al.* 1987) and may be treated briefly here. There is a northward-moving current, the Benguela current, occupying a zone about 200 km wide off the west coast of southern Africa from Cape Town north to about latitude 15° S. It is the eastern part of the South Atlantic gyre and the driving force is the wind associated with the South Atlantic anticyclone. Since the sys-tem is in the southern hemisphere, a wind from the south blowing parallel with the coast gives rise to offshore Ekman transport. The tem-perature anomaly produced by the associated upwelling delineates the area occupied by the system, which is seen to be larger in winter than in summer (Fig. 5.18). The differences in shape of the SST patterns at the two seasons are largely attributable to the fact that the upwelling centred on Luderitz and on Cape Columbine (near Cape Town) reach their maxima in summer, while the very major upwelling at Cape Frio reaches its maximum in winter.

However, as we saw in connection with the California current, the mean temperature anomaly obscures the complexity of the system. A satellite picture taken at any instant shows a complex pattern of whirls and jets, many of them determined primarily by the bottom topography. In some places the cold upwelled water is carried far out into the At-lantic while at other places the upwelled water is confined to a relatively

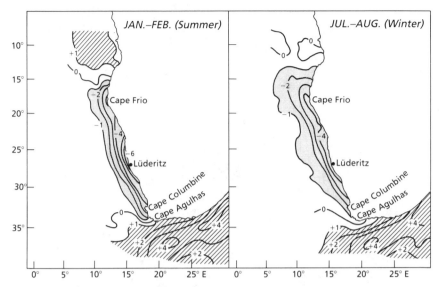

Fig. 5.18 Average sea-surface temperature anomalies in the Benguela current system, in summer and winter. From Parrish *et al.* (1983).

narrow band close to the coast. The picture is further complicated by variations in wind strength on a scale of a few days.

As in other upwelling systems world-wide, *Sardinops* and *Engraulis* are the most abundant fishes. The history of the fishery off southern Africa has parallels with the California fishery. Beginning in the 1950s, there was an intensive fishery for pilchard, *Sardinops ocellata*, with peak landings of over 1.3 million tonnes in 1968. The fishery then declined sharply and has not recovered. Attention was then turned to the anchovy, *Engraulis capensis*, which is now the most abundant species in the South African purse-seine fishery.

The stocks may be roughly divided into those breeding in the Cape area and those breeding further north. For the southern stocks, the main spawning ground for both anchovy and pilchards is the inshore water off the southern tip of Africa, centred on Cape Agulhas and extending to Cape Town in the west and Cape Infanta to the east. The area is not particularly productive of plankton, but during the breeding season has a layer of warm Indian Ocean water at the surface which confers stability in an otherwise turbulent area. Starting at the western end of this spawning area is a frontal jet, the 'Good Hope Jet' which runs northward along the shelf edge for more than 250 km (Fig. 5.19). Shelton and Hutchings (1982) showed that anchovy eggs and young larvae are carried in the jet, and appear as recruits in the upwelling system north of Cape Town. Note that this arrangement, where the anchovy spawning

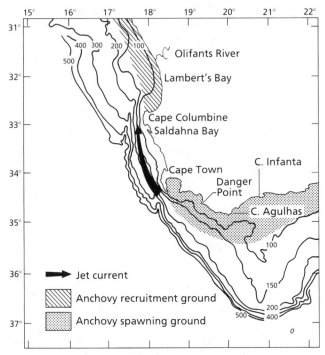

Fig. 5.19 Spawning and recruitment grounds of the South African anchovy, and the jet current that connects them. From Shelton and Hutchings (1982).

ground is 'upstream' of the highly productive upwelling area, is the reverse of that found in southern California where the spawning area is downstream of the intensive upwelling.

In this review of coastal upwelling systems we have seen three categories of fish stocks: (i) stocks poleward of summer upwelling maxima; (ii) stocks equatorward of summer upwelling maxima; and (iii) stocks equatorward of winter upwelling maxima. The anchovy stocks in the Cape Town area, those off the Columbia River in the northern part of the California current system and the most southern (Talcahuano) stock in the Peru system are poleward of summer upwelling. The anchovy stocks in the southern California bight, various stocks in the Canary current system, and the Arica stock in Peru are equatorward of summer upwelling, their spawning grounds being located in bights with weak offshore transport. It is the most northern stocks off Peru, between San Juan and Chimbote that are unique in lying equatorwards of winter upwelling maxima, and in spawning in a region of strong Ekman offshore transport. Historically, these stocks have been the largest in the world. Various explanations have been offered for the ability of the larvae to flourish in these conditions. The continental shelf is much wider than at other upwelling centres (section 5.4) and, being closer to the equator than the others, the Coriolis force is less, so it has a large Rossby radius and a greater Ekman depth. Hence, it is suggested, for a particular wind stress, the offshore water velocity is lower than for other upwelling areas, and since the whole system extends further offshore, the larvae have more time to grow to an actively swimming stage before being advected out of the favourable upwelling system. Parrish *et al.* (1983) observed that for stocks poleward of upwelling areas the maximum biomass has been in the range 0.1–0.4 million tonnes, for those equatorwards of summer upwelling areas it has been 0.8–2.0 million tonnes, but the stocks that are equatorward of winter upwelling have exceeded 10 million tonnes. It is not clear whether the location of the stocks in relation to upwelling centres is the key determinant of productivity, or whether latitude, with its attendant differences in solar radiation and Coriolis force may not have more to do with the matter.

5.7 SOME SMALLER SCALE UPWELLING SYSTEMS

5.7.1 Summer upwelling off Nova Scotia

Upwelling is thought to occur sporadically on the western margins of the major oceans, but few examples have been well documented. A study of summer upwelling off Nova Scotia, Canada, suggests that it is a major mechanism for transporting nutrients onto the shelf from deep water (Petrie *et al.* 1987). First reports of coastal upwelling in this area go back to Hachey (1935). He found that strong winds from the southwest were correlated ($R^2 = -0.7$) with lower than normal sea surface temperatures (SST) such that a wind of about $0.9 \ m \ s^{-1}$ would produce a drop in SST of 1 °C. The observations of Petrie *et al.* (1987) are derived from infrared satellite images. Thirteen of them were available for the period 24 June to 6 August 1984. On 7 July most of the water over the Scotian shelf was around 16 °C except that there was a narrow band of cold water close to the coast. A week later there was a substantial area of water of 7–8 °C extending about 20 km from the coast on average with plumes extending up to 60 km from the coast. By 21 July the average width of the cold water zone was 20–30 km and plumes extended up to 85 km offshore. Four days later there were plumes extending up to 175 km from the coast. Between 25 July and 21 August the zone of cooler water disappeared.

The authors examined records of alongshore wind stress and found that the mean for the period 27 June–27 July was 0.03 Pa. They numerically solved the equations of motion for a two-dimensional, two-layer model of the shelf which included time-varying wind stress, representative bathymetry, bottom and interfacial friction, and the effects of varying layer depths. They chose an initial upper layer depth of 20 m, and a density difference of 3 kg m^{-3}. They began with zero wind stress on 27 June, forced the flow by the alongshore wind stress, set the interfacial stress initially at zero and the horizontal eddy viscosity at 100 m^2 s^{-1}. The model predicted that surface waters would move offshore and the thermocline would reach the surface after 8.1 days. The infrared images were spaced about 7 days apart, but the upwelling first appeared on the image of 7 July, 10 days after the start of the winds. This was taken as good agreement with the model.

The pycnocline was found to upwell at a rate of 20 m in 10 days, over a coastal strip about 10 km wide and 500 km long. This gives a vertical transport of about 1×10^5 m^3 s^{-1}. At a mean nitrogen concentration of 10 μg-at N l^{-1} the authors calculated that the upwelling of nutrients amounted to 10 μg-at N s^{-1} per 1-cm-wide strip of shelf normal to the coast. This is approximately equal to previously calculated nutrient requirements of the phytoplankton. It was therefore concluded that coastal upwelling is an important mechanism supporting primary production on the Scotian shelf.

The satellite images also showed that nearly stationary wave-like features with scales of 50–75 km appeared to grow rapidly in the upwelled temperature front. Results of modelling exercises suggested that that these were caused by baroclinic instability. The properties of fronts will be dealt with in detail in Chapter 6.

5.7.2 Summer upwelling on the west coast of Spain

The northwest corner of Spain, near Cape Finisterre is near the northern limit of the upwelling associated with the Canary current system (Wooster *et al.* 1976). Blanton *et al.* (1984) calculated the Bakun upwelling index off Cape Finisterre and showed that in most years winds from the north predominate, causing strong offshore Ekman transport. The authors argue that offshore Ekman transport lowers the sea level at the mouths of the coastal inlets (rias) setting up a seaward pressure gradient that forces the water above the pycnocline out of the rias. Oceanic water upwelled onto the continental shelf then enters the ria close to the bottom. Figure 5.20 shows data taken from the Ria de Vigo for 1977. The upper histogram shows monthly averages of upwelling indices. The offshore transport in the summer months (shaded) is clearly seen. When this occurs, cold (<12 °C), nutrient-rich (>5 μm NO$_3$) water is brought into the rìa at 20–50 m depth. Blanton *et*

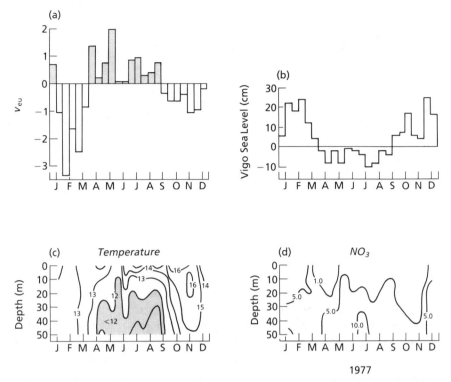

Fig. 5.20 Physical and chemical data from the mouth of the Ria de Vigo, Spain (1978). From Blanton *et al.* (1987): (a) monthly averages of upwelling index. Positive values (shaded) imply offshore transport; (b) monthly averages of mean sea level; offshore transport corresponds with low sea level; (c) annual pattern of temperature distribution with depth (note strong stratification June–October); and (d) annual pattern of nutrient distribution with depth, with high surface-nutrient levels during winter mixing.

al. (1984) also showed that the topography of Cape Finisterre had the effect of intensifying the upwelling.

Tenore *et al.* (1982) have documented the enormous biological productivity of these Spanish rias, especially in the form of cultured mussels. Two thousand 20×20 m rafts in Ria de Arosa produce annually over 100,000 tonnes total weight of mussels, *Mytilus edulis*. The upwelling and subsequent translocation of nutrient-rich water into the rias is believed to be the secret of this high level of productivity. Blanton *et al.* (1987) used atmospheric pressure maps to calculate the upwelling indices for a point off the western coast of Spain for the period April–September in each of the years 1973–1983. They found that the condition index of the mussels (measured as % solids in the meat) was well correlated with the upwelling index. ($R^2 = 0.66$, Fig. 5.21). In the best year, 1977, the meat had 15.8% solids, while in the worst year, 1983, it had only 9.5% solids. The authors suggested that if a prediction could

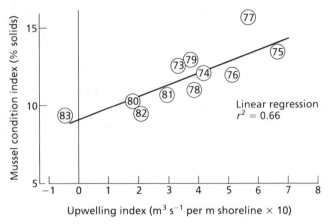

Upwelling index (m^3 s^{-1} per m shoreline $\times$ 10)

Fig. 5.21 Correlation between mussel condition index (% solids in meat) and upwelling index. Numbers indicate years. From Blanton *et al.* (1987).

be made in the early spring about the most probable intensity of up-welling in the ensuing summer, the mussel culture could be managed more economically. They pointed out that the winds off Spain are influenced by the relative positions of the high pressure atmospheric cell off the Azores (the 'Azores high') and the low pressure cell off Greenland (the 'Labrador' or 'Icelandic low'). Long range forecasts of synoptic-scale weather patterns could be used to provide the appropriate forecasts of pressure gradients and hence of wind strengths, Ekman transports and the estimated carrying capacity of the rias for the ensuing season.

It seems highly probable that the productivity patterns of a wide range of coastal habitats are affected by the incidence of wind-induced coastal upwelling. We have evidence that it can occur on the western sides of ocean basins as well as in eastern boundary currents. The extent of the phenomenon appears to be a fruitful line of enquiry.

5.8 PRODUCTIVITY DEDUCED FROM HYDROGRAPHIC AND CHEMICAL FIELDS

Minas *et al.* (1986) attempted to estimate 'net community production' (NCP) in three major upwelling areas from hydrographic and chemical fields rather than from *in situ* measurements of ^{14}C uptake, arguing that the latter are the subject of much debate about their validity, and are difficult to interpret in a situation where the water masses are in constant vertical as well as horizontal motion. The authors used data sets from Peru (15° S), northwest Africa (21° N) and the Benguela system

(20° S). The principle of their model, originally devised by Broenkow (1965), is illustrated in Fig. 5.22(a). Suppose point A represents the salinity (S) and oxygen content (O_2) of the source water that is being upwelled from depth, and B represents the same for offshore surface water. We may assume that water in various parts of the upwelling system is some mixture of these two types. If there were no change in oxygen content of water as a result of photosynthesis, respiration or uptake from the atmosphere, any mixture whose salinity lay intermediate between those of A and B would have an oxygen concentration corresponding with points along the mixing line AB, which has the equation:

$$\{O_2\}\text{mix} = \{O_2\}A \times (S_{\text{mix}} - SB)/(SA - SB) + \{O_2\}B$$
$$\times (SA - S_{\text{mix}})/(SA - SB).$$

Samples taken in the upwelling area had the values shown in the figure; departure from the mixing line (vertical distance) must be due to photosynthesis, respiration or uptake from the atmosphere. To distinguish the biological effects from atmospheric effects, a parallel calculation may be made for nutrients (Fig. 5.22(b)). Any departure from the mixing line indicates biological uptake of nitrate. Using an appropriate equivalence between nitrogen uptake and oxygen evolution, the authors calculated how much of the oxygen departure is attributable to biological processes, i.e. photosynthesis minus community respiration, and attributed the balance to atmospheric input. Calculation of the rate of production from oxygen production and nutrient consumption requires determination of the time scale for the observed changes. Minas *et al.* (1986) did

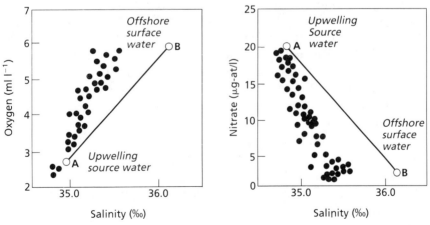

Fig. 5.22 Mixing lines and sample values of salinity with (A) oxygen and (B) nitrate in an upwelling area. For explanation, see text.

this by using Bowden's (1977) procedure. The first step is to calculate the temperature increase ΔT in the upwelled waters. This is done by creating a mixing line on a temperature–salinity diagram and determining the vertical distance between this line and the location of each sample on the diagram. The total heat gain ΔQ is calculated by integration. To calculate the time interval Δt, in days, appropriate to the heat gain, the net daily solar heat input Q_0, taken from published charts, is inserted in the formula

$$\Delta t = \Delta Q/Q_0 \, .$$

When this procedure was applied to various upwelling systems it was found that for a data set off northwest Africa the time to effect the observed heat increase was only 6.64 days, while for a transect off southwest Africa it was 63.1 days. Two transects from Peru gave intermediate values. When the average rates of net community production were calculated for the various upwelling systems it was found that, on the days when the data sets were taken, NCP averaged $0.59 \, \mathrm{g\,C\,m^{-2}\,d^{-1}}$ off Peru, 1.06 off southwest Africa and 2.51 off northwest Africa.

Minas *et al.* (1986) then went on to discuss the fact that off northwest Africa there was good agreement between observed average ^{14}C uptake $(2.31 \, \mathrm{g\,C\,m^{-2}\,d^{-1}})$ and the calculated figure of 2.51, but off Peru there was a wide discrepancy, $2.76 \, \mathrm{g\,C\,m^{-2}\,d^{-1}}$ observed, but 0.59 calculated. There was evidence of a much larger biomass of herbivores off Peru than off northwest Africa at the time of the observations. This would have two consequences: (i) the respiration of the herbivores would depress the apparent NCP, which results from the oxygen evolved in photosynthesis minus the respiration of the plants and the respiration of the animals; (ii) the herbivores would provide substantial amounts of regenerated nitrogen, which would boost the ^{14}C uptake but not the NCP calculated by this method, which is only the 'new' production. The authors further pointed out that when ^{14}C incubations are performed in small bottles the herbivores are excluded. This allows the phytoplankton biomass to increase beyond what happens in nature, and hence may inflate the ^{14}C uptake measurement. Finally, they pointed out that net community production per unit area is not the best measure of the productivity of an upwelling system. For example, the Peru system extends much further from the coast than the other two, and productivity per kilometre of coastline is greater for Peru than for northwest Africa.

Since the method just described gives an estimate of 'new' production based on nitrate uptake, Minas *et al.* (1986) combined their estimates with the estimates of Smith and Whitledge (1977) for ammonium utilization in the same waters, and concluded that the f ratio for northwest Africa was 0.64, indicating that upwelled nitrogen was responsible for about two-thirds of the total production in that upwelling system. The authors also pointed out that differences in grazing pressure on the phytoplankton were reflected in differences in rate of nutrient uptake. In situations such as that seen off Peru where an abundance of herbivores restricted the phytoplankton biomass and depressed the rate of utilization of nitrate, the system exported a higher proportion of its nutrients to offshore water, whereas in a system with low herbivore biomass the nutrients are taken up relatively rapidly by the phytoplankton and export is in the form of phytoplankton biomass.

Looking back over this approach by Minas *et al.* (1986) we see that it provides a fresh point of view on the biological processes occurring in an upwelling system. However, it suffers from the limitation that each calculation applies only to one section at one time. Since very few sections have been studied in sufficient detail to permit the calculations, we have a rather limited view of the activities of a particular upwelling system. Furthermore, it measures net community production, which is difficult to interpret without detailed knowledge of the populations of heterotrophs present in the water column.

5.9 CONCLUSIONS

Reviewing this chapter, we see that major anticyclonic gyres in the atmosphere associated with similar gyres in the four major ocean basins — North and South Atlantic, North and South Pacific — give rise to equatorward winds parallel with the coast along the eastern margins of those basins. The combination of wind stress and the Coriolis effect determine that Ekman transport will take place, i.e. surface waters will move away from the coast. Deeper waters upwell along the coast to take their place, but the intensity of vertical transport is not uniform. It appears to be influenced by topographic features such as headlands and canyons, so that satellite images show complex patterns of intense upwelling which in the course of time give rise to complex plumes, jets, squirts and gyres that penetrate varying distances from the coast into the main stream of the boundary current.

This complexity is further compounded by temporal variation in wind strength. Off Peru the process may be continuous more or less year round but off Oregon, for example, it is confined to the summer season. This seasonality is in part explained by the tendency for the major anticyclones to move poleward in summer and equatorward in winter. Superimposed on this seasonality is 'event'-scale variability in which wind stress and the associated upwelling event tend to build and decline over a period of 5–10 days.

Rather surprisingly, it seems that the event-scale variability tends to enhance biological productivity rather than the reverse. There is evidence from several locations that growth of phytoplankton populations (and also feeding by early larval stages of fish) occurs most rapidly when wind stress is low and the water column is well stratified. There seems to be a parallel with the 'Gran effect' discussed in section 3.3.2. In temperate waters it is the alternation between winter mixing and spring stratification that provides conditions required for the spring bloom. In upwelling areas the period of maximum wind stress provides for the upwelling of nutrients and the periods between provide well-stratified conditions in which the phytoplankton is held in or near the euphotic zone, permitting high photosynthetic rate, rapid uptake of nutrients, cell division and population growth.

We also saw in Chapter 3 that during well-stratified periods there is a tendency for the phytoplankton to accumulate in a chlorophyll maximum just above the pycnocline. It seems that these aggregations of phytoplankton cells are necessary for some kinds of zooplankton and for herbivorous fish to feed at their optimal rate. Hence, the times of low wind stress in upwelling areas are the times when strong chlorophyll maxima are formed and when the first-feeding larvae of sardine and anchovy encounter sufficient food for them to make the difficult transition from passive, yolk-absorbing creatures to active herbivores. The timing of their life histories is adapted to the average timing of the upwelling events. Any major disruption of the timetable leads to poor larval survival and hence a poor year-class of fish.

The fish stocks of upwelling areas world-wide are dominated by clupeids such as sardines, pilchards and anchovies, with mackerel and hakes also being abundant. The clupeids, for the most part, tend to avoid the areas of strong Ekman transport at spawning time. Those in higher latitudes tend to spawn upstream of the major upwelling areas, those in lower latitudes tend to spawn in sheltered bights with relatively little upwelling, located downstream of the major upwelling areas. Historically, the lower latitude stocks have had the higher biomasses. The lower latitude stocks off Peru have had the largest biomass of all, and these appear to be able to spawn right in the plumes of the major upwelling areas. It has been suggested that it is possible because in the Peru upwelling system, relative to the others, the Ekman transport is less intense, less variable, and extends further from shore.

6

Fronts in Coastal Waters

6.1 INTRODUCTION

It is obvious from the preceding two chapters that continental shelves are mosaics of water masses with different properties. The estuaries and estuarine plumes often have lowered salinities as a result of fresh water run-off, while adjacent waters may be fully saline. Some areas are strongly mixed by tidal currents while the next water mass is stratified. At the boundaries between them there are normally sharp horizontal changes in temperature and other properties.

The waters over the continental shelves are usually less saline than the open ocean, but more prone to be cold in winter and warm in summer. The shelf-break front marks the point of transition. Almost all fronts are regions of enhanced plankton production and this leads to higher fish production. Mooers *et al.* (1978) showed that the shelf-break front off eastern North America moves considerable distances under the influence of winds and the meanders of the Gulf Stream. A long-term plot of the positions occupied by the shelf-break front delineates an envelope within which are found most of the highly productive fisheries of the region. An aerial survey of the distribution of fishing vessels on the east coast of Canada showed the highest concentrations in the vicinity of the shelf-break front (Fournier 1978).

In the coastal upwelling situation that we described in the last chapter, upwelled water moves away from the coast by Ekman transport and eventually arrives at a region where the surface water will be pushed back no further, so the upwelled water sinks under it (Fig. 6.11). This is a highly productive frontal region, and one to which large numbers of tuna are attracted.

There is no agreement about the classification of fronts, but a partial listing of those occurring in coastal waters would be: tidal fronts (also known as shallow-sea fronts), shelf-break fronts, upwelling fronts, estuarine fronts, plume fronts, and fronts associated with geomorphic features such as headlands, islands and canyons. In this chapter we investigate the distinctive physical and biological properties of each.

6.2 THE PHYSICS OF FRONTS

6.2.1 Definitions, examples and causes

In the ocean, variables such as temperature, salinity and density do not vary gradually with horizontal distance. Instead we find large regions where horizontal variations are small, bounded by narrow regions where horizontal gradients are large. The narrow high-gradient regions are called fronts following the meteorological custom of naming regions of high horizontal temperature gradient in the atmosphere. Fronts in the ocean come in many sizes. The fronts associated with the large western-boundary currents such as the Gulf Stream, for example, can be easily traced for thousands of kilometres, exhibiting a temperature change of 10 °C in 50 km near the surface and extending with diminishing contrast to the bottom at 4000 m. At the other end of the scale small transient fronts are generated and destroyed by tidal flow in the upper layers of coastal waters every day. These fronts are typically a few kilometres long, confined to the upper layers and separate waters which differ by only 1–2 °C.

The infrared satellite image of the western North Atlantic Ocean shown in Plate 2 contains a number of examples of fronts. The warmest water in the picture is the orange and yellow water at the latitude of Cuba and Florida which flows north in the Gulf Stream, that is, the long crooked finger of yellow lying close to the coast past Cape Hatteras to the latitude of Chesapeake Bay where it turns eastward. On both sides of the flow, and especially to the west or north, the colour changes rapidly to green and blue, indicating a rapid drop in temperature. From shipboard measurements, we know these temperature fronts are intimately tied to similar changes in salinity and density. Thus a map of one variable is similar to the map of the others.

Across fronts like the one associated with the Gulf Stream, properties are distributed as illustrated in Fig. 6.01(a). Density is used in the illustration because it is the important variable when considering the dynamics of the flow. On either side of the current, in regions A and C, the isopycnals are flat and no horizontal gradient exists. In section B, where the geostrophically balanced current flows perpendicular to the page, the isopycnals rise from one side of the current to the other creating a horizontal gradient in density. Both the vertical and horizontal density gradients diminish with depth as the strength of the current decreases.

In Fig. 6.01(b), for comparison, is an illustration of the idealized density distribution across the typical tidal front discussed in section 4.6.1. In the tidally-mixed front the density of the shallow-mixed water is constant with depth and intermediate between the surface and bottom density of the stratified side which lies in greater depths. The isopycnals therefore diverge across these fronts; some sloping up and some sloping down.

The front associated with the Gulf Stream is not the only one visible in Plate 2. In the area between the Gulf Stream and Nova Scotia, where the cold red-coloured water lies, there are a number of situations

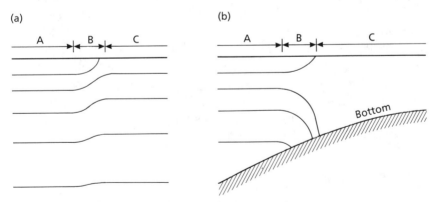

Fig. 6.01 (a) Density section through a geostrophically balanced current in which the flow is out of the page (in the northern hemisphere) and decreases with depth. The current is confined to the region B where there exists a horizontal density gradient. In regions A and C, where the isopycnals are flat there is no current. (b) Density section through a front created by an increase in the turbulent stirring in the shallower water. Region A is vertically stratified while region C is completely mixed by the turbulence. Horizontal density gradients exist in region B, the 'tidal front'.

where the colour jumps from one fairly uniform value to another in a narrow band. Some of these features are tied to permanent currents which are guided over the continental shelf by the bathymetry. Others are signatures of the large-scale turbulent motions which slowly stir the waters of different temperatures being brought together in the region. The density structure in these fronts is similar, except for the size, to that found in the large ocean currents illustrated in Fig. 6.01(a).

Images similar to Plate 2 are now available from all parts of the world ocean and from these it is clear that fronts are a ubiquitous feature of the sea surface. Some are created, as mentioned, by the large currents and tidal mixing, others by large turbulent eddies, or by the coastal upwelling discussed in Chapter 5. Tidal currents have also been found to cause fronts at the edge of the continental shelf where the flow of water interacts with the rapidly changing bathymetry. The large internal waves which are generated affect the temperature of the surface layer. This type of front, because it is closely associated with the tides, is discussed in the next chapter.

Fronts are important to us because higher biological activity is often associated with them, as has been documented by Pingree *et al.* (1975) and Le Fèvre (1986) who found dense aggregations of phytoplankton and zooplankton along tidal fronts in the English Channel. Our purpose in this section is to explore the physical processes that may account for this increased production. We are looking specifically for processes which can transport nutrients into the stratified euphotic zones where biological activity is fast enough to maintain growth. Our task, however, is not an easy one. The field is new, exciting, changing rapidly and there is no consensus yet on what the dominant processes are in the different situations. Some of the processes advocated today will be shown to be unimportant in a few months as new observations provide a better understanding. Part of the difficulty is that fronts are relatively narrow features with high gradients. Observing the changes in these fields at the necessary time and length scales requires a large repertoire of highly sophisticated *in situ* instrumentation coupled with accurate navigation and powerful data processing.

6.2.2 Tidal fronts

These fronts, as we discussed in section 4.6.1, sometimes occur over the continental shelves where the stratified water characteristic of the deeper areas lies beside unstratified water characteristic of shallower areas. The strong mixing found on the unstratified side of the front is a result of the high levels of turbulence generated at the bottom by the tidal currents which tend to be faster in the shallower regions. A useful illustration of a typical tidal front and some of the processes associated with it is shown in Fig. 6.02. In this section the existence of the front is assumed and the problem is to find and explain the mechanisms which transport nutrients into the front to maintain the higher observed levels of primary production.

One mechanism often cited for this purpose and first suggested by Pingree *et al.* (1975) is peculiar to tidal fronts. It arises because the speed of tidal currents varies over the lunar month (28 days) between the spring and neap tides (Chapter 7). As the current speed increases the energy available for mixing increases. This presumably causes the level of turbulence generated at the bottom to increase. The depth of water where the turbulence is energetic enough to break down the stratification will increase. Thus the boundary between the mixed and stratified water will move towards deeper water, decreasing the area of stratified water. On the other hand, when the tidal currents decrease, the turbulence declines and the front moves towards shallower water, allowing the area of stratified water to increase again.

The newly stratified water behind the advancing front will contain nutrient levels characteristic of the previously mixed water. Observations of nutrients across these features (Simpson and Hunter, 1974)

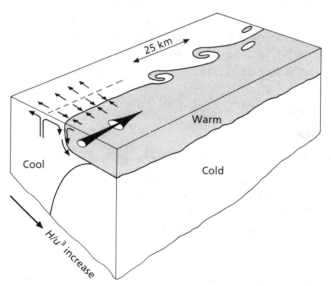

Fig. 6.02 Diagram of frontal structure and circulation, after Simpson (1981). Note strong along-front mean flow, convergence and downwelling at the front, upwelling on the well-mixed side, and frontal eddies, some of which close on themselves to form isolated patches.

show that nutrient concentrations tend to be higher in the mixed region than in the upper stratified layers, where presumably they have been depleted by biological activity earlier in the year. Thus the newly formed part of the front should contain higher concentrations of nutrient than the water which has been stratified for along time.

A simple calculation of the nutrient flux into the stratified side of the front by the tidal variation has been used by Loder and Platt (1985). They proposed that if the front moves L km between the neap and spring tides and the depth of the upper mixed layer (D) remains constant throughout the advance, then the amount of nutrient captured by the advancing front is just $L \times D \times \Delta C$, where ΔC is the increase in the nutrient concentration from the stratified to the mixed side of the front. If the front advances in a time T then the rate at which the nutrient is captured per unit length of front and per unit time is

$$F_m = L \times D \times \Delta C / T, \tag{6.01}$$

where F_m is in units of mg-at $m^{-1} s^{-1}$.

The effectiveness of this process critically depends on the length of the frontal excursion (L). This is examined by Simpson and Bowers (1981) who created a model to predict the position of the tidally-mixed front at various times throughout the summer season depending on the variations in the current speed. Their first attempt predicted a lateral spring-neap movement three times the observed distance. This discrepancy was thought to be due to the fact that the efficiency with which the turbulence mixes the water changes with the degree of stratification. As the stratification increases it takes more mixing energy to stir the water up. Incorporating this effect decreased the excursion in the model closer to the observed values.

6.2.3 Eddies

The line on the surface of the water which indicates the presence of a front tends to be irregular in shape rather than straight, as is obvious in Plate 2. These irregularities often appear to be eddies which twist the cold and warm waters together in a circular fashion. Such features are thought to be created by instabilities in the front which allow small departures from the geostrophic balance to grow at the expense of the potential and/or kinetic energy in the front. These are the same processes generating the cyclonic lows (eddies) in the mid-latitude atmosphere which provide the variable weather and storms of those latitudes. The frontal eddies are of interest to us because it is often suggested (Pingree 1978; Garrett and Loder 1981; Loder and Platt 1985) that they cause water to be exchanged across the front and thereby contribute a significant flux of nutrients.

In an attempt to examine the potential effects of eddies on the biology of fronts Pingree (1979) analysed a number of eddies in the front found at the Celtic Sea shelf break. His analysis showed the eddies had a diameter of 20–40 km and may have been created by baroclinic instability. His suggestion, which has been taken up by others, is that the eddies transport nutrients across the front and probably contribute to the enhanced biological productivity in the front. Building on the theoretical work of meteorologists he derived the equation

$$F_E = \gamma (gD\Delta\rho/\rho)^{1/2}D\Delta C, \tag{6.02}$$

to estimate the flux, by the eddies, of nutrient across the front. F_E is the flux of the nutrient C which exhibits a difference ΔC across the front, ρ is the density, $\Delta\rho$ is the density difference across the front, D is the depth of the upper layer, g is the acceleration due to gravity and γ is a constant derived from theoretical considerations and equal to 0.0055.

An alternative method of calculating the nutrient flux due to eddies was suggested by Garrett and Loder (1981). In their scheme the nutrient flux is just the number of eddies times the eddy volume times the difference in the nutrient concentration, ΔC. The radius of the eddies is assumed to be four times the Rossby deformation scale, R_i (section 5.2.3), because this is the length of the fastest growing perturbation in baroclinic instability theory. The volume of an eddy of depth D is then $16\pi R_i^2 D$ and the supply of nutrient is $16\pi R_i^2 D \times \Delta C$. If eddies are exchanged across the front every $16R_i$ in time T the nutrient flux is

$$F_E = \pi R_i \times D \times \Delta C/T. \tag{6.03}$$

We shall return to this equation in section 6.3.3 where the relative magnitudes of the various processes are discussed.

6.2.4 Vertical eddy diffusion

The increase in the rate of vertical eddy diffusion from the stratified side to the mixed side of a tidal front may also be an important factor in the higher biological activity found in the front. The increase is caused by the increasing intensity of the turbulence towards the mixed side which parallels the decrease in the vertical density gradient. Tett (1981), in a mathematical model, used this horizontal change in the vertical eddy diffusivity to explain the increased chlorophyll concentrations in the Ushant front observed by Pingree *et al.* (1975). An outline of the situation is shown in Fig. 6.03.

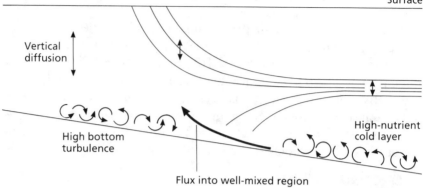

Surface

Vertical diffusion

High bottom turbulence

High-nutrient cold layer

Flux into well-mixed region

Fig. 6.03 Density section through a tidally-mixed front with vertical arrows indicating the increase in the rate of vertical eddy diffusion from the stratified side to the mixed side. At the bottom higher levels of turbulence help to transport nutrients from the cold, high-nutrient layer into the fully-mixed region.

On the left the water is homogeneous because it is kept well mixed by the high vertical diffusion created by the tidal stirring. On the right the water stays stratified because the vertical diffusion across the pycnocline is low. In the front the vertical diffusion varies between the two extremes and will be higher than where the stratification is fully developed. If the nutrient concentration increases with depth as observed by Pingree *et al.* (1975), the higher vertical diffusion rate in the front results in a greater upward flux of nutrient there than on the stratified side.

The model produces encouraging results by predicting chlorophyll distributions with many of the observed characteristics. It works because the vertical diffusivity is high enough in the front to supply the required nutrients but it is also low enough to keep the phytoplankton cells in the euphotic layer long enough for them to multiply.

Vertical diffusion in the near-bottom waters is also likely to be an important consideration in the nutrient balance near tidally-mixed fronts, especially those which have cold nutrient-rich lower layers on the stratified side (Le Fèvre 1986). These layers which are remnants of the winter conditions tend to contain higher nutrient concentrations than in either the fully mixed region or the upper layer on the stratified side. The turbulence created by the high tidal currents is most energetic at the bottom and at the front is thought to stir the water enough to mix some of the cold high-nutrient layer into the fully-mixed region where it gets distributed throughout the water column (Fig. 6.03). Such a process occurs during each tidal cycle but may increase with the increasing currents of the spring tides. New investigations with instruments capable of measuring levels of turbulence in the water coupled with detailed nutrient distributions will be required to quantify this flux and estimate its importance.

6.2.5 Convergences and downwelling

From a ship the first indication of a front is usually a long narrow band of seaweed, plastic garbage bags, etc., on the surface and feeding birds in the air. The water in the band is also quite smooth compared to the surrounding waters indicating a concentration of floating organic matter on the surface which damps the small waves. These observations indicate the surface waters are converging towards the front, bringing the floating material together, before sinking beneath the front. A well-documented visit to a front with a strong convergence is given by Pingree *et al.* (1974).

They described a rather violent tidally-mixed front between Guernsey and Jersey in the English Channel where direct observations both at the surface and underwater showed that the water on one side of the front was flowing towards then underneath the water on the other side of the front. The authors concluded from physical measurements that the front was not in geostrophic balance. This implies that forces other than those due to the rotation of the earth and the pressure gradient controlled the flow. This more complicated situation makes it harder to understand the physics of the front without more sophisticated measurements.

6.2.6 Upwelling due to friction in a nearly geostrophic current

Garrett and Loder (1981) demonstrated that friction in the water, small as it is, could produce an important vertical component to the flow in fronts which are to a first approximation geostrophically balanced. Under the right conditions the vertical circulation could deliver water containing a higher nutrient concentration to the euphotic layer in the centre of the front.

Consider first the ideal front in which a pressure gradient perpendicular to the front drives a current parallel to the front. If there is no friction in the water, the pressure gradient is exactly balanced by the Coriolis force. The process described by Garrett and Loder (1981) stems from the fact that the Coriolis force and the pressure gradient are not exactly balanced. This occurs because friction causes the velocity to be slightly less than would be achieved in a perfect geostrophic balance. The force due to the pressure gradient therefore is slightly greater than the Coriolis force and a small component of velocity is generated down the pressure gradient. This component is perpendicular to the geostrophic flow and tends to be high in the centre of the front where the geostrophic flow is high. We end up with the situation shown in Fig. 6.04 where the small cross-stream velocities u_f are shown increasing from each side of the front towards the centre. These cross-stream flows create a slight horizontal convergence to the left of the centre and a horizontal divergence to the right of the centre. Because the water is incompressible such convergences and divergences in the horizontal are compensated for by vertical flows. These are expected to be as indicated in the figure; up on the left and down on the right. The vertical flow has been estimated near the centre of a large front to be about 1 mm d^{-1} or about 1 litre d^{-1} up through each square metre. If the upwelling water contains 5 mg-at l^{-1} of nitrate the upwelling delivers $\simeq 5 \text{ mg} - \text{at m}^{-2} \text{d}^{-1}$.

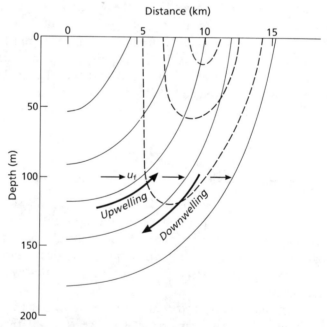

Fig. 6.04 Isopycnals (solid) and velocity (dotted) contours through a front. The small cross-stream velocity, u_f, due to friction creates a divergence and upwelling on the left of the centre and convergence and downwelling on the right.

6.3 THE BIOLOGY OF FRONTS

6.3.1 Tidal (or shelf-sea) fronts

A concise description of the mechanism underlying the occurrence of shelf-sea fronts was given by Simpson (1981):

> Over the continental shelf tidal flows exert strong frictional stresses on the seabed. In stress terms, these tidal streams are equivalent to hurricane-force winds in the atmosphere blowing regularly twice a day. As well as playing an important part in the tidal dynamics, these stresses serve to produce turbulent kinetic energy which is responsible for vertical mixing.
>
> Variations in the level of tidal stirring divide the shelf seas, during the summer regime, into well mixed and stratified zones separated by high gradient regions called fronts.

The details of tidal waves and currents will be discussed in Chapter 7, but we may note in passing that most of the dissipation of tidal energy occurs on continental shelves, and the extensive shelf seas of northwest Europe are believed to dissipate one-eighth of the world's tidal energy, about 19×10^{10} J s^{-1} (Miller 1966). One of the earliest systematic studies of a shelf-sea front was Simpson's (1971) study of a persistent front in the Irish Sea. His record of the profile is shown in Fig. 6.05. As the diagram shows, there were tidally-mixed waters at

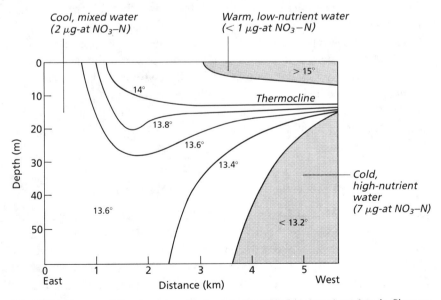

Fig. 6.05 Diagram of the structure of a front in the Irish Sea, based on data in Simpson (1971) and Simpson and Hunter (1974).

13.6 °C on the eastern side of the front and stratified waters (>15 °C at the surface, < 13.2 °C near the bottom) on the western side. As one passed over the front there was a change of surface temperature of 1.4 °C within about 2 km. In a later study of the same front, Simpson and Hunter (1974) reported that in August the nitrate-nitrogen content of the tidally-mixed water was about 2 μg-at l^{-1} while that of the surface water in the stratified region was <1 μg-at l^{-1}. On the other hand, the deep water in the stratified region contained 7 μg-at l^{-1}. They estimated that the chlorophyll content of the mixed water was about 1.5 μg l^{-1} while that in the surface water of the stratified layer was only 0.5 μg l^{-1}. A moment's reflection on these results (see Fig. 6.05) suggests that if there was a mechanism for mixing the nutrient-rich deep water from the stratified region into the mixed region, or a mechanism for mixing the somewhat nutrient-enriched water from the mixed region into the stratified surface layer, primary production could be expected to increase.

During the early 1970s biologists had been noticing that shelf-sea fronts were often the sites of persistent high densities of phytoplankton and zooplankton. Le Fèvre and Grall (1970), investigating the distribution of *Noctiluca* (a phytoplankter which, in high abundance, imparts a blood-red colour to the water, known as a 'red tide') noted that red tides occurred in a region off western Brittany characterized by convergence of surface waters with tidally-mixed waters on the landward side and stratified waters on the seaward side. It was later shown that this frontal structure extends northwards towards the coast of England, across the mouth of the English Channel and is now known as the Ushant front. Pingree *et al.* (1975) reported that there was high phytoplankton biomass along it, and that the most abundant organism in the phytoplankton was *Gymnodinium aureolum,* a species also noted for red tide formation.

Since the northwest European shelf area was the site of these pioneering studies of shelf-sea fronts, we may make this area the subject of our first case study.

6.3.2   Biological production at northwest European shelf-sea fronts

After the work of Le Fèvre and Grall (1970), the next biological study on the Ushant front was that of Pingree *et al.* (1974). They worked on a section of the front between Jersey and Guernsey and drew particular attention to a strong accumulation of floating seaweed, accompanied by numerous crustaceans, fish and birds, along a line that was obviously a site of convergence and sinking. The authors drew attention to the possibility that high biomass of phytoplankton at a shelf-sea front might be as much due to convergence as to stimulated growth of phytoplankton at the front. The physics of this situation was discussed in section 6.2.5.

On the next cruise Pingree *et al.* (1975) made a series of observations across the Ushant front in the mouth of the English Channel and to the west of Brittany. On each crossing of the front they encountered a chlorophyll peak. A detailed section (Fig. 6.06) showed that there was a persistent chlorophyll maximum from the surface down to the pycnocline on the stratified side of the front, and a subsurface chlorophyll maximum in the pycnocline some distance behind the front. The suggested explanation was that during the spring tide period of maximum tidal currents the front is pushed back into the previously stratified area, so that nutrients are brought to the surface. When the tidal currents

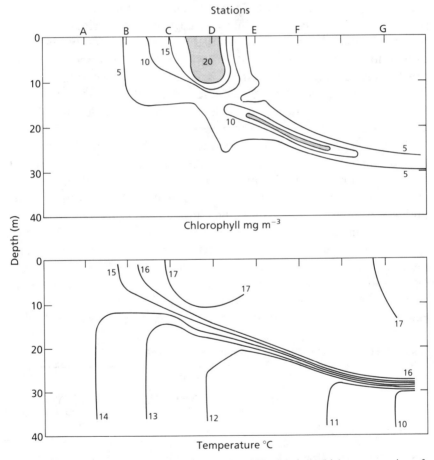

Fig. 6.06 Section through the Ushant front, July 1975, showing a high concentration of chlorophyll near the surface on the stratified side of the front and another high concentration lying just above the thermocline. After Pingree *et al.* (1975).

relax during the neap tides, the waters close to the front that have previously been tidally mixed become stratified and there is a burst of phytoplankton growth. Simpson and Pingree (1978) obtained similar results. They also obtained records of internal wave activity in the stratified region and explained the subsurface chlorophyll maximum in the pycnocline just behind the front as resulting from vertical transport of nutrients across the pycnocline during the passage of internal waves.

In the earlier study (Pingree *et al.* 1975) measurements had been made of the parameters of primary production. In addition to chlorophyll, they measured the light extinction coefficient, the assimilation index (rate of carbon fixed per milligram chlorophyll-a), and integrated water column primary productivity, which they expressed relative to a value of unity for the vertically-mixed regime (Table 6.1). In the stratified waters away from the frontal region they found that primary production below the thermocline was negligible, but primary production within the thermocline accounted for at least 50% of the total carbon fixed. In the frontal region with a weak thermocline, water column production was 6.5 times greater than in the tidally-mixed area but it was achieved by having more than 40 times the concentration of chlorophyll, which had a relatively low assimilation index.

Subsequently Tett (1981) constructed a one-dimensional model based on Pingree *et al.* (1975). While acknowledging that several mechanisms for the enhancement of primary production might be at work, he singled out vertical mixing as the dominant factor. He modelled the upper layer of the stratified region as nutrient-limited, and the lower layer as light-limited. In the tidally-mixed area he considered that the phytoplankton were light-limited because they experienced too low an average level of illumination during their vertical excursions. In the frontal region, where the pycnocline shallows, he considered that the phytoplankton were subject to minimal turbulent transport and were held in a region of moderate illumination from above and moderate rates of nutrient supply by diffusion from below. Model parameters were taken from relevant values in

Table 6.1 Primary production data from Pingree *et al.* (1975).

			Stratified		
			Wind-mixed		Bottom-mixed
	Mixed	Frontal	layer	Thermocline	layer
Chlorophyll-a (mg m^{-3})	0.45	19.0	0.55	9.5	0.66
Extinction coeff. k (m^{-1})	0.12	0.34	0.10	0.1–0.3	0.10
Assimilation index (AI) [mg C h^{-1} (mg chl-a)$^{-1}$]	4.2	1.9	2.2	5.7	0.2
Relative production rate (P)	1.0	6.5	0.6	0.7–1.1	0.05

the literature, except that vertical diffusivity was computed from observed vertical temperature gradients and appropriate values for the mean downward heat flux in summer and physical conditions were set to resemble the situation at a front near the Scilly Isles.

The model was used to simulate changes in stratified, mixed and frontal stations for 40 days, by which time a steady state had been reached and it was found that the distribution of nutrients and chlorophyll closely resembled the natural situation at the front. In discussion, the author emphasized that his aim was only to show that it is *possible* to explain much of the distribution of chlorophyll at a front as a result solely of vertical mixing of nutrients and chlorophyll. He made no claim that this mechanism was the only process at work.

Savidge (1976) studied the Celtic Sea front and showed that as he steamed across it there was a change in surface temperature of about 4 °C within 6–7 km, and that there was a sharp peak of chlorophyll associated with it. He suggested that waters from each side of the front might each be deficient in some important ingredient, not necessarily the same one, and that mixing of water types through the frontal structure might lead to enhanced primary production. By means of shipboard experiments he confirmed that when water from either side was mixed with a relatively small quantity of filtered water from the opposite side, production rate doubled. The hypothesis has been expressed in the phrase 'complementation of waters of inherently different properties induces fertility'. It is rather more complex than the hypothesis of Pingree *et al.* (1975) in which the simple addition of nutrients to stratified water was enough to stimulate phytoplankton production.

On another cruise to the Celtic Sea front, Savidge and Foster (1978) encountered a more complex structure. They passed first through the main front. Just beyond this was a zone of upwelling marked by a temperature minimum. Further away from the front lay a lens of warm water separated from the main mass of cool, tidally-mixed water by minor but distinct thermal gradients. There were three thermal gradients in all, and chlorophyll maxima were associated with each one of them. On a later crossing of the front, strong winds had developed and the pattern had broken up into numerous patches of dense chlorophyll, each about 1.2 km across. As with the earlier cruise (Savidge 1976) they interpreted the chlorophyll peaks as resulting from 'complementation of waters' by cross-frontal mixing.

Recalling the physical mechanisms discussed in section 6.2.3, we see (Fig. 6.02) that it is now understood that a typical shelf-sea front has a strong mean flow parallel with the front, giving rise to baroclinic eddies which often close in on themselves and break free, forming patches of cool water on the warm side, and vice versa. These are readily seen in satellite images. The results of Savidge and Foster (1978) are consistent with their having run a sampling transect through the front and through one of these isolated water masses. If such a water mass, originating from the cool side, eventually mixes with the nutrient-deficient water on the warm side, or vice versa, there will be a net transport of nutrients. This is one possible mechanism for enhancement of primary production.

In a review of the biological implications of fronts on the northwest European continental shelf, Holligan (1981) pointed out that the distribution of chlorophyll and primary production at the Ushant front varies seasonally (Fig. 6.07). In late April and early May the Ushant front marks the landward limit of the spring bloom taking place in stratified waters offshore. As the season progresses, those diatoms that are not removed by grazing tend to sink and are replaced by flagellates (mostly dinoflagellates) which flourish in the boundary zone between the warm, wind-mixed layer and the cool, tidally-mixed water. Finally, by late September, phytoplankton are again most abundant in the surface water on the stratified side of the front. The highest phytoplankton standing crops of the year are found when the flagellates bloom close to the stratified side of the front in July and August.

Holligan (1981) also pointed out that the depth of the mixed water on the Ushant front is 60–100 m. As a consequence, the growth of phytoplankton is light-limited and inorganic nutrients are never fully depleted. Fronts which form in shallow water have a different biological dynamic, for the phytoplankton in the tidally-mixed water may not be light-limited, but more probably nutrient-limited. Pingree *et al.* (1978a) identified shallow fronts in the North Sea off Flamborough Head and in the area between the Scottish Island of Islay and the northern tip of Ireland. The tidally-mixed areas appeared to have higher nutrient concentrations than the stratified regions, presumably because nutrients were being regenerated by the benthos. The enhanced chlorophyll concentrations associated with these fronts were once again ascribed to the alternation of mixing and stratified regimes present at certain places as a result of the alternation between neap and spring tides.

After presenting this review, Holligan and his colleagues made several intensive studies of the Ushant front (Holligan *et al.* 1984a,b). In late July and early August 1981 a large interdisciplinary team described the vertical distribution of the plankton at three stations representative of stratified, frontal and vertically-mixed regions. The phytoplankton was most dense in the frontal regions ($26.5\,\mathrm{g\ C\ m^{-2}}$)

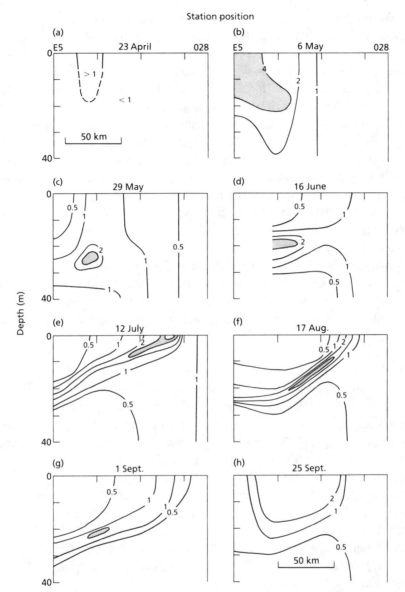

Fig. 6.07 Distribution of chlorophyll-a (mg m^{-3}) across the Ushant tidal front on dates given. From Holligan (1981).

and was dominated by the dinoflagellate *Gymnodinium aureolum*. At the stratified station there was a relatively small biomass (0.42 g C m^{-2}) of small naked flagellates that were concentrated at a subsurface chlorophyll maximum. The biomass at the tidally-mixed station was intermediate in value (7.91 g C m^{-2}) and consisted mainly of diatoms.

We may note in passing that these results are explicable in terms of properties discussed in earlier chapters. Diatoms, which tend to be larger than flagellates, have a marked tendency to sink, and flourish best under turbulent conditions which bring them back to the surface from time to time. The small flagellates in the stratified waters form a sub-surface chlorophyll maximum close to the thermocline, where there is a small but significant steady vertical transport of nutrients, possibly supplemented by bursts of increased vertical transport under the influence of internal waves. It is also possible that the flagellates make use of their locomotory powers to migrate down through the nutricline to take up nutrients and back up again to reach improved illumination (Holligan 1981; see also Kamykowsky *et al.* 1988). Fogg (1985) reported a similar distribution of phytoplankton types in waters adjacent to a front in the western Irish Sea.

When turnover rates were calculated for the phytoplankton, it was found that the small flagellates in the stratified, nutrient-limited surface waters were turning over rapidly, estimated at 1.17 doublings per day, the diatoms in the mixed region were achieving only 0.16 doublings per day and the flagellates in the frontal region were found to have a negative index of population growth. The authors estimated the vertical diffusion coefficient in the stratified region and the frontal area from observations on heat flux, then applied it to the observed nitrate gradient to determine the nitrate flux through the thermocline. Their results suggested that in the stratified region the small flagellates were obtaining about half their nitrogen needs from upwelled nitrate and the other half from recycling of ammonia. Results in the frontal region suggested that the vertical flux of new nitrogen was sufficiently high to have accounted for the accumulation of high biomass of *Gymnodinium* earlier in the season. They stated that the precise mechanisms that permit the accumulation of large biomasses of dinoflagellates in the frontal region are not yet clear, but suggested that one factor is that they appear to be subjected to low grazing mortality compared with other types of phytoplankton.

At each of the three stations studied, the standing stock of hetero-trophs was between 2.3 and 3.2 g C m^{-2}, bacteria comprising 10–30%. At the stratified station, though not at the others, the biomass of consumers exceeded that of the phytoplankton. Respiration data indicate that the meso-zooplankton were responsible for less than 10% of the oxygen consumption, and micro-heterotrophs were responsible for the greater part of biological energy dissipation.

6.3.3 What causes dense phytoplankton patches on shallow-sea fronts?

From what we have seen so far, there are two possible mechanisms to account for the high densities of phytoplankton commonly found on shallow-sea fronts: advection as a result of the convergent flows that have been observed at these fronts, or *in situ* growth of phytoplankton populations stimulated by some mechanism or transport of nutrients into the mixed layer of the stratified zone adjacent to the front. The role of advection is still an open question, but the mechanisms of cross-frontal transport of nutrients are becoming better understood.

We have seen that the high biomass of phytoplankton is usually found in surface waters on the stratified side of the front in spring and autumn, but in the thermocline in summer. In Chapters 3 and 4 we saw that a spring bloom in surface waters is normally associated with the shallowing of the mixed layer to a critical depth (the Gran effect). This will account for the bloom formation over the whole of the stratified region of the continental shelf but not for the elevated phytoplankton biomass and productivity adjacent to the front. To achieve this by *in situ* growth requires that the supply of nutrients be augmented. There are two possible routes for the nutrients to come, either horizontally across the front from the tidally-mixed region or vertically from the nutrient-rich water mass below the thermocline. Let us first examine mechanisms of cross-frontal transport.

(a) Spring-neap tidal cycle

The mechanism most often cited by Pingree and his co-workers is the movement of the front during the spring-neap tidal cycle (see section 6.2.2). During spring tides the tidal currents are at their maximum strength which means that tidal mixing extends to regions with deeper water, i.e. there is an erosion of the stratified region. During this phase, nutrients from below the thermocline will be brought to the surface in the area covered by the excursion of the front. As the tidal currents relax towards the neaps, the area of tidally-mixed water decreases, permitting stratification to return in that same area covered by the excursion of the front. Figure 6.08 from Simpson and Pingree (1978) shows the movements of the Ushant front relative to the Brittany coast between neap and spring tides.

Simpson and Bowers (1981) documented the movement of fronts as recorded by satellite images. Much of the observed variability in position is due to tidal advection at the semi-diurnal frequency, so this was

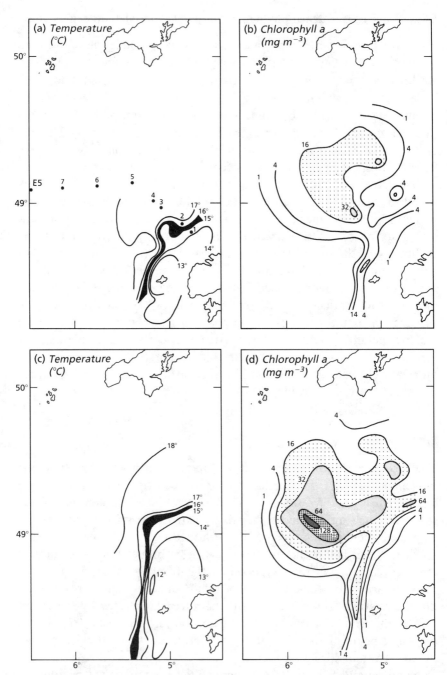

Fig. 6.08 Temperature and chlorophyll distribution at the Ushant front. (a),(b) neap tide conditions, 27–28 July 1976; (c),(d) spring tide conditions, 31 July–2 August 1976. Note that the high concentration of chlorophyll is further west at spring tides. From Simpson and Pingree (1978).

first removed. The residual indicated that the average spring-neaps excursion for eight tidal fronts lying west of Great Britain was 4 km.

Loder and Platt (1985) made a rough calculation of the rate at which nitrate might be captured by the upper layer on the stratified side as it advances into the tidally-mixed region. They used a variant of Eq. 6.01 and substituted 4 km for the tidal excursion of the front (Simpson and Bowers 1981), 20 m for the depth of the mixed layer, and 20 mg-at m^{-3} for the increase in nutrient concentration between the stratified and the mixed side of the front. From this they calculated the average supply of nitrate to the frontal zone as 0.12 mg-at m^{-1} s^{-1}. This value will later be compared with calculations for other methods of cross-frontal transport.

(b) Baroclinic eddies

It was shown in section 6.2.3 that the strong net flow parallel with a shelf-sea front gives rise to baroclinic eddies which are known from satellite observations to break free. A cool-water eddy introduced into the stratified side of a front will carry nutrients with it. Loder and Platt (1985) used two methods of calculation to arrive at an estimate of the rate of cross-front transport of nutrients by this mechanism. The first method follows Pingree (1979), who adapted a semi-empirical formula developed for poleward heat transport by baroclinic eddies in the atmosphere (Eq. 6.02). Taking $\Delta\rho/\rho = 6.7 \times 10^{-4}$, $\Delta C = 2$ mg-at m^{-3} and $\gamma = 0.0055$, the average rate of supply of nitrate by baroclinic eddies, per unit length of front is given by $F_e = 0.08$ mg-at m^{-1} s^{-1}.

The second used Eq. 6.03 in which the nutrient flux transported by eddies breaking free from the front is given by the number of eddies times the eddy volume times the difference in nutrient concentration, ΔC. For the two methods to yield the same result, T in Eq. 6.03 would have to be set at 60 days. This is an unreasonably long interval between the generation of eddies, judging by satellite images. Loder and Platt (1985) therefore suggested that the value of γ used (a constant derived from theoretical considerations) was much too low, and that the value of 0.08 mg-at m^{-1} s^{-1} for cross-frontal transport of nitrate may also be low.

(c) Residual currents

As we saw in section 6.2, the residual currents in a tidal front are primarily along the front, but a weak cross-frontal circulation induced by internal friction is also expected. Garrett and Loder (1981) provided formulae for this circulation and Loder and Platt (1985) evaluated them for the Ushant front. They arrived at a figure for nitrate transport of less than 0.01 mg-at $m^{-1} s^{-1}$, which is insignificant compared with the transport by the two mechanisms previously described. As we shall see later, work on the tidal front of Georges Bank leads to the view that the role of residual currents in bringing about a cross-frontal flux may have been underestimated.

In addition to the three methods we have just discussed, by which nutrients may pass horizontally across a front to fertilize the stratified low-nutrient water, there remains the possibility of vertical transport of nutrient-rich water from below the thermocline into the mixed layer above. Loder and Platt (1985) attempted a quantification of this term for comparison with the cross-frontal terms.

Vertical transport

Possible mechanisms for vertical transport in the vicinity of a tidal front were described in section 6.2.4. Loder and Platt used the equation

$$F_v = K_v(\Delta C/\Delta z)L_w, \tag{6.04}$$

where K_v is the vertical eddy diffusivity and L_w is the cross-frontal distance along which there is significant vertical transfer. Taking the value of $K_v = 10^{-4}$ m^{-2} s^{-1} which had been obtained by Pingree and Pennycuick (1975) from a stratified station, and $L_w = 10$ km they calculated the vertical nitrate flux as 0.2 mg-at $m^{-1} s^{-1}$.

Summarizing to this point, the Loder and Platt (1985) calculations suggested that reasonable estimates for the various types of transport of nitrogen into the mixed layer of the stratified water adjacent to the Ushant front were: spring-neap tidal cycle 0.12 mg-at $N m^{-1} s^{-1}$, baroclinic eddies 0.08 mg-at $N m^{-1} s^{-1}$, residual currents 0.01 mg-at $N m^{-1} s^{-1}$ and vertical transport 0.20 mg-at $N m^{-1} s^{-1}$. From this they concluded that all the mechanisms considered might bring 0.40 mg-at $N m^{-1} s^{-1}$, which might be approximately equated to a carbon flux of $0.28 g C m^{-2} d^{-1}$ in a 10 km wide frontal zone.

It is instructive to compare this calculation with those made by Holligan *et al.* (1984b). They concerned themselves with the vertical diffusion of nitrate up through the bottom of the thermocline from the mixed layer below. They first assumed that the vertical diffusion coefficient, K_v, was the same for heat and nitrate and constant throughout the thermocline. Thus the flux of heat, F_H, and nitrate, F_N, through the thermocline can be expressed by the equations

$$F_H = -\rho c\, K_v \Delta T/\Delta Z, \tag{6.05}$$

$$F_N = -K_v \Delta N/\Delta Z, \tag{6.06}$$

where ρ is the density of the water, c is the heat capacity of the water, and ΔT and ΔN are the differences in temperature and nitrate across the thermocline of thickness ΔZ.

From vertical profiles of temperature obtained at three-month intervals the authors estimated the downward flux of heat (F_H) across the thermocline and derived a value for K_v from Eq. 6.05. Substituting this value of K_v into Eq. 6.06 allowed them to calculate vertical fluxes of nitrate from measurements of the nitrate gradient.

Their main argument is that biological depletion of nitrate in the mixed layer and thermocline causes an increase in the vertical gradient of nitrate at the base of the thermocline, i.e. in the nutricline. This higher gradient must, if the vertical diffusion coefficient stays the same, lead via Eq. 6.06 to a greater upward flux of nitrate.

They performed this calculation at a number of stations. At one station a chlorophyll maximum attributable to flagellates was observed in the middle of the thermocline and the gradient of nitrate in the nutricline was not particularly steep. As a result, the calculated vertical transport of nitrate was relatively low, 3.22 mg nitrate-N $cm^{-2} d^{-1}$. On another occasion there was a chlorophyll maximum attributable to the dinoflagellate *Gymnodinium aureolum* near the base of the thermocline producing a high vertical gradient in nitrate. This higher gradient led to an estimated vertical flux that was much greater than in the first case, 22.4 mg nitrate-N $cm^{-2} d^{-1}$.

Similar calculations were carried out for observations in a tidally-mixed front which resulted in a maximum value for the vertical flux of nitrate of 39.9 mg nitrate-N $cm^{-2} d^{-1}$, equivalent to 'new' carbon production in excess of $3\,g\,C\,m^{-2} d^{-1}$. They concluded that dinoflagellates can increase the vertical nutrient gradient and hence the vertical nutrient flux; however, their flux values are about 10 times those calculated by Loder and Platt (1985). This may be because their estimate of K_v is an average value based on temperature changes in the bottom mixed layer over 90 days while the nitrate gradient was obtained over about 1 hour. Direct measurements indicate (Gargett 1984) that K_v is not a constant for it must vary continuously with the level of turbulence and stratification. The value of K_v concurrent with the high nutrient gradient observed by Holligan et al. (1984) could have been much lower than the 90 day estimate and unable to provide the high nutrient flux the authors calculate.

We shall now return to the question of the role of residual currents in cross-frontal transport. In particular there have been studies at the

Georges Bank front, which is the most extensively studied frontal system after the northwest European shallow-sea fronts.

6.3.4 The Georges Bank frontal system

Georges Bank is a major submarine bank lying off the eastern coast of North America between Cape Cod and Nova Scotia. Its characteristics have been extensively reviewed in Backus and Bourne (1987). Over most of the bank the water depth is 50–60 m and the tidal currents are sufficiently strong to keep the water column vertically mixed throughout the year. Around the bank the combination of fresh water run-off and solar heating causes persistent stratification throughout the spring and summer. As a consequence, a tidal front develops around the central bank, between the mixed and stratified water. (There is, in addition, a clockwise gyre driven primarily by rectification of the strong tidal currents over the sides of the bank, but this will be treated in Chapter 7.)

Biological productivity of the waters over the bank is high, and there is intensive fishing of both finfish and shellfish. Since there is a substantial loss of nitrogen associated with removal of commercial catches, it is important to understand the mechanisms for supply of 'new' nitrogen to the top of the bank from deeper water. Work on this topic up to 1988 has been reviewed by Horne *et al.* (1989).

The position of the front is in rough agreement with the prediction made by Loder and Greenberg (1986) using Simpson and Hunter's (1974) formulation (see Chapter 4). From observations made in 1985 it is deduced that the front on the northern and northwestern side of the bank makes a twice-daily excursion of 10–15 km as a result of changes in tidal current velocity. Records from a depth of 33 m at a fixed station in the frontal zone showed semi-diurnal fluctuations not only in current velocity but in temperature (approximately 7–15 °C on tides of maximum amplitude). From other data sets it was found that there was a good correlation between nitrate concentration and temperature, and it was possible to construct a time series for variations in nitrate concentration with time, indicating semi-diurnal fluctuations between about 1 and 10 mg-at m^{-3} on spring tides at the 33 m-deep fixed station.

The current meters at the fixed 33 m station showed the expected gyral current parallel with the front, but also a cross-frontal current averaging 0.05 m s^{-1} at 13 m depth and 0.03 m s^{-1} at 33 m depth, both directed towards the tidally-mixed area on the top of the bank. Use of this information in conjunction with the fluctuating nitrate concentrations enabled Horne *et al.* (1989) to calculate a depth-integrated flux of nitrate through the front towards the centre of the bank of

$\simeq 12$ mg-at N m^{-1} s^{-1}. We note that this is about 30 times the flux calculated by Loder and Platt (1985). Horne *et al.* (1989) then went on to consider what physical mechanisms might be responsible for this flux.

They first considered shear-flow dispersion. This would work approximately as follows. A parcel of nutrient-rich water moving onto the bank with the tidal flow and returning half a tidal period later might have been subjected to vertical mixing with nutrient-poor surface water and return with its nutrient concentration reduced. Examination of the data showed that this was unlikely to be a major component of the observed flux.

It is instructive to compare the estimated nitrogen flux to the centre of the bank (12 mg-at N m^{-1} s^{-1}) with the observed biological requirement for nitrogen. During a cruise in August 1985 *in situ* incubations of phytoplankton were carried out in the presence of either ^{14}C (as HCO^{3-}) or ^{15}N (as NO^{3-} or NH^{4+}), and used to calculate rates of photosynthesis and nitrogen uptake. Measurements were made of nutrients, chlorophyll and physical characteristics of the water column. It was found that, because of the semi-diurnal movement of the front mentioned earlier, some of the stations on the bank or in the frontal zone were in fact stratified. Grouping stations according to their physical characteristics rather than their position led to the conclusion that the frontal stations had the highest carbon-fixation rates (265 mg C m^{-2} h^{-1}) and nitrate-uptake rates (1.3 mmol N m^{-2} h^{-1}), the stratified stations had the lowest (160 mg C m^{-2} h^{-1} and 0.65 mmol N m^{-2} h^{-1}) while the mixed stations were intermediate. Nitrate-based 'new' production accounted for 60% of the total at frontal stations, 41% at the mixed stations and only 27% at the stratified stations. The nitrate demand of the phytoplankton populations was 0.36 mg-at m^{-2} s^{-1} at the frontal stations, 0.09 mg-at m^{-2} s^{-1} at the mixed stations and 0.018 mg-at m^{-2} s^{-1} at the stratified stations. At an oceanic station far from the bank the demand was only 0.02 mg-at m^{-2} s^{-1}.

Using reasonable approximations for the area of the mixed zone on the top of the bank, and of those parts of the frontal zone on the bank side of the mooring site, the nitrate requirement of the mixed zone was estimated at 2.5 mg-at N m^{-1} s^{-1} and that of the frontal zone between the bank and the mooring as 2.9 mg-at N m^{-1} s^{-1}. Hence the calculated supply, 12 mg-at N m^{-2} s^{-1}, is roughly twice the total demand by the phytoplankton. Bearing in mind that these calculations are rough approximations, they seem to indicate that at the time of study crossfrontal transport by residual flow was sufficient to meet the observed needs of the phytoplankton, without there being any requirement for other mechanisms.

In comparing results from Georges Bank with those from the north-west European area, it must be remembered that the front associated with the gyre on Georges Bank may well show very different properties from the shallow-sea fronts off the coast of Europe. Measurements to date suggest that the enhanced primary production at the Georges Bank front is made possible by a strong cross-frontal transport of 'new' nitrogen, and that the principal physical mechanism involved is unknown. At the European fronts it appears that primary production is enhanced by vertical transport of nitrate and by a variety of mechanisms of cross-frontal transport.

6.3.5 Heterotrophic activity at shelf-sea fronts

An integrated multidisciplinary study of several kinds of heterotrophic activity at a shallow-sea tidal-mixing front in the Irish Sea has been reviewed by Fogg (1985). The investigators recognized three water masses as distinct ecosystems: surface stratified water, SSW, bottom stratified water, BSW, and mixed water, MW.

After thermocline formation the SSW system was occupied by an expanding phytoplankton population which gradually used the available nutrients then reached a plateau of biomass that was maintained throughout the summer, with the greatest concentration of chlorophyll being on the pycnocline. In parallel with this was development of heterotrophic populations of bacteria and zooplankton which expanded until photosynthesis was balanced by heterotrophic activity. At this stage, nitrogen turnover, as measured by urea turnover rates, was rapid and there was a high degree of interdependence between the food web components that persisted until destabilization of the water column in the autumn.

In the bottom stratified water, by contrast, both photosynthesis and heterotrophic activity was at a low level. In the mixed water, where the phytoplankton is light-limited rather than nutrient-limited, bacterial and zooplankton biomasses, heterotrophic activity and nitrogen turnover rates were lower than in the SSW.

Fogg (1985) reported that, like many others before them, the investigators had found patches of high chlorophyll density in the SSW adjacent to the front and on the pycnocline away from it. However, he said that if chlorophyll concentrations were averaged down to 30 m, the frontal waters were scarely distinguishable fron the rest of the SSW. In other words, the distribution was different, but not the total amount in the water column.

They found that there was no significant increase in bacterial or zooplankton biomass at the front, but a marked increase in physiological activity. In June and July the levels of cellular ATP were about twice those found elsewhere in the SSW, and there were patches of high glucose uptake activity and urea uptake activity adjacent to the front. The tentative explanation put forward by Fogg (1985) is interesting from several points of view. He suggested that eddies formed along the front (see Fig. 6.02) might, by their rotation, induce vertical circulation and bring to the surface concentrations of phytoplankton from the isopycnal region of the SSW. This phytoplankton would be immersed in nutrient-poor water and would be expected to be active photosynthetically but unable to divide. Under these circumstances large quantities of DOM would be liberated and this might explain the high levels of heterotrophic activity. Fogg (1985) also suggested that this mechanism could account for the appearance of high concentrations of chlorophyll at the surface near the front. This is yet another potential mechanism to add to those discussed in the previous two sections.

When the distributions of zooplankton were examined species by species it was found that the front demarcated distinct communities. *Temora longicornis* and the nauplii of *Acartia clausi* were absent from SSW, whereas *Calanus finmarchicus* and *Membranipora membranacea* were absent from MW. *Oithona similis* was concentrated around the pycnocline, while *Microcalanus pusillus* was found only in the BSW.

On the stratified side of the front most of the zooplankton were found above the thermocline and made limited vertical migrations. When zooplankton found themselves in patches of high concentrations of phytoplankton they ceased vertical migration. Fish eggs, larvae and post-larvae were more abundant in the frontal zone than elsewhere. Birds, such as Manx Shearwaters, were more numerous on the stratified side of the front than elsewhere, especially in areas of chlorophyll maxima. The general impression that frontal waters are regions of enhanced primary and secondary productivity is sustained, even though the mechanisms responsible are still a matter of debate.

Benthic productivity

It has been observed (Glémarec 1973; Creutzberg 1985) that in the benthic areas corresponding with a frontal change from tidally-mixed waters to summer stratified waters there is a change in benthic community structure. Creutzberg (1985) showed that on a transect across a front off the Dutch coast in the North Sea (Fig. 4.13) the tidally-mixed area to the south was marked by sandy sediments, while the stratified area to the north had muddy sediments. The boundary between the two has remained remarkably constant for at least 80 years, and just north of the boundary is a zone about 15 km broad characterized by high organic matter content and a high biomass of benthic animals.

Two hypotheses have been put forward to explain it. The simplest is that the fast-moving tidal streams of the southern region scour the bottom and carry a heavy load of organic and inorganic sediment. Further north, where tidal streams are slower, there is reached a point at which the waters deposit their load of organic and inorganic sediment, creating ideal conditions for the growth and reproduction of benthic organisms. A more complex explanation invokes the presence of a front with its enhanced phytoplankton biomass in the waters above. Sinking of the phytoplankton leads to the accumulation of organic matter in the sediments below.

Creutzberg (1985) made observations on the position of the front and of its chlorophyll distribution on eight occasions between 1982 and 1984. He discovered that there was a chlorophyll maximum which remained stationary over the enriched benthic zone even when the front itself migrated 50–60 km north. His conclusion was that the first of the two explanations was the correct one, namely that the advection of organic matter and its deposition, over a long period, in a position determined by the speed of the tidal currents has led to the development of an enriched benthic community. He suggested that rapid release of nutrients from the enriched benthic community was the cause of the chlorophyll maximum over the top of it. This is the converse of the view that holds that deposition of phytoplankton from chlorophyll maxima associated with the front has led to the enrichment of the benthic community. He did not go so far as to suggest that release of nutrients from the benthos was the sole cause of chlorophyll maxima at fronts, but it is worth remembering that the bottom stratified water below fronts may be receiving unusually large amounts of nutrients across the benthic boundary layer.

6.4 SHELF-BREAK FRONTS

Along the outer continental shelf of almost all of eastern North America in winter is to be found a front marking the transition between the colder and fresher shelf water mass and the warmer and more saline slope waters. Data from Flagg and Beardsley (1975) showing the situation south of Rhode Island in early April 1974 (Fig. 6.09) show that the shelf water was well mixed at a temperature of 5–10 °C and 32.6–34.4‰ salinity, while the offshore waters were mildly stratified

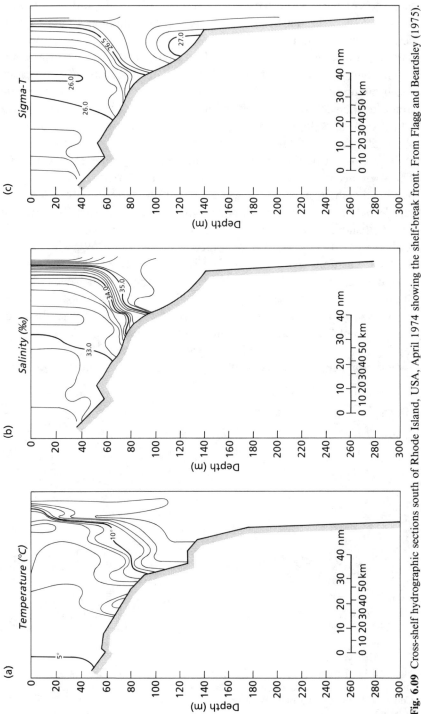

Fig. 6.09 Cross-shelf hydrographic sections south of Rhode Island, USA, April 1974 showing the shelf-break front. From Flagg and Beardsley (1975).

(not shown) with a surface temperature of 14–16 °C and about 36‰ salinity. When the density is considered, these differences tend to compensate for one another so that the σ_t difference across the front was only 0.4 units. In various parts of the eastern seaboard of North America similar changes occur over distances of 7–40 km and a vertical depth ranging from 15 to 60 m (Mooers *et al.* 1978). The front normally straddles the shelf break, intersecting the bottom near the 100 m isobath (60–120 m) and sloping up from the bottom in an offshore direction for about 50 km (0–100 km).

The coming of the spring warming drastically changes the picture. The shelf becomes stratified and below the thermocline, on the deeper part of the shelf, a parcel of water of reduced salinity with a temperature less than about 10 °C becomes isolated from the warmer, shallower waters inshore and the more saline waters offshore. This was first described in northwest Europe by Vincent and Kurc (1969), who called it the 'cold cushion' (*bourrelet froid*) (Fig. 6.10). The effect is to maintain a weak front near the shelf edge below the thermocline, even through the period of summer stratification.

As was mentioned earlier, Mooers *et al.* (1978) have described how the position of the shelf-break front changes under the influence of wind forcing, by the impact of Gulf Stream rings (see Chapter 8) and in association with the 'calving' of lens-shaped parcels ('bubbles') of shelf

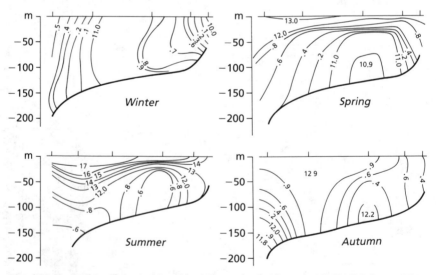

Fig. 6.10 Formation of the 'cold cushion' (*bourrelet froid*) on the shelf off Brittany. The structure is visible in spring and becomes warmer and smaller as the seasons advance. Compiled from the data of Le Magueresse and of Vincent and Kurc, by Le Fèvre (1986).

water of scale 10–20 km, which break through the frontal structure and move offshore. The net result is that the position of the shelf-break front off the east coast of North America may vary in distance from shore by as much as 200 km. The movement is least off the coast of Florida and Georgia to the south, and reaches a maximum at the latitude of Cape Cod. The envelope in which this front moves corresponds with the position of highly productive fisheries, from the Grand Banks off Newfoundland, south to Cape Hatteras.

6.4.1 Biological production at shelf-break fronts

There have been various observations of nutrient enrichment and/or enhanced phytoplankton and zooplankton biomass associated with shelf-break fronts (Fournier 1978; Pingree and Mardell 1981; Le Fèvre and Frontier 1988). For example, Fournier (1978) reported that the average value for chlorophyll-a across the Scotian Shelf (in March 1977) was less than 1 mg m^{-3}, while over the shelf break chlorophyll reached a maximum of 4.2 mg m^{-3}. As explanation, he suggested that shallowing of the mixed layer in the vicinity of the front allowed enhanced phytoplankton production, even in winter.

The observations of Herman *et al.* (1981) on the shelf break south of Nova Scotia can be interpreted as supporting the views of Le Fèvre and Frontier (1988). Using an undulating 'batfish' sampler which measured salinity, temperature, depth, chlorophyll-a and copepods, they found that plant production and copepod abundance were much higher at the front than in surrounding shelf and slope waters. Chlorophyll concentrations were $\simeq 1.5$ mg m^{-3} in shelf water, peaked to $\simeq 6$ mg m^{-3} in the front and were 3–4 mg m^{-3} in the slope water. The corresponding abundances of copepods in the shelf, front and slope waters were $\simeq 30,000$ m^{-2}, 70,000 m^{-2} and 10,000 m^{-2}, respectively. The high concentrations of copepods occurred above the thermocline in the frontal region, and these copepods did not migrate down through the thermocline during the day, while those in the shelf waters often did so. Having considered the various factors that might account for high copepod abundances at the front, Herman *et al.* (1981) concluded that convergence of surface waters at the front was the most plausible mechanism. However, at that time the mechanism of fertilization of surface waters by internal tidal waves (Le Fèvre and Frontier 1988) had not been proposed.

Pingree and Mardell (1981) reported on a series of studies of the Celtic Sea shelf break and pointed out that in summer the edge of the shelf is characterized by a band of water about 1–2 °C cooler than water on either side of it and about 100 km broad. At times the band can be traced for 800 km along the Armorican and Celtic Sea slopes. In it, the levels of inorganic nitrate and chlorophyll are significantly higher than in adjacent water. They noted that there seemed to be an association between the patches of increased chlorophyll and the bottom topography. Although some had invoked upwelling as the mechanism responsible for the presence of cooler, nutrient-rich water at the surface (e.g. Heaps 1980), Pingree and Mardell (1981) were inclined to favour tidally-driven internal waves interacting with the bottom topography as the important mechanism.

This idea was developed further by Mazé (1983), Mazé *et al.* (1986) and by Pingree *et al.* (1986) to the point where Le Fèvre and Frontier (1988) could write that 'a reasonably safe picture of what is taking place has now emerged'. Details of the physics of internal waves will be given in Chapter 7, but in summary the story goes approximately as follows. The incoming oceanic barotropic tidal wave is rather small, of the order of 2 m. In stratified water the interaction of the tidal wave with the shelf break gives rise to an internal baroclinic tide with an amplitude which reaches a maximum of about 60 m at the very edge of the shelf. From here it propagates in two directions, towards land and away from it, gradually damping out. Propagation towards the open sea is more or less sinusoidal, but propagation onto the shelf is distorted by the barotropic tidal currents.

There has been much discussion about whether an internal tide would produce the observed reduced temperature and elevated nutrient concentrations in surface waters. It is now thought that there is an interaction between the internal waves and the wind mixing of the layer above the thermocline. Assuming some sort of equilibrium between the depth of the thermocline, the temperature of the mixed layer and the prevailing wind conditions, a forced upward displacement of the thermocline leads to a disruption of that equilibrium. Additional energy is injected into the mixed layer, which leads to an increase in the depth of the mixed layer with upward mixing of cool water from below the thermocline. This carries with it additional nutrients, so that primary production is enhanced. Another way of looking at it is to consider that the depth of the mixed layer is a function of the buoyancy provided by surface warming and the strength of winds causing mixing. If the thermocline is pushed upwards, the mixed layer will tend to return towards its original thickness by mixing upwards some of the cooler, nutrient-rich water. The frequency of this occurrence is that of the M_2 tide, i.e. twice daily.

Le Fèvre and Frontier (1988) pointed out that augmentation of the nutrients on a daily basis provides a more-or-less continuous increase in phytoplankton productivity to which the copepod population can respond. This is in contrast to the situation at a tidal-mixing front, where the cycle of enhanced production follows the fortnightly cycle of spring and neap tides, leading to a short burst of phytoplankton production every two weeks. It seems unlikely that zooplankton populations could increase and decrease in such a way as to exploit this production to the full. Hence shelf-break fronts should show that copepod populations increase in parallel with phytoplankton, while tidal-mixing fronts should not. The limited field data available support this view. Results given by Le Fèvre and Frontier (1988) for the Armorican shelf-break front show increased copepod abundance in the frontal region, suggesting that a classical copepod herbivore food chain is functioning. On the Ushant tidal-mixing front the dominant zooplankter was the pteropod mollusc *Limacina,* which is known to feed on micro-zooplankton. They suggested that much of the phytoplankton at this front remains ungrazed by copepods and becomes senescent, giving support to a food web involving bacteria and ciliates, which in turn are taken by *Limacina* . This line of reasoning is supported by the high levels of heterotrophic activity observed by Fogg (1985) and others as characteristic of tidal-mixing fronts.

In summary, the most recent view of the mechanism controlling biological productivity at shelf-break fronts is that tidally-generated internal waves add energy to the mixed layer, causing a deepening with incorporation of nutrient-rich water from below the nutricline. This process varies in intensity with the M_2 tide so that it produces a twice-daily augmentation of nutrients, a fairly constant elevation of primary production levels and a corresponding increase in secondary productivity. This mechanism contrasts with that believed to be operative in tidally-mixed fronts, where the maximum enrichment of nutrients in primary production coincides with the neap tides at fortnightly intervals. This stimulation of primary productivity is considered to be too infrequent to support an elevated population of macro-zooplankton. Instead, there is a burst of phytoplankton production, followed by a period of decline and decomposition, which supports a dense population of bacteria and micro-zooplankton.

6.5 UPWELLING FRONTS

As we saw in Chapter 5, there are many places around the world where wind-induced offshore transport of surface waters occurs on a major scale, leading to coastal upwelling. While the forcing winds continue to blow, cool, nutrient-rich waters from below the pycnocline continue to rise to the surface and move offshore, displacing the warmer, less nutrient-rich waters that preceded them. The interface between the up-welled waters and the offshore waters constitutes an upwelling front. A section through such a system (Fig. 6.11) reveals that the front is the place where the pycnocline intersects the surface of the ocean.

During an upwelling event, the front at first forms close to shore, then moves offshore, but at some point an equilibrium is reached at which the front ceases to move offshore and the upwelled waters, as they continue to move offshore, are driven downwards beneath the off-shore waters (Fig. 6.11). The front lies at a prograde angle. At the surface there is a convergence of waters and a tendency for buoyant biological material to accumulate. When the wind stress decreases and upwelling ceases, the front retreats landward and may disappear.

In spite of the large amount of biological study of upwelling systems (see Chapter 5), there is a surprising lack of detailed information about upwelling fronts. The modelling studies of Thompson (1978) suggested that part of the sinking water is entrained in a shoreward-flowing deeper current, but there may be a one-celled or a two-celled circulation, de-pending on the wind stress, stratification, current shear and bottom relief. In any event, the presence of the front is normally marked by a sharp temperature transition from the cold, nutrient-rich upwelled water inshore to the warmer, less nutrient-rich stratified water offshore. Clearly, any mechanism causing a mixing of these waters is likely to stimulate primary productivity.

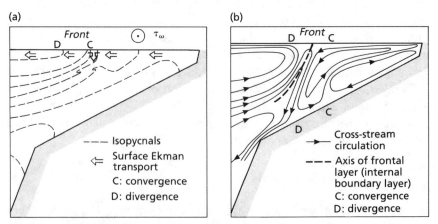

Fig. 6.11 Conceptual model showing (a) density field and (b) cross-shelf circulation in the vicinity of an upwelling front. From Mooers *et al.* (1978).

Pearcy and Keene (1974) studied the upwelling system off Oregon by airborne remote sensing and found abrupt changes in colour coincident with changes in temperature which they thought denoted enhanced biological activity at upwelling fronts. Ryther (1967) found large concentrations of a ciliate, *Cyclotrichium meuneri*, which coloured the water red, on the inshore edge of an upwelling front off Peru, and Packard *et al.* (1978), after studying the biology of a related ciliate, *Mesodinium rubrum,* in the California upwelling system, showed that large concentrations formed on the inshore edges of upwelling fronts. Peterson *et al.* (1979) studied zooplankton distribution in the Oregon upwelling system and found that very high concentrations were retained shoreward of the front, and were carried below the pycnocline when the upwelling relaxed. Wroblewski (1982) modelled the movement of the various stages of the copepod *Calanus marshallae* in the Oregon upwelling system and showed how vertical migration may interact with the physical oceanography to cause the organisms to move over a spiral path on the continental shelf, moving offshore at the surface, but returning in deeper water after sinking at the front. At the highest trophic level, there is evidence that albacore tuna, which make trans-Pacific migrations to feed close to the California upwelling systems, tend to concentrate in the vicinity of the upwelling fronts (Laurs *et al.* (1977). On the whole, the biological information that we have about upwelling fronts is rather fragmentary, and shows that the study of upwelling fronts is much less advanced than that of tidal-mixing fronts.

One of the most recent field studies of an upwelling front is that carried out by Armstrong *et al.* (1987) across the southern Benguela system north of Cape Town. They emphasized throughout their account that the frontal zone is dynamic and subject to rapid change, especially since they encountered on the outward boundary of the front a strong north-flowing jet current, with speeds up to almost 1 m s^{-1} and extending to a depth greater than 200 m. However, they found during their cruise that the greatest biomass and productivity of the phytoplankton, when integrated over the euphotic layer, was in the frontal region (Table 6.2). Inshore of the front very high concentrations of nutrients and chlorophyll occurred in surface waters, but penetration of light was limited and integrated water-column productivity was correspondingly low. Offshore of the front the nutrient levels in the euphotic zone were extremely low, so that photosynthesis was limited to that level that could be sustained by nutrient recycling. In the frontal region itself, however, relatively high biomasses of phytoplankton were well distributed through the euphotic zone and production rates were high. It was inferred that there was an enhanced supply of new nitrogen into the

Table 6.2 Station number, water type, sea surface temperature (SST), 1% light depth ($D_{1\%}\simeq$depth of euphotic zone), integrated phytoplankton production rate $0-D_{1\%}$, mean phytoplankton production rate $0-D_{1\%}$, integrated chlorophyll-a and bacterial production rates $0-D_{1\%}$ and the percentage ratio of bacterial to phytoplankton production. From Armstrong et al. (1987).

Station	Water type	SST (°C)	$D_{1\%}$ (m)	Phytoplankton production		Chlorophyll-a (mg m^{-2})	Bacterial production	B/A (%)
				A (mg C m^{-2} h^{-1})	Mean (mg C m^{-3} h^{-1})		B (mg C m^{-2} h^{-1})	
2–03	Inshore	15.3	9	112.8	12.5	17.5	39.2	34.7
4–04	Frontal	16.3	16	80.9	5.1	58.9		
3–03A	Frontal	16.2	19	258.9	13.6	108.5		
4–05A	Frontal	15.7	14	187.0	13.4	73.4	21.4	11.4
2–07	Offshore	18.1	33	65.4	2.0	20.3		
4–10	Offshore	18.5	38	81.0	2.1	15.5	54.5	67.3
1–16	Offshore	18.0	27	27.3	1.0	13.1		

euphotic zone of the frontal region. The mechanism suggested was interleaving of water masses of differing salinities, temperatures and nutrient contents but similar densities, and there was some evidence for this from the CTD data. However, as we have seen, mechanisms of cross-frontal transport are not well understood and are under active investigation.

6.6 RIVER- AND ESTUARINE-PLUME FRONTS

At the edges of river or estuarine plumes (see Chapter 4) where water of reduced salinity meets more saline coastal water, fronts are formed which are conspicuously marked by lines of foam and by sharp change in colour of the water. As mentioned in Chapter 4, the less saline, lighter water lens has an elevated surface and is carried by gravity seawards to ride over the top of the more dense coastal water, forming a retrograde front.

An early study of the physics of an estuarine-plume front was that carried out by Garvine and Monk (1974) on the Connecticut River plume in Long Island Sound. They showed that adjacent to the front the layer of brackish water was only about 1 m deep, and that the pycnocline sloped up to intersect the surface over a distance of about 50 m. There was a vigorous convergence of surface waters from both sides of the front and locally intense sinking at the front itself.

There has been little work on the biological significance of plume fronts, but the evidence available suggests that they are sites of enhanced biological activity and that fish tend to aggregate there. Pearcy and Keene (1974), who studied the coastal waters of Oregon by airborne remote sensing found evidence of increased phytoplankton biomass at the Columbia River plume front, and Owen (1968) found that catches of albacore tuna were higher at that front than at other stations nearby. In Japan, Tsujita (1957) found that the spawning and hatching of Japanese sardine off western Japan was consistently located close to an estuarine-plume front.

In the Chesapeake Bay area, there are several places where the rivers emptying into the bay form plume fronts, and Tyler and Seliger (1978, 1981) have shown that high concentrations of various phytoplankton species, including the red-tide dinoflagellate *Prorocentrum* are found in these frontal regions especially in winter and early spring.

6.7 OTHER TYPES OF FRONTS

6.7.1 Estuarine fronts

Relatively small-scale fronts commonly appear in estuaries, forming and dispersing with every tide. They are similar in some ways to the tidal-mixing fronts of coastal waters. At certain states of the tide the shallow water at the edges of the estuary is tidally-mixed while the deeper water in the middle remains stratified under the influence of the freshwater run-off. Such a front has a marked salinity gradient, as well as a temperature gradient, at the surface. Examples are given by Bowman and Iverson (1978). The influence of these fronts on biological production is not well understood, but striking examples of concentration of pollutants in the surface convergence have been noted.

6.7.2 Effect of geomorphology

When tidal or other currents interact with irregularities in the sea bed, or coastline, it is usual to find consistent patterns of eddies and associated fronts. Pingree *et al.* (1978b) reviewed the formation of fronts at coastal headlands, and Bowman *et al.* (1986) reviewed the interactions of tidal currents with islands and features of the sea floor. There is scattered evidence of enhanced phytoplankton growth, for example around islands, off headlands and above sea mounts, but the mechanisms involved, and the role of fronts in these mechanisms, are as yet not at all clear.

Wolanski and Hamner (1988) recently reviewed the distribution of zooplankton in relation to topographically-controlled fronts formed by headlands, islands and reefs. When a tidal current streams past one of these structures there may be formed a front (often called the dividing streamline, Fig. 6.12) which separates the normal tidal flow from the more turbulent eddies on the downstream side of the object. If the water is fairly shallow the turbulent eddies interact with the bottom giving rise to convergence and sinking at the front. As a result, surface-dwelling planktonic organisms tend to aggregate there, especially those that are buoyant. In the vicinity of coral reefs, where many species of coral tend to spawn synchronously, very dense aggregations of buoyant coral eggs are formed at these fronts, which can then be readily identified from the air.

Alldredge and Hamner (1980) documented the formation of dense aggregations of zooplankton in the turbulent area downstream of a headland when tidal streaming was at its maximum, and Oliver and Willis (1987) reported that high concentrations of coral eggs which formed at the topographic fronts at certain stages of the tide, remained as coherent linear patches as they drifted away from the places where the fronts had formed. It is clear that calculations of the encounter rates between planktonic organisms around coral reefs, islands and headlands must not be made on the basis of average density in the water mass, but

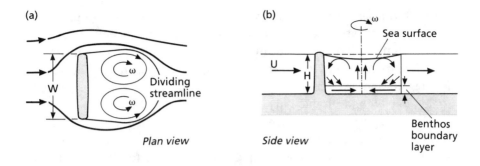

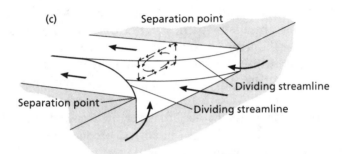

Fig. 6.12 (a) Plan of a commonly occurring flow pattern around a reef in shallow coastal water, showing the front (labelled dividing streamline) separating the area of turbulence in the wake from the less turbulent area beyond. (b) Vertical section showing upwelling produced by the clockwise (southern hemisphere) gyre. (c) Formation of turbulent-mixing region as a tidal stream enters a reef passage. For details see text. From Wolanski and Hamner (1988).

must take into account the way in which topographically-controlled fronts bring about massive aggregations.

It is worth noting in passing that in the situation depicted in Fig. 6.12(c) in which a tidal current streams through a reef passage, nutrient-rich water from depth is upwelled from in front of the reef by Bernoulli suction and stimulates algal production on the reef (Wolanski *et al.* 1988).

6.8 SUMMARY

More effort has gone into the study of shelf-sea (tidal-mixing) fronts than into any other kind. It is now clearly understood that tidally-induced currents cause mixing from top to bottom of the water column in many shelf areas. The strength of the tidal currents varies with the diurnal tidal cycle, but also with the fortnightly alternation between spring and neap tides. At places where the depth in relation to the tidal currents is too great for mixing to occur from top to bottom of the water column, thermal stratification occurs in summer. Shelf-sea fronts mark the discontinuity between the tidally-mixed region and the stratified region and are frequently found to be the sites of enhanced planktonic biomass and productivity.

Early observations on such fronts emphasized the convergence of surface waters and downwelling that occurs there. Any organisms buoyant enough to resist the downwelling would be expected to aggregate at the front. Thus, it was thought that passive advection was one explanation of the concentrations of planktonic organisms found at fronts.

An alternative explanation invoked the concept of enhanced *in situ* production, made possible by particularly favourable conditions of light and nutrients. In the upper mixed layer of stratified water in summer, phytoplankton growth is usually nutrient-limited because the presence of a pycnocline restricts the upward movement of nutrients from the lower regions where they are more plentiful. Any mechanism that causes a transfer of nutrients to the surface layer on the stratified side of a front is likely to lead to enhanced biological production. A number of mechanisms have been proposed.

1   The spring-neap tidal cycle. At a fixed position relative to the bottom, the water may be tidally mixed at one stage of the tidal cycle and stratified at another. Nutrients are brought up during the mixing phase and utilized in the upper mixed layer during the stratified phase.

2   Cross-frontal transport. A variety of mechanisms has been proposed for transferring nutrients from the tidally-mixed side of the front, where phytoplankton is often light-limited but not nutrient-limited, to the stratified side of the front. Baroclinic eddies, in which parcels of nutrient-rich water break through to the stratified side, have been observed by remote sensing.

3   Vertical transport. Some evidence suggests that conditions in the frontal zone are favourable for the vertical transport of nutrients through the front to the stratified water above. This could result in enhanced phytoplankton in the immediate vicinity of the front.

The relative importance of these three types of mechanism has yet to be evaluated. It is probable that different mechanisms dominate at various times and places. Only a sophisticated multidisciplinary approach using state of the art technology to yield physical and biological data on a range of scales from microns to kilometres will clarify the mechanisms involved.

Shelf-break fronts occur where winter cooling and wind mixing lead to cooler, less saline water on a continental shelf with warmer, more saline water offshore. A front forms at the interface between the two water masses and production of phytoplankton, zooplankton and fish is particularly high. The mechanism most widely accepted as causing the enhanced production is the generation of tidally-driven internal waves at the shelf edge. The consequent rise and fall of the nutricline leads to vertical transport of nutrients into surface water. The process is seen as being of daily occurrence, thus supporting a relatively constant level of enhanced phytoplankton production on which a food web involving zooplankton and fish can be built. This situation has been contrasted with that found at tidal-mixing fronts. At these, the mechanism of enhanced biological production appears to be linked to the neap-spring tidal cycle, so that pulses of high phytoplankton biomass occur at fortnightly intervals. This is too far apart for zooplankton populations to be able to exploit them effectively. As a result, there is accumulation, death and decay of phytoplankton, with the release of dissolved and particulate organic matter, which is exploited by bacteria, ciliates and specialized macro-zooplankton, rather than copepods and fish (see section 6.4.1).

Upwelling fronts form at the interface between normal shelf water and the cool, nutrient-rich water brought to the surface during wind-driven coastal upwelling. Since upwelling-favourable winds are not constant, but increase and decrease to give recognizable upwelling events, upwelling fronts are also variable in time and space. At the interface, upwelled water is driven down under the warmer shelf water, forming a prograde front. There have been few systematic studies of the biology of upwelling fronts, but the information that is available suggests that planktonic organisms tend to aggregate on the coastal side of these fronts and that tuna and other pelagic fishes occur there in particularly large numbers.

At places where plumes of fresh or brackish waters run into saline coastal waters it is common to find plume fronts, where the less saline waters ride over the top of the saline waters, forming retrograde fronts and where surface convergence and downwelling occurs. Once again we find that there are few systematic studies of the properties of such fronts, but there is fragmentary evidence suggesting that aggregations of plankton form at or near the convergence and that good catches of pelagic fish are often obtained there.

Within estuaries, local changes in depth give rise to small-scale variations in the degree of tidal mixing, and to the formation of fronts between mixed and stratified waters. Such fronts are often formed parallel to the shores of estuaries, separating the tidally-mixed shallow waters from the deeper waters stratified under the influence of fresh water run-off. Bathymetric features also give rise to a variety of eddies and fronts in coastal waters. Whenever a tidal current impinges on a geomorphic feature such as a submerged canyon, an island or a headland, transient fronts are found which form and disperse with the daily or twice-daily tidal cycle. Little is known about the biological significance of these small-scale fronts.

7

Tides, Tidal Mixing and Internal Waves

7.1 INTRODUCTION

Tides are created by the gravitational pull of the moon and the sun and are most familiar as a rise and fall in the level of the sea twice a day. This changing water level leads to interesting patterns of zonation of intertidal organisms. They have been well studied over a long period and we shall not review them here. The tides also generate currents in the water which interact with the bottom to produce turbulence. This bottom-produced turbulence tends to mix the lower layers of the water and if the currents are sufficiently strong the turbulence may prevent any stratification and a whole area may be permanently tidally mixed. It has recently been shown that the breeding grounds of a number of stocks of herring are each located in a different, discrete area of strong tidal mixing (Iles and Sinclair 1982), and in this chapter we explore the possible significance of this observation.

In situations where tidal mixing is less strong and the water column becomes stratified, the interaction of the tidal currents with the bottom topography may lead to the formation of internal waves on the thermocline. These in turn propagate horizontally and cause vertical mixing

255

and the redistribution of nutrients, with important effects on phytoplankton production. The surface of the ocean often shows slicks and
areas of convergence associated with the passage of internal waves, and
these can have a strong influence on the distribution of zooplankton and
larval fishes.

As tidal currents move on and off shallow banks, they interact with
bottom topography to generate unidirectional currents which form gyres
round those banks. The combination of tidally-mixed water over the top
of a bank and a gyre round its periphery is thought to provide conditions particularly suitable for the eggs and larvae of fish. In this chapter
we shall explore some of these interesting consequences of tidally-
induced water movement.

7.2 THE PHYSICS OF TIDES

7.2.1 Tide-generating forces and the equilibrium tide

Tides in the ocean result from a slight imbalance between two forces;
the first is the gravitational pull of the moon and sun and the second is
the centripetal force which is required to keep the ocean's water moving
along with the rest of the earth in a circular path through space. We
examine these forces and motions in this section to show how the main
features of the tide are generated.

We first examine the effect of the gravitational force of the moon on
the world's ocean in Fig. 7.01. Here the earth and the moon are seen
from above the north pole of the earth and both are assumed to be
stationary except for the earth's rotation. Also, the earth in this case has
no continents but is covered with a uniform layer of water. If the moon
wasn't there the oceans would cover the earth to a constant depth. The
moon however exerts a gravitational attraction on each particle of water

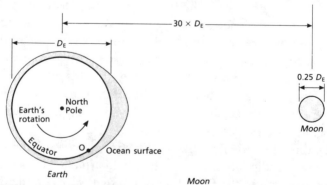

Fig. 7.01 The earth and moon viewed from above the north pole showing the ocean
(shaded) pulled by the moon's gravity into a tidal bulge under the moon. An observer is
at 'O' on the equator.

in the ocean and causes it to pile up under the moon. An observer at the equator at 'O' rotating with the earth through the day will notice that the depth of the ocean rises and falls once a day. The high tide will always be under the moon and the low tide will be on the side opposite to the moon. This elementary model produces a tide but it is not a very good model because at most places on the earth there are usually two high and two low tides during the day.

One needed improvement in the model is found by considering the rotation of the earth–moon pair once every $\simeq 29\frac{1}{2}$ days. To keep a body moving in a circle, a force must be directed towards the centre of rotation otherwise the body would continue in a straight line according to Newton's first law of motion. This phenomenon is of course well known to anyone who has swung a ball on the end of a string and let go. The force which maintains the circular motion is called the centripetal force and in the case of the earth–moon system it is supplied by the gravitational force between the two bodies. If the average gravitational pull over the whole earth was not equal to the required centripetal force, the distance between the earth and moon would have to change until they were equal.

If the earth and the moon each had the same mass they would rotate about a common centre of mass which would lie half way between the two. But because the earth's mass is roughly 80 times that of the moon the common centre of rotation is about 1600 km or about one-quarter of the earth's radius inside the earth. The rotation of the two bodies about this point is illustrated in Fig. 7.02(a). As the moon moves around this centre of mass, the earth also moves around it in a circle with radius equal to the distance between the centre of mass of the earth and the centre of mass of the two bodies. The curved lines drawn through the centre of the earth and the point 'A' illustrate that every point in the earth moves in a circle of similar radius. Now, because all particles in the earth are moving in circles of identical size the centripetal force must be the same for every particle. The direction of this force is always towards the moon which is also towards the centre of the circle described by the particle (Fig. 7.02b).

The moon's gravitational pull on the earth decreases with distance from the moon. Particles on the side closest to the moon experience a greater gravitational pull than do particles on the side away from the moon. The average gravitational pull on particles in the earth is found at the centre of the earth and must equal the required centripetal force but for all other points there is an imbalance in the forces. This imbalance is illustrated in Fig. 7.03 which shows the earth and the moon again from above the north pole. Along the centre line perpendicular to the line joining the earth and moon the gravitational force on each

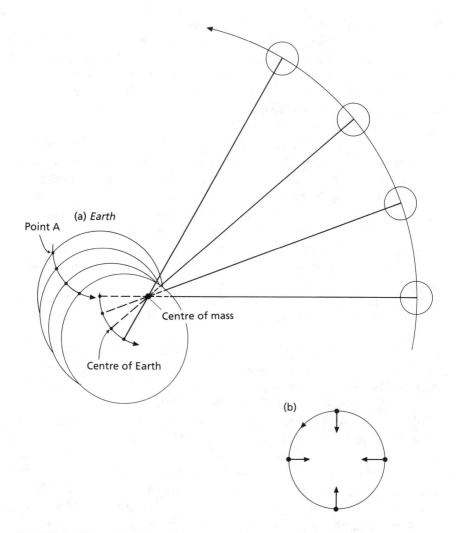

(a) *Earth*

Point A

Centre of mass

Centre of Earth

(b)

Fig. 7.02 (a) The earth and moon in mutual rotation about the common centre of mass. The curves through points 'A' and 'C' illustrate the curves traced out by typical points on the earth. (b) The circular path of a typical particle on the earth through one cycle of the motion illustrated in (a) and the direction of the centripetal force required to keep the particle travelling in the circle.

particle supplies the required centripetal force. On the half closest to the moon the gravitational force is greater than the centripetal force which causes the ocean to pile up under the moon. On the half of the earth away from the moon the gravitational force is not strong enough to supply all the centripetal force required to keep the particles moving in the required circle, and the particles tend to move farther away. This creates the tidal bulge on the side of the earth away from the moon. The

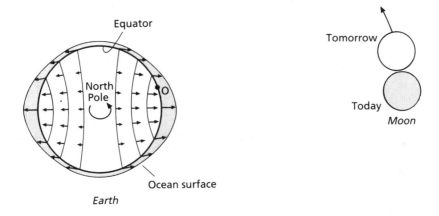

Fig. 7.03 The small arrows on the earth represent the net force due to the imbalance between the gravitational pull of the moon and the centripetal force which leads to 'tidal bulges' on two sides of the earth. The moon is shown at two positions one day apart to illustrate the delay of the tides by $\simeq 50$ min each day as seen by the observer at 'O'.

observer at 'O' now observes two high tides and two low tides every day as the earth rotates.

So far we see that the centripetal and gravitational forces of the earth–moon system account for the fact that there are two tides every day. Figure 7.03 may also be used to explain why the tides do not occur at the same time every day. The earth and moon rotate around each other in a lunar month which is roughly $29\frac{1}{2}$ days. As our observer at 'O' rotates through exactly one day back to 'O' the moon has moved about $12°$ further around in its orbit to the position marked 'tomorrow' in Fig. 7.03. The high tide which is under the moon will still be under the moon tomorrow but it will be observed later in the observer's day. Since the earth spins on its axis at about 4 min per degree the tides will appear about 50 min later each day.

For the next refinement of these qualitative arguments consider Fig. 7.04 in which the earth and moon are viewed from the side. Instead of having the moon directly over the equator, as in the previous figures, it is shown about $25°$ north of the equator. This is not an unusual situation and arises because the moon's orbit around the earth is tilted at an angle to the equator. The moon can be found at various angles to the north and south of the equator up to a maximum of $35°$ depending on the season and the time of the lunar month. The aim of the diagram is to show that the tidal bulges are no longer on the equator as they

Moon

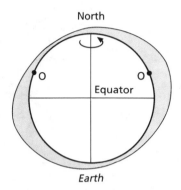

North

Equator

Earth

Fig. 7.04 Viewed from the side (north up) and with the moon north of the equator the tidal bulges are asymmetric about the equator resulting in the diurnal inequality in the heights of the high and low tides.

were imagined in Figs 7.01 and 7.03 but now one lies north of the equator under the moon while the other lies on the opposite side of the earth, to the south of the equator. Our observer at 'O' now notices that there are two high and two low tides but that the high tides are of unequal height and similarly the low tides are of unequal height. This difference is called the diurnal (daily) inequality.

This elementary view of tides on a world with a moon but no sun, without continents, and in an ocean that is always in equilibrium with the forces which are acting is called the equilibrium tide. It cannot predict the tide at any particular location but it does explain some of the main features of the tide such as the diurnal variation, the diurnal inequality in the height of the tide and the daily delay of $\simeq 50$ min in the time of the highs and lows. The next improvement in the model is introduced by considering the effects of the earth–sun system on top of the effects of the earth–moon system.

The mass of the sun is 27×10^6 times the mass of the moon but its distance from the earth is 400 times that of the moon. Because of this great distance, the gravitational attraction on a particle of water on the earth due to the sun is about one-half that due to the moon. So we can construct a diagram like that shown in both Figs 7.03 and 7.04 for the tide due to the sun, but the height of the tide will only be half that due to the moon. The important effect of the tide due to the sun arises because its tidal bulge moves relative to the moon's tidal bulge throughout the lunar month. When the two tidal bulges coincide they add together to create the extra high tides called the spring tides. When the tidal bulges are opposed their effects tend to cancel each other creating the neap tides. Thus the equilibrium tide model can also qualitatively explain the fortnightly inequality in addition to the other effects.

7.2.2 Tides in the real ocean

The equilibrium tide helps us to understand some of the main principles, but when it comes to predicting the tide in the real ocean this theory is of little help because the water, that is raised up as the tidal bulge, has to move around a world which is spinning on its axis and which is cluttered with continents. It is possible to formulate a rigorous dynamical theory to predict the tides but the way that the tidal wave moves in an ocean basin is very dependent on the shape of the basin, and this is not amenable to simple mathematical description. Such theories can be used in simple situations which have elementary geometries, for example in bays with flat bottoms and straight vertical sides (Bowden 1983), but generally the tide must be predicted by extrapolating from existing measurements.

This is normally done by measuring the height of the tide for at least a month, then decomposing the record into sinusoidal constituents. There are three main categories of constituents (Pond and Pickard 1983): (i) semi-diurnal, period about 12 h; (ii) diurnal, period about 24 h; and (iii) long period, greater than 24 h. Although 20 or more constituents may be required to predict the tidal height accurately, the four most important constituents are:

the lunar semi-diurnal	M_2	Period = 12.42 h
the solar semi-diurnal	S_2	Period = 12.00 h
the luni-solar diurnal	K_1	Period = 23.93 h
the principal lunar diurnal	O_1	Period = 25.82 h

The M_2 constituent is roughly twice the amplitude of the other three.

The form of the tide, or the pattern of the water's rise and fall, is not the same at all locations around the oceans but varies according to the relative importance of the different constituents. There are generally four main classifications of the form of the tides and these are illustrated in Fig. 7.05. The top record from Immingham, England, illustrates a tide with two high and two low tides every day. Both of the highs are about the same height and both the lows are about the same

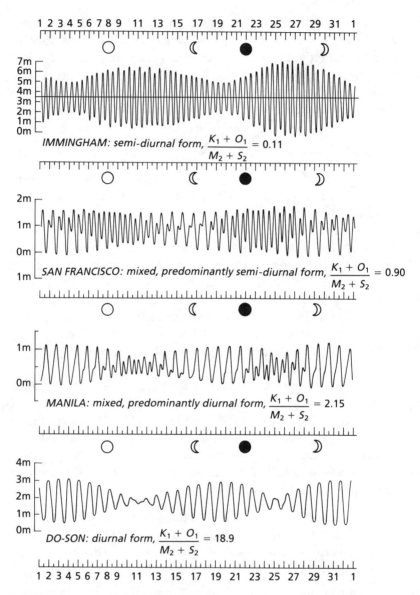

Fig. 7.05 Tidal records through March 1936 at four coastal stations illustrating variations in the amplitudes of the semi-diurnal ($M_2 + S_2$) and diurnal constituents ($K_1 + O_1$). Adapted from Defant (1958).

height. Such a tide is called semi-diurnal in form because there are two per day and both are about the same height. In this case the semi-diurnal constituents dominate the diurnal ones. This fact is often quantified with the ratio $F = (K_1 + O_1)/(M_2 + S_2)$, where each letter stands for the amplitude of the constituent. If F is small (0.11), as in the upper record, the sum of the amplitudes of the diurnal constituents $(K_1 + O_1)$ is small relative to the sum of the semi-diurnal ones $(M_2 + S_2)$.

The four records in Fig. 7.05 show a marked decrease from top to bottom in the amplitude of the semi-diurnal tidal oscillation relative to the diurnal oscillation which is confirmed by the increase in F from 0.11 to $\simeq 19$. At San Francisco there are always two tides per day but they are usually of unequal amplitude. At Manila there are two tides per day during the neap tides but only one rise and fall per day during spring tides. At Do-Son there is only one rise and fall of the tide per day throughout the month which is a purely diurnal form of tidal oscillation. This is the rarest form of tide. The changing form of the tide between locations is due partly to the shape of the ocean basin in which the tidal wave is contained and partly to the latitude (Hendershott 1981).

7.2.3 Moving the tidal bulge over the earth: Kelvin waves

The tidal bulge shown in Fig. 7.04 travels over the surface of the rotating earth as a very long wave. And because the length of the wave is very much greater than the depth of the water it is called a 'shallow water' wave similar to the waves that can be set up in a bathtub by making the water rock back and forth. The other kind of wave, the 'deep water' wave, is the one normally seen generated by the wind on the ocean's surface. One important feature of shallow water waves which sets them apart from the deep water waves is that the velocity of the wave motion is constant throughout the depth of water while the motion in the wind waves dies out a few metres below the surface. Thus the velocity in the tidal wave is approximately constant throughout the depth of the ocean. Such flow is sometimes referred to as a barotropic wave, more specifically a barotropic Kelvin wave.

Because the tidal waves cause the water to move relative to the earth for a long time the Coriolis force is an important feature of the motion. The effect of the Coriolis force is to push the water to the right in the northern hemisphere. Figure 7.06 shows a Kelvin wave travelling south, with the coast to the west of it. The Coriolis effect causes the amplitude of the wave to increase towards the shore and leads to the expression that the wave is 'trapped' against the shore. Such a trapped Kelvin wave causes the water particles to move back and forth parallel to the coast as the wave goes by.

The horizontal scale of the wave, or the approximate distance from the coast to where the sea level is undisturbed by the wave, is estimated by the Rossby radius. This is the same scale we used in estimating the width of the coastal upwelling regions in section 5.2.3 but in the present case we use the formula

$$R_e = (gh)^{1/2}/f, \tag{7.01}$$

where g is the acceleration due to gravity, h is the depth of the water and f is the Coriolis parameter. In our earlier example we were concerned with two layers of different density and the g was replaced with the reduced gravity (g') and the h represented the depth of the upper layer. For the Kelvin wave in Fig. 7.06 we put $g = 10$ ms^{-2}, $h = 4000$ m and $f = 10^{-4}$ s^{-1} to get $R_e = 2000$ km. This is commonly called the external Rossby radius or deformation scale as opposed to the internal Rossby radius derived in section 5.2.3. The external deformation scale is roughly 100 times the internal one.

7.2.4 Tidal currents

We see from Fig. 7.06 that a coastally-trapped Kelvin wave will retain its integrity only if it is moving in an anti-clockwise manner round the coastal margin of a basin. The Coriolis force ensures that it will have a high amplitude close to the coast and a low amplitude far from the coast, typically at a distance of 2000 km. For the North Atlantic Basin,

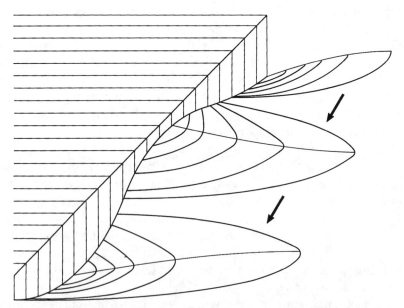

Fig. 7.06 A Kelvin wave travelling from the upper right to the lower right with the coastal wall to the right of its direction of travel. Note that the height of the wave decreases with increasing distance from the coast.

therefore, a pattern similar to that shown in Fig. 7.07 develops. In mid-Atlantic there is an amphidromic point at which there is zero rise and fall with the tides, and the tidal amplitude increases as one moves towards the coast. High tide appears successively at points along the coast, proceeding in an anti-clockwise direction. Similar patterns are found within smaller basins, such as the North Sea.

In the deep ocean the vertical range of the tide, as suggested by the illustration in Fig. 7.07, is only a few centimetres and tidal currents tend to be only a few centimetres per second. Over the continental shelves however the currents can be in the metres per second range as they are associated with much higher tidal amplitudes.

The direction of tidal currents varies greatly and depends on the way in which the tidal wave propagates in the local area. Along a straight coast or in a confined channel the currents tend to be parallel to the shore. In open areas the tidal wave is not constrained to be rectilinear and currents in general will have both north–south (v) and east–west components (u). Each of these components may be decomposed into the same sinusoidal tidal constituents that were found in the vertical excursion of the tide at the coast.

One method of presenting these data is shown in Fig. 7.08(a) which shows an ellipse representing the currents over one cycle due to the M_2 tidal constituent. The ellipse is constructed by plotting the u component against the v component and is commonly called a hodograph. The variations of the u and v components through time are further illustrated by the curves in the upper left and lower right of the figure. The

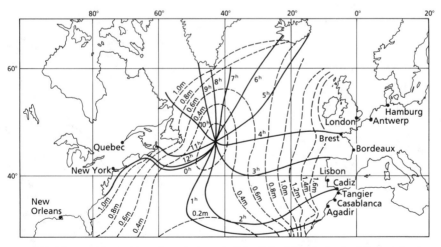

Fig. 7.07 The amplitude (dotted) and phase (solid) of the M_2 tide in the North Atlantic Ocean. The amplitude is in metres and the phase is in hours after the passage of the moon over the prime meridian. Adapted from Defant (1958).

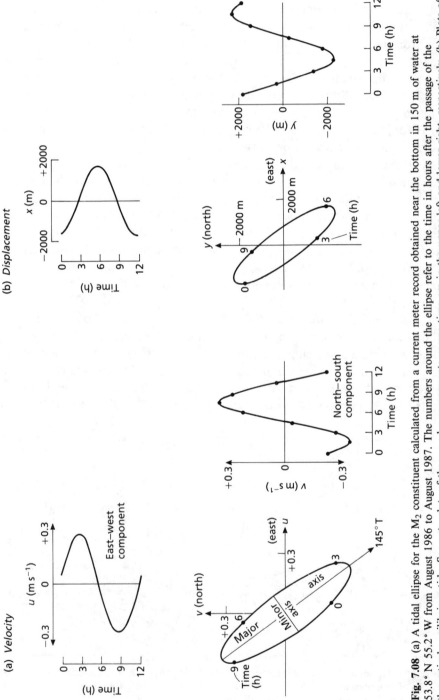

(a) Velocity

(b) Displacement

East–west component

North–south component

Fig. 7.08 (a) A tidal ellipse for the M_2 constituent calculated from a current meter record obtained near the bottom in 150 m of water at 53.8° N 55.2° W from August 1986 to August 1987. The numbers around the ellipse refer to the time in hours after the passage of the theoretical equilibrium tide. Separate plots of the u and v components versus time are in the upper left and lower right, respectively. (b) Plots of the particle displacement due to the M_2 tidal currents shown in Fig. 7.08(a).

one in the upper left represents the flow along the east–west axis while the one in the lower right is the v component. It is quite common to specify the M_2 tidal flow by the length of the major and minor axes of the ellipse plus the orientation of the ellipse relative to north. Using this method the M_2 tidal ellipse in Fig. 7.08(a) has amplitudes along the major and minor axes of $\simeq 0.4$ m s^{-1} and 0.1 m s^{-1}, respectively, and the major axis lies along 145 °T.

By noting the velocity of water along two axes at right angles at a fixed point throughout a tidal cycle it is possible to calculate the total distance that a parcel of water has moved — the tidal excursion. Movement is usually highly variable and even oscillatory, so the displacement over a tidal cycle is calculated by dividing the records of u and v into segments so short that each can be assumed constant, then summing them. A shorter method, shown in Box 7.01 gives the same result.

It should be remembered that the estimation of displacement developed in Box 7.01 does not use velocity observations obtained at the location of the moving particle but uses data obtained at one location which is sometimes far from the moving particle. The calculation assumes, for lack of better information, that there are no horizontal variations in the velocity field. This is not a bad assumption over short distances in regions where horizontal uniformity is likely, but near the shore and over rough shallow bathymetry the velocity field can change significantly in a few kilometres and extrapolating particle excursions from velocities at one position should only be done with care.

The current and displacement ellipses in Figs 7.08(a) and (b) only specify the currents and displacements associated with the M_2 constituent and each of the other constituents may be plotted in the same manner. The other constituents tend to be of smaller amplitude and the main features of the tidal flow can usually be approximated by the M_2 tidal ellipse alone. At some locations the tidal ellipse appears more open than those in Fig. 7.08, while at others the ellipses collapse into a straight line. These variations are due to differences in the amplitudes and phases between the u and v components of velocity. These can be easily demonstrated by plotting figures similar to Fig. 7.08(a) after changing the values of the amplitudes and phases in the functions for u and v in Eqs 7.03.

7.2.5 Internal waves

Tidal currents often cause internal waves to be generated on the pycnocline. These waves were mentioned in section 3.2.4 where it was shown that when the pycnocline is displaced vertically a restoring force

Box 7.01 Estimating tidal displacements mathematically

We start with the u and v components of velocity displayed in Fig. 7.08(a). These are single frequency oscillations and can be expressed by sine functions, so mathematically the components can be written in the form

$$u = A_u \sin(\omega t + \phi_u),$$
$$v = A_v \sin(\omega t + \phi_v),$$

(7.02)

where A_u and A_v are the amplitudes of the oscillations in metres per hour, ω is the frequency of the oscillation in radians per hour which for the $\simeq 12$ h M_2 constituent is $2\pi/12 = 0.52$ rad h^{-1} (or 30° h^{-1}), t is the time in hours and ϕ_u and ϕ_v are the phase angles to indicate where the zero of the sine function is relative to the time zero. For the u and v components shown in Fig. 7.08(a) the amplitudes (A) are 0.25 m s^{-1} and 0.34 m s^{-1} respectively. In metres per hour these numbers become 900 m h^{-1} and 1224 m h^{-1}. The phases are 10° or 0.17 rad and 220° or 3.84 rad respectively and we can write the u and v components as

$$u = 900 \sin(30t + 10)$$

and

(7.03)

$$v = 1224 \sin(30t + 220),$$

where we have used degrees to indicate the phase inside the brackets because it is a more familiar unit than radians.

The displacement, D_x, due to the oscillating velocity component, u, is given by the integral of the velocity as

$$D_x = \int A_u \sin(\omega t + \phi_u)\, dt = -A_u/\omega \cos(\omega t + \phi_u),$$

(7.04)

where the symbols are the same as in Eqs 7.02. The amplitudes of the displacements due to the u and v velocity components are $A_u/\omega = 900/0.52 = 1719$ m and $A_v/\omega = 1224/0.52 = 2338$ m, respectively. And the displacement functions in the x and y directions are therefore

$$D_x = -1719 \cos(30t + 10)$$

and

(7.05)

$$D_y = -2338 \cos(30t + 220),$$

where again the phase of the argument is in degrees. Outside the brackets ω must be in radian measure. These functions are plotted in the upper left and lower right of Fig. 7.08(b). By plotting the displacement in the x direction at a given time against the displacement in the y direction the ellipse in the lower left results. This is the same type of plot as the velocity hodograph and results in a similar ellipse except that it describes the time sequence of a particle displacement rather than the time sequence of its velocity. Comparing the ellipses in Figs 7.08(a) and (b) shows both to have the same shape and orientation but the phase is different in the two which we already knew because sine (velocity) and cosine (displacement) functions with the same arguments are 90° out of phase.

is generated which pushes the pycnocline back towards its undisturbed position. This force leads to the possibility of a vertical oscillation and of waves being propagated on the interface. Some features of such a wave on a sharp density interface are illustrated in Fig. 7.09.

The interface in the figure has been placed close enough to the sea surface that some effects of the internal wave motions are felt at the surface. The lower layer is assumed to be infinitely deep. The wave is propagating from left to right and causes the water particles to move

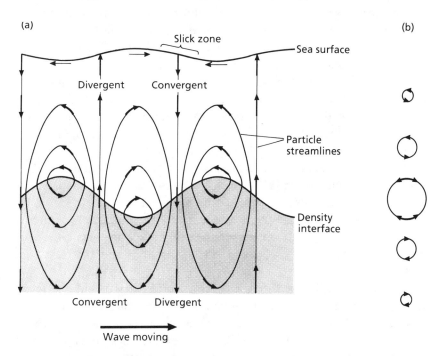

Fig. 7.09 (a) A wave progressing from left to right on the density interface showing the instantaneous particle streamlines, the resultant surface wave, the zones of convergence and divergence and the region where slicks are likely to form on the surface. (b) The particle orbits over one cycle due to the passing internal wave. The diameter decreases away from the interface and the direction of the trajectory reverses across the interface as indicated by the double arrows on the central orbit.

along the particle streamlines. These lines indicate the instantaneous paths of the particles while the arrows within the streamlines suggest the magnitudes of the particles' velocities which diminish with distance away from the interface. As the wave progresses, the pattern of streamlines moves along with it but the individual particles of water remain in their original locations moving in wave-induced circular orbits. The examples of these orbits in Fig. 7.09(b), illustrate that the orbit diameters, like the particle velocities, decrease with distance from the interface.

Interesting biological consequences of the internal wave motion are created by regions of convergent and divergent flow in the upper layer. Above the pycnocline the thickness of the upper layer clearly varies along the length of the wave. As the wave advances, water must flow from the thin regions above the wave crests to the thicker regions above the wave troughs. The water must therefore converge behind the wave crests and diverge behind the wave troughs as noted in Fig. 7.09(a). These regions of convergence and divergence exist up through the water column to the sea surface with diminishing intensity but their strength is enough to create a wave in the sea surface. This surface wave has the same length as the interface wave but it is 180° out of phase with the internal wave and has a much smaller amplitude.

On the sea surface the convergent zones cause floating organic matter to accumulate in bands which are ahead of but parallel to the wave crests. The organic material increases the surface tension of the surface layer which reduces the amplitudes of the smallest surface ripples. This gives the appearance of a band of water that is smoother than the water over the divergent zones where the small waves continue to exist. These smooth patches are often called 'slicks'. The contrast in the roughness of the sea surface can also be detected by a special instrument called the synthetic aperture radar. When these instruments are carried in aircraft or satellites they allow mapping of the sea surface features over large areas, as demonstrated by New (1988) for an area near the Celtic Sea shelf break.

Below the pycnocline the particle trajectories (Fig. 7.09b) due to the internal wave are opposite in direction to those above the interface. Also the regions of divergence and convergence below the interface are displaced by one-half of a wave length from the ones above the interface. These features make sense because the thickness of the lower layer varies along the wave in the opposite sense to the thickness variations above the interface. The reversal in the particle orbits across the interface leads to a shear across the interface which is especially strong at the wave crests and troughs. Half way between the crests and troughs the motion due to the wave is predominantly vertical as indicated by the vertical streamlines in Fig. 7.09(a).

The interface waves of Fig. 7.09 are a special form of internal wave. Throughout most of the ocean's depth the density of water increases gently and smoothly rather than rapidly across thin layers. The larger-scale density stratification also supports internal waves but they tend to be oscillations of the whole water column rather than waves confined to the region near an interface. For a description of these large-scale oscillations and a more thorough examination of the interface waves introduced here see Pond and Pickard (1983), Roberts (1975), New (1988) or Apel (1987). More information of the biological significance of internal waves is given in section 7.4.1.

7.2.6 Generating internal interface waves: lee waves

One of the more common mechanisms producing internal waves on the pycnocline occurs when the water flows over an obstacle. Long (1954) investigated this process in a long tank filled with two layers of unequal density (Fig. 7.10). The denser layer fills the lower third of the channel. When a small smooth obstacle is placed in the bottom of the channel and the lower layer is forced to flow from left to right lee waves are formed on the interface between the fluids downstream of the obstacle. These lee waves are moving at the same speed but in the opposite direction as the water. Waves that are shorter than the ones shown travel too slowly for the speed of the water and are swept downstream. The longer and faster waves propagate upstream away from the obstacle. Thus the observed lee waves are the interface waves that are selected by the speed of the stream. These lee waves are usually associated with steady flows but often tidal flows generate such waves on the lee side of shallow ridges and when the tidal stream slows down the waves continue to exist but moves away from the obstacle through the more slowly moving water.

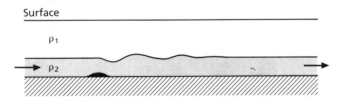

Fig. 7.10 The interface in a two layer fluid in which the lower layer is moving from the left to right over a smooth obstacle whose height is less than the thickness of the lower layer. Adapted from Long (1954).

Another type of lee wave generated by the same mechanism but over the period of the flood or ebb tide is sometimes created at the edge of continental shelves. An elementary illustration of its generation is given in Figs 7.11(a) and (b). We first assume that the tidal flow through section A in the deep water is the same as that through section B in the shallow water and that far away from the shelf edge the currents are the same from top to bottom, that is, the currents are barotropic. On the flood tide, when the water is streaming onto the continental shelf, there must be more water passing through the upper layer in section B than through the upper layer in section A. This occurs because the layer is the same thickness in each location and the water through section B is going faster than through A. In the lower layer there is more water passing through section A than through B. The deficiency in the upper layer coupled with the surplus in the lower layer causes the interface between the two to rise in the vicinity of the shelf break. On the ebb flow the opposite occurs and the interface between the layers is forced down as indicated in Fig. 7.11(b).

The raising and lowering of the pycnocline can be tens of metres and represents a large fraction of the depth of the upper layer. These large undulations propagate as waves both into deep water and over the shelf but because of their large amplitudes non-linear effects tend to dominate their existence. The fate of the wave which is called the internal or baroclinic tide is outlined Fig. 7.12. The wave when it is first formed is in the form shown in stage 1 and travels towards the left over the shallower water. The front of the wave (to the left) begins to steepen because the troughs of the waves travel faster than the peaks. As the waves get steeper (stage 3) shorter waves begin to form which grow with time until only high amplitude but short waves

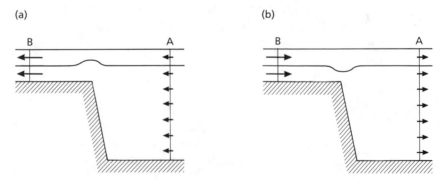

Fig. 7.11 (a) A section through the ocean in the vicinity of a shelf break illustrating the higher currents of the flood tide over the shallow water at B than in the deep ocean at A and the resultant rise in the pycnocline at the shelf break. (b) The same situation as in (a) except the tide is ebbing and the pycnocline is depressed at the shelf break.

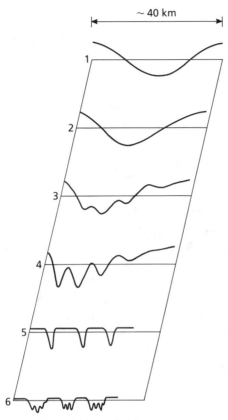

Fig. 7.12 Six stages in the life of the internal tide as it progresses to the left over the shelf, first becoming steeper at the leading edge then forming shorter waves that become solitary waves which decay into turbulence.

remain (stage 5). These short waves, because they are quite far apart relative to their wavelength, are sometimes called solitary waves or solitons. They usually travel in groups however and are not really solitary but they may be treated mathematically as isolated phenomena.

As these packets of solitary waves propagate further onto the shelf their amplitude decreases as the high shears within the waves lead to Kelvin–Helmholtz instabilities (section 4.4.2) and turbulence (stage 6). The waves are dissipated within about 30 km of the generation region and it is the conversion to turbulence in this 30 km strip of wave energy that is thought to provide the mixing which increases the nutrient flux up into the euphotic zone and the subsequent biological enhancement (Sandstrom and Elliott 1984; Pingree and Mardell 1985).

7.2.7 Tidal rectification

The bathymetry of continental shelves is often uneven, being marked by shallow banks and basins. An example of this changing bathymetry in the Gulf of Maine is shown in Fig. 4.12. Oscillating tidal currents washing back and forth over the banks sometimes generate mean currents at the edge of the bank where the depth is increasing rapidly. This process of extracting energy from an oscillating tidal current to produce a mean uni-directional flow is called tidal rectification. Two places where it is thought to play an important role are around Georges Bank in the Gulf of Maine (Loder 1980) and around large sandbanks in the North Sea (Huthnance 1973).

The rectification of the oscillating flow depends on the fact that the tidal excursion is larger over the bank than in the deep water. This is illustrated in Fig. 7.13(a) in a cross section of the ocean through the edge of the bank similar to Fig. 7.11 but without density stratification. The tidal flow through the vertical section 'A' is the same from top to bottom and if the same amount of water is to pass through section 'B' as section 'A' the speed of the flow in the shallow water must be larger than in the deep water. The same phenomenon is illustrated in Fig. 7.13(b) by the comparison of the M_2 tidal ellipses over the bank and in the deep water.

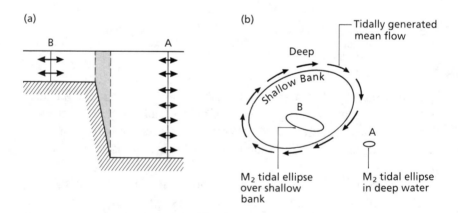

Fig. 7.13 (a) A section through the ocean across the edge of a bank such as Georges Bank in the Gulf of Maine illustrating that the tidal current is higher in the shallow water than in the deep water. The shaded region over the sloping bottom indicates where the currents are accelerated and decelerated leading to 'tidal rectification' which generates the mean flow around the bank. (b) A view down on the bank shows the large tidal ellipse over the shallow water relative to the ellipse in the deep water and the tidally-generated clockwise flow around the bank.

The change in speed of the tidal flow between the deep and shallow water occurs in a narrow band over the sloping edges of the bank and creates non-linear forces associated with the acceleration and deceleration of the water. The rapidly changing speed also affects the magnitude of the forces due to the Coriolis effect and bottom friction which both depend on the speed of the flow. It turns out that in this region of rapidly changing forces the balance between them is disturbed in such a way as to produce the clockwise mean flow around the bank as illustrated in Fig. 7.13(b). The details of the mechanisms are not easily described outside the elegant analytic formulations of Loder (1980) and Huthnance (1973) and numerical models of Greenberg (1983).

7.3 TIDAL MIXING IN THE WATER COLUMN

We saw in section 4.6 that tidally-driven vertical mixing changes the seasonal pattern of phytoplankton production. There is also evidence that tidally-mixed areas are favoured as spawning grounds by herring. This section explores the evidence for these ideas.

7.3.1 Tidal mixing and plankton production

If tidal currents are strong enough to mix the water column all year, there is a continuous supply of nutrients from near-bottom waters up to the euphotic zone, which permits phytoplankton production to continue at a good level throughout the summer. This is in contrast to the situation in stratified waters, where the supply of nutrients tends to become depleted after the spring bloom. As a result, tidally-mixed areas like Georges Bank, near the Gulf of Maine, and the Dogger Bank in the North Sea, have levels of primary production considerably higher than adjacent stratified areas of the shelf. It has been found that these sites are often selected as breeding grounds by commercially important fish stocks, and it is of interest to explore the connection between tidal mixing and the early life histories of fish.

7.3.2 Tidally-mixed areas and the spawning of fish

Fisheries biologists have long been puzzled by the ability of a single species such as herring to divide into a number of discrete breeding stocks, each with a characteristic place and time of breeding, and often with recognizable small differences in body structure. The adults of the various stocks are often found mixed together in the same place but at breeding time they segregate and return to the place where they were hatched. In this way they maintain genetic differentiation of the stocks. How do they recognize those breeding grounds, and what defines their limits? For some stocks, the breeding ground is readily identified as a group of fjords (e.g. the Norwegian coast) or estuaries (as on the Maine coast), but in other cases there is no clear physical boundary to the spawning ground or the nursery area in which the young fish develop. Iles and Sinclair (1982) noticed that the breeding grounds of a number of herring stocks are located in areas of vigorous tidal mixing, bordered in summer by a tidal front, and they proposed that tidally-mixed areas act to define and delimit the breeding and nursery areas of a number of stocks of herring on both sides of the North Atlantic.

The early work on predicting the occurrence of tidally-mixed areas was done in waters around the UK (Section 4.6.1). When tidally-driven currents moving across the sea floor are sufficiently strong to generate turbulence that breaks down any stratification in the water column above, that area is said to be tidally mixed. As we showed in earlier chapters, Simpson and Hunter (1974) and Pingree *et al.* (1978a) showed how to predict the occurrence of tidally-mixed areas on Georges Bank, Nantucket Shoals, in the mouth of the Bay of Fundy and off the southwest coast of Nova Scotia.

Iles and Sinclair (1982) showed (Figs 7.14 and 7.15) that there was a remarkable similarity between the distribution of larval herring and the occurrence of tidal mixing, on both east and west coasts of the North Atlantic. In the Gulf of Maine area larval herring are found on Nantucket Shoals, Georges Bank and in the tidally-mixed areas off New

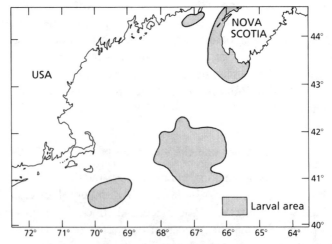

Fig. 7.14 Larval distribution areas for several populations of Atlantic herring. From Iles and Sinclair (1982).

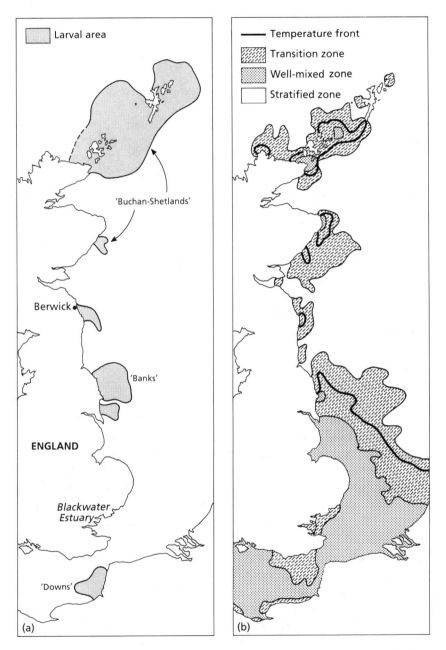

Fig. 7.15 (a) Distribution of early larval stages of Atlantic herring in the North Sea. (b) Location of tidally-mixed areas. From Iles and Sinclair (1982).

Brunswick and Nova Scotia. In the North Sea, the spawning grounds of the 'Downs', Banks' and 'Buchan-Shetlands' stocks of herring are in tidally-mixed areas. Even in the Gulf of St Lawrence, where the occurrence of tidally-mixed areas had been predicted by Pingree and Griffiths (1980), five out of six major herring spawning grounds are in tidally-mixed areas.

The question then arises: what is it about these tidally-mixed areas that makes them suitable as herring spawning grounds? Is the food-production process particularly favourable or could it be that the strong tidal currents remove fine particles and leave behind a gravel particularly suited to herring spawning? In general (see Chapter 4), tidally-mixed areas have a relatively uniform biomass of phytoplankton throughout the growing season, and the zooplankton biomass tends to peak late in the year. Since the herring in these areas may be spring spawners or autumn spawners, there appears to be no critical pattern of food production to which they are adapted. In fact, off southwest Nova Scotia there is an autumn-spawning stock that occupies the tidally-mixed area principally in winter, when food production is minimal. There remains the possibility that the young herring make particular use of the high productivity associated with tidal fronts, but this has not been demonstrated.

Iles and Sinclair (1982) suggested that the important feature of these tidally-mixed areas is that the boundaries act as natural barriers to the dispersion of the larvae during early development, thus ensuring the geographical discreteness of the stock. They pointed out that as tidal currents flow over the edges of banks like Georges Bank and the waters surrounding Grand Manan in the Gulf of Maine, the changing topography causes rectification of the currents and anti-cyclonic gyres are formed (Loder 1980; Loder and Wright 1985; Smith 1989; and see section 7.2.7). These persist even in winter when the front between stratified and tidally-mixed waters disappears, and might well be the physical barrier that reduces dispersal of the young fish. In general, the idea that tidally-mixed areas assist in the retention of young stages of fish during development, thus defining the geographical limits of the breeding grounds of the stock, has come to be known as the larval-retention hypothesis. An important aspect of the hypothesis is the suggestion that the size of the retention area sets an upper limit on the size of the stock that breeds there. Where the breeding ground is in a fjord or an estuary, the land forms a natural boundary of the breeding area. Otherwise, the boundaries of the tidally-mixed area serve the same purpose. At one site in the Gulf of St Lawrence that is an important herring breeding ground but is not tidally mixed, there is a geographically fixed gyre that might well serve as a retention area.

Cushing (1986) argued that the Iles and Sinclair hypothesis probably was not applicable to the waters around the British Isles. He pointed out that herring eggs in tidally-mixed areas often hatch outside the period of stratification, at a time when tidal fronts are not present. He also pointed to the wide spatial separation of spawning grounds and nursery grounds for various North Sea stocks, and argued that it would be necessary for the larvae to begin drifting towards the nursery grounds soon after hatching, rather than experience larval retention over the spawning grounds. Finally, he suggested that since herring deposit their eggs on gravel of a particular size range, it might be that this gravel was mainly found in tidally-mixed areas.

Sinclair (1988) countered the first point by suggesting that tidally-mixed areas are distinct from adjacent areas even during winter when there is no stratification. To the second, he argued that the information used by Cushing when discussing larval drift was based on very simple older models of physical oceanography, and that the contemporary understanding of the physical oceanography of the North Sea would lead to a quite different interpretation. He also stated that displacement of larval distributions from the location of spawning was not necessarily inconsistent with the hypothesis. The essence of the hypothesis was that a persistent physical oceanographic feature was present to define a spawning area.

More recently, Townsend *et al.* (1986) have produced some evidence from the Bay of Fundy–Gulf of Maine area that herring larvae may leave the tidally-mixed area in large numbers. They identified herring spawning grounds in the tidally-mixed area west of Grand Manan Island, and in September 1983 located concentrations of newly hatched larvae in the same general area. However, the distribution of larger larvae indicated a drift westwards, out of the tidally-mixed area, and towards waters where copepods were considerably more abundant. The data, though of a preliminary nature, tended to favour the more orthodox view that fish are normally spawned upstream of their feeding grounds and get carried by the mean currents towards sites that are productive of the zooplankton that the fish need to thrive and grow.

The question was pursued further by Chenoweth *et al.* (1989). They carried out three intensive surveys along the New Brunswick–Maine coast in September and October 1986, and confirmed that large numbers of larvae were produced in the tidally-mixed area southwest of the Isle of Grand Manan. Larvae hatched in August were retained for several weeks in the tidally-mixed area, but after the breakdown of the offshore stratification in late September, the larvae from Grand Manan moved rather rapidly westward along the Maine coast. Chenoweth *et al.* (1989) speculated that herring larvae do not drift passively, but exert some control over their horizontal movements by making vertical migrations into water masses moving at different speeds and in different directions, as Fortier and Leggett (1983) had demonstrated for herring larvae in the St Lawrence Estuary (Chapter 4).

Brown's Bank, off southwest Nova Scotia, has a permanent clockwise gyre over the western cap, interrupted only when wind-forced currents break its circulation for short periods. It is also tidally mixed, except when pulses of low-salinity water produce temporary vertical stratification (Smith 1989). O'Boyle *et al.* (1984) noticed that in ichthyoplankton samples over the Scotian shelf there was a marked tendency for both eggs and larvae of cod to appear in considerable numbers in the same samples taken on Brown's Bank. They interpreted this as evidence in support of the larval retention hypothesis, arguing that if there were not a retention mechanism the larvae would have drifted away from the site of egg deposition.

Suthers and Frank (1989) reported on surveys of cod larvae and pelagic juveniles over Brown's Bank and adjacent waters, carried out in three successive years. Their first point was that in any one year, larval stages and pelagic juveniles tended to be found in the same samples, supporting the arguments advanced by O'Boyle *et al.* (1984) that there seems to be a retention mechanism that keeps the different stages in the same general area. However, their second point was that the spatial distribution of those larvae and pelagic juveniles varied greatly from year to year. In 1985 and 1986 the numbers of both larvae and juveniles were greater close to shore than on the offshore bank but in 1987 the reverse was true. They speculated that the gyre on Brown's Bank served as a retention mechanism, but that at times when wind forcing breaks it, the larvae and juveniles are carried by prevailing currents towards the coast. On the basis that high concentrations of developing cod were found close to the coast in two years out of three, they suggested that the coastal waters are important nursery grounds. The debate continues.

Both here and in section 9.5.2 it is very noticeable that the argument does not proceed in a rigorous 'hard science' manner from one testable hypothesis to another, but consists of fragmented observations on different aspects of a very complex system. Ecologists who choose to investigate the properties of large ecosystems face a dilemma. They may isolate a small part of the system, find a 'manageable' problem, and formulate a rigorously testable hypothesis. But the total system, functioning in an integrated manner, is much more than the sum of its parts and at the present stage of development of ecosystem science one often has to resort to the process of collecting evidence which appears to support or not support a particular hypothesis.

7.4 THE BIOLOGICAL SIGNIFICANCE OF INTERNAL WAVES

7.4.1 Internal waves as nutrient pumps

We may first recall what was said about shelf break fronts in section 6.4.1. It has often been observed that the shelf break is an area of enhanced biological productivity. Fish are particularly abundant there (Fournier 1978) and it is very common to find greater amounts of phytoplankton and zooplankton than in adjacent areas. Fournier (1978) reported that in March 1977 the average value for chlorophyll-a across the Scotian shelf was less than 1 mg m^{-3} while over the shelf break it reached a value of 4.2 mg m^{-3}. Herman *et al.* (1981) found that chlorophyll-a was $\simeq 1.5$ mg m^{-3} over the shelf, rose to $\simeq 6$ mg m^{-3} at the front and was 3–4 mg m^{-3} in the slope water. The corresponding abundances of copepods in the water columns of shelf, front and slope waters were 30,000, 70,000 and 10,000 m^{-2}, respectively. The high concentrations of copepods occurred above the thermocline in the frontal region, and these copepods did not migrate down through the thermocline during the day, while those in shelf waters often did so.

Pingree and Mardell (1981) reported on a series of studies of the shelf break adjacent to the Celtic Sea in the eastern North Atlantic. In summer this region is characterized by a band of water about 100 km broad that is 1–2 °C cooler than water on either side of it. At times the band can be traced for about 800 km south from the Celtic Sea. The levels of inorganic nitrate and chlorophyll are significantly higher in this band than in adjacent water. They suggested that tidally-driven internal waves interacting with the bottom topography were the cause of the phenomenon. During 1983 and 1984 Pingree and his colleagues continued to investigate the phenomenon, in collaboration with French oceanographers (Mazé 1983; Pingree *et al.* 1986; Mazé *et al* 1986). Using both fixed-current metres and an undulating towed sampler they showed that during the off-shelf streaming phase of the barotropic tide the isotherms in the upper shelf region were depressed. When the tide slackened, the depression separated into on-shelf and off-shelf propagating internal tides. At large spring tides the crest-to-trough height of the internal tide was 50–60 m at the point of origin. The off-shelf waves had a wavelength of about 46 km and propagated at about 1.03 m s^{-1} while the on-shelf waves had a wavelength of 31 km and propagated at 0.7 m s^{-1}. The latter had a more rapid decay rate, indicating that they caused more vigorous mixing in shelf waters than offshore. As the field data show (Fig. 7.16) the large amplitude waves have relatively high frequency internal waves associated with them and concentrated in the troughs. This is a clear example of the phenomena described in section 7.2.5 and Fig. 7.12.

According to New (1988) these short waves produce a roughening of the sea surface that is often visible as a series of 'rips', many kilometres long and spaced about 1 kilometre apart. As mentioned earlier, these are detectable from space by synthetic aperture radar imaging and illustration was given of the occurrence of several groups of these waves over the shelf edge in the Bay of Biscay. Each group of waves indicates the position of a depression of the thermocline in one of the long-period waves.

New (1988) also investigated the properties of such a system by means of a linear numerical model that described the topographic generation of the internal tides. He was able to show that there was good agreement between his model and field observations, and then went on to investigate variations in the Richardson number (Eq. 4.09). Since a steady, stratified shear flow is thought to become unstable to small perturbations (subcritical) when $Ri < 0.25$, New (1988) was able to calculate the position and duration of regions of potential instability and the probability that these would cause upward mixing of nutrient-rich water through the thermocline. He found that at spring tides during summer stratification, mixing was highly probable both on the shelf and beyond

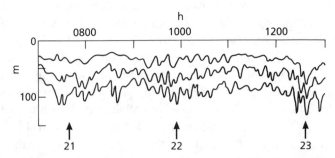

Fig. 7.16 Isopycnals along a line between points 50 and 150 km off the shelf break of the Bay of Biscay. Isopycnals are plotted at densities of 1026, 1026.5 and 1027 kg m^{-3}. 21–23 indicate tidal troughs with packets of short-period waves. From New (1988).

the shelf break. At neap tides the Richardson number was not critical at any location, so no mixing was predicted.

7.4.2 Internal waves concentrate and transport planktonic organisms

It has been observed many times (e.g. Halpern 1971; Chereskin 1983) that as the tidal current generated by an ebbing tide flows over the edge of the continental shelf, or over the edge of a reef or bank, one or more depressions of the thermocline are formed on the downstream side, and these are known as lee waves. When the ebbing tide slackens off, the lee waves generate a train of internal waves which propagate shoreward (see section 7.2.6).

The water movement associated with these internal waves, when they are close enough to the surface, is as shown in Fig. 7.09. There are alternating zones of upwelling with divergences and downwelling with convergences. Ewing (1950) drew attention to the fact that convergences may be visible at the surface as smoother areas, or slicks, on a lightly rippled sea. He pointed out that a surface film of organic matter is naturally present on biologically productive waters and that in the convergence zones this film becomes thicker and is able to damp out small surface ripples. The internal waves are quite long, so that slicks may be hundreds of metres apart, and they move slowly shoreward ($\simeq 20$ cm s^{-1}) with the internal waves. There is a tendency for buoyant material such as floating seaweed or amorphous organic matter to aggregate at the convergences, and for planktonic organisms to become associated with the aggregations. This suggests that the shoreward transport by these slicks may have significance in the adaptations of organisms inhabiting them.

Shanks (1983) made daily collections of the pelagic megalopa larvae of the crab *Pachygrapsus crassipes* from the end of the 320-m-long pier at the Scripps Institution of Oceanography, California. He found that the daily abundance fluctuated with fortnightly maxima that occurred about 5 days before the spring tide. Noting the occurrence of surface slicks, and their relationship to fortnightly tidal cycles, he used styrofoam cups weighted with sand as small drogues and deployed 50 cups 20 m apart on a line at right angles to the lines of the slicks. On three of the five occasions that he did this, the cups were not transported by the slicks, but on the other two occasions the cups were caught by the slicks and carried shoreward. At the end of the experiment over 90% were concentrated in two slicks and had been carried 1–2 km shoreward. Associated with these slicks were dense accumulations of flotsam.

In parallel with these physical experiments standard neuston net tows were taken in the slicks and in the rippled water between the slicks, to sample the larvae of invertebrates and fishes. On the days when the drogues were concentrated, but not on the other days, the larvae of invertebrates and fishes were concentrated 6–40 times more densely in the slicks than in the water between them. While the author had no data on the differences between days when concentration and shoreward transport occurred, and days when it did not, he suggested that the slicks might be a common method of onshore transport for larvae.

If larvae are to make effective use of slicks induced by internal waves, they must have behavioural adaptations that will take them to the surface and enable them to stay there in downwelling conditions. Shanks (1985) found that when megalopa larvae of *Pachygrapsus crassipes* were individually released by a diver at 1–5 m depth, they all swam to the surface. Laboratory experiments showed that they swam towards a light source, away from the direction of gravity and in the direction of decreasing pressure. They also showed a tendency to cling to objects in their vicinity. Under a variety of experimental conditions their average swimming speed was 9.5 cm s^{-1}, which was considered more than enough to enable them to swim against the downwelling observed in the vicinity of slicks. Having arrived at the surface, they could then hold their position by clinging to floating objects.

A New Zealand crab, *Munida gregaria* has bright red larvae which aggregate near the surface of the sea in daylight. Zeldis and Jillett (1982) carried out aerial surveys and published beautiful photographs showing the larvae concentrated on what appeared to be mid-shelf internal wave slicks, as well as on river plume fronts and headland fronts. While their report lacked the experimental evidence provided by Shanks, it appears to be strong confirmatory evidence for the role of these physical mechanisms in transporting crab larvae shoreward.

Satellite observations show that tidally-generated internal waves are refracted by bottom topography, and that some coastlines are more suitable for the production of internal waves than others (Apel *et al.* 1975). Shanks and Wright (1987) thought that the differential pattern of shoreward transport of slicks might account for some of the patchiness in barnacle distribution on rocky shores. They repeated the experiments of Shanks (1983) using styrofoam cups as drogues and found that barnacle larvae were indeed concentrated in slicks associated with the internal waves, and that settlement of barnacle larvae was more abundant in places where the drogues tended to accumulate. In addition to the barnacle larvae, they found that two kinds of crab larvae and gammarid amphipods were at times concentrated in the slicks.

Kingsford and Choat (1986) paid particular attention to internal waves and larval and juvenile fish, at a site off the coast of New Zealand. They consistently found greater abundances in the slicks than in the rippled areas. They also found that fish were more abundant in the vicinity of drift algae. When the internal waves and their slicks were not present, the algae and the fish were in scattered patches at different distances from shore. When the slicks formed, the algae and the fish were arranged in lines and carried shoreward.

The implications of these findings are that quantitative studies of the larval abundances of shallow-water and intertidal organisms must take into account the non-random distribution of those larvae in coastal waters and of the physical mechanisms that may carry them to the coastal shallows from far out on the continental shelf. Since the occurrence of internal waves and slicks is dependent on the tides, it will not be surprising if there is a fortnightly tidal rhythm in inshore transport and settlement. Further studies of the relationship between internal wave packets and bottom topography will be expected to throw light on the patchy distribution of many coastal organisms.

7.5 TIDAL CURRENTS AND ISLAND STIRRING

Ever since Doty and Oguri (1956) described enhanced phytoplankton biomass and production in the vicinity of Hawaii and coined the phrase 'island mass effect', it has been suspected that the disturbance of flow caused by the presence of an island may lead to upwelling of water from below the thermocline and hence nutrient enrichment of surface waters. However, there has been a lack of simultaneous physical and biological measurements to give firm support to the concept. Looking at this question more closely, Simpson and Tett (1986) drew attention to the evidence of enhanced phytoplankton biomass around two groups of islands on the continental shelf adjacent to the British Isles. They suggested that the cause was an interaction of the tidal currents with the islands.

Both the Scilly Isles (off the southwest tip of England) and St Kilda (west of the Hebrides, western Scotland) have surface temperatures around the islands about 3 °C lower than the surface layer of the summer stratified region offshore. In part this may be explained by tidal mixing in the shallower water close to shore, but other effects are at work. For example, as tidal currents flow in a curved path round the island mass they are accelerated to a velocity about twice that in the far field, and this leads to centrifugal displacement of surface waters and a compensatory upwelling of nutrient-rich deeper water. The biological productivity stimulated by these nutrients gives rise to increased phytoplankton biomass that is advected away from the islands by the residual currents. To the west of the Scilly Isles a plume of chlorophyll exceeding 4 mg chl-a m^{-3} sometimes extends for about 50 km into water that otherwise contains less than 0.5 mg chl-a m^{-3}. At St Kilda, where the tidal currents are less strong, the zone of enhanced chlorophyll-a is at a rather lower concentration, 2–3 mg m^{-3}, but it extends over a much larger area, about 5000 km^{-2}. A numerical model based on the interaction of tidal currents with the Scilly Isles gave results that were in good agreement with the field observations.

Both island groups are nesting grounds for large numbers of sea birds (about one million on the Scilly Isles alone), and the enhancement of primary production by the interaction of the island mass with the tidal currents is thought to be a factor in their success. That is not to say that this mechanism is the only one operating. There may be a positive feedback in which the enhanced phytoplankton production gives rise to a richer population of planktonic and benthic animals close to the islands, which in turn stimulate the recycling of nutrients so that total productivity is increased both by the 'new' nitrogen upwelled from depth and by the nitrogen regenerated locally.

7.6 SUMMARY

In this chapter we have chosen not to review the enormous body of literature on life between tidemarks. The diurnal or semi-diurnal rise and fall of the tide creates on the shore a vertical pattern of physical zonation in which the lower regions have less exposure to the air, less tendency to dry out and a more equable temperature, while the upper regions have converse properties. This creates a delightfully interesting stage on which organisms enact complex biological plays that are only now beginning to be understood. Since this zone is accessible for study by both naturalists and experimental scientists, there is copious literature on the subject that would require a separate volume to do it justice.

Instead we have concentrated on the effects of the tides on biological processes remote from shore. We have seen that the vertical mixing in the water column, caused by tidal currents, creates conditions in which nutrients are supplied to the phytoplankton all summer long, creating conditions favourable for the growth of larval and juvenile fishes. It has been noted that a very large proportion of the known herring spawning grounds are in areas that are tidally mixed all year round. Many of these are on offshore banks, where the water shallows enough to permit tidal mixing throughout the water column.

The interaction between tidal currents and offshore banks also tends to create a strong circular flow around the perimeter of a bank, and this feature is thought by some to create a barrier to the dispersal of fish larvae, retaining them in the biologically productive waters over the banks long enough for them to learn to recognize the bank as a site to which they should return for spawning. It has been suggested that this is an important mechanism in the evolution of distinct stocks within a single species of fish.

Tidal currents moving over the bottom in stratified water also tend to generate internal waves at the pycnocline. Particularly large waves, 50–70 m in amplitude and of the order of 30 km wavelength, are generated where strong tidal currents flow over the shelf break. These waves are thought to mix nutrient-rich water up into the nutrient-depleted mixed layer, thus stimulating biological production near the shelf edge. Details of the food webs have not been worked out, but we know that the shelf-break region of the North Atlantic is particularly productive of fish.

On a smaller scale, as internal waves approach shore they create convergences at the surface in which organic films accumulate and produce visible slicks. These convergences also accumulate planktonic organisms and appear to be important mechanisms for carrying aggregations of planktonic larvae towards shore. It is thought that various crabs, fish and corals may rely on this mechanism for the completion of their life histories.

Finally, we have seen that as tidal currents interact with islands they cause vertical mixing which brings nutrients to the surface and stimulates biological productivity around the islands. It has been suggested that this mechanism accounts for the large populations of sea birds that find various islands the best places for rearing their young.

PART C
PROCESSES ON A SCALE OF THOUSANDS OF KILOMETRES

8

Ocean Basin Circulation: the Biology of Major Currents, Gyres, Rings and Eddies

8.1 INTRODUCTION

In this chapter we begin to consider the ocean basins as wholes. We look at the major gyres which circulate anticyclonically in the subtropical regions of the North and South Atlantic and Pacific basins. We note that they are driven by the major global winds, which in turn are driven by the inequalities in the solar energy flux between the equator and the poles. North of the subtropical gyres in the North Atlantic and North Pacific are subpolar gyres which rotate in the opposite sense, i.e. cyclonically. Some of the most productive waters in the world are in the southern halves of these subpolar gyres, but the seasonal patterns of production are quite different in the two ocean basins.

Along the western margins of all these gyres are particularly intense currents; for example the Gulf Stream and the Kuroshio Current in the northern subtropical gyres, the Labrador Current and the Oyashio Current in the sub-Arctic gyres. Organisms within these currents are transported

291

long distances very rapidly, and we find that a variety of commercially important marine organisms, such as salmon, eels and squid migrate in these currents in the course of completing their life histories.

There is a tendency among biological oceanographers to take these intense western boundary currents as given natural phenomena, without delving too deeply into the mechanisms that drive them and cause them to be different from eastern boundary currents. However, as we enter an age in which human activity appears to be capable of altering the climate of the planet, and with it the circulation of the ocean, it is appropriate to try to understand the underlying mechanisms, in order to try to predict the consequences of climate change. In section 8.2.1 the theory of what drives boundary currents is explored. The story is not a simple one, and physical oceanographers do not have all the answers, but a study of the theory sheds new light on the functioning of whole ocean basins, and on the way in which they may respond to climate change.

Satellite images of boundary currents show that along some parts of their length they begin to meander. Sometimes the meanders grow so large that the ends are cut off to form isolated rings of water, of the order of 100–300 km in diameter. These rotating water masses, which are particular forms of eddies, retain their distinct identity for several months and have their own special biological characteristics. If they are formed between the boundary currents and the coast they may impinge on the continental shelf. Their enormous kinetic energy has the effect of drawing large volumes of water off the shelf on one side of the ring, while adding water on the other. The biological consequences for juvenile fish populations inhabiting the shelf water are often quite drastic.

It is now known that eddies are present throughout the ocean, though their energy often appears to decrease with distance from major ocean currents. It seems possible that there is a cascading of energy from the rings associated with the major currents, down to eddies remote from those currents. Since cyclonic eddies have the potential to cause upwelling in their centres, they may well be important in the global primary production budget.

Finally, in this chapter we note that the centres of the major subtropical gyres are permanently stratified and are thought to be biologically relatively unproductive. However, recent work suggests that these areas may have transient episodes of upwelling of nutrients which stimulate bursts of primary production. If this is true, we need to know more about them, for the subtropical gyres occupy a large fraction of the world's ocean, and their fixation of carbon dioxide is an important factor in the budget for carbon dioxide in the atmosphere.

8.2 THE WINDS AND THE WIND-DRIVEN CIRCULATION

All the major surface currents in the oceans are created by the drag of the wind on the surface of the water. The winds, in turn, are created because the earth's surface is heated unevenly by the sun, making the tropical regions warm and the polar regions cold. If the system were simple, the cold and dense polar air would flow under the warm and light tropical air while the tropical air moved north over the top of the cold polar air. But the fact that it all happens on a spinning sphere complicates the patterns. A very simplified diagram of the two major wind systems, the trades and the westerlies, is given in Fig. 8.01.

The trade winds arise because the warm air in the equatorial regions rises and is replaced by air flowing towards the equator in both the

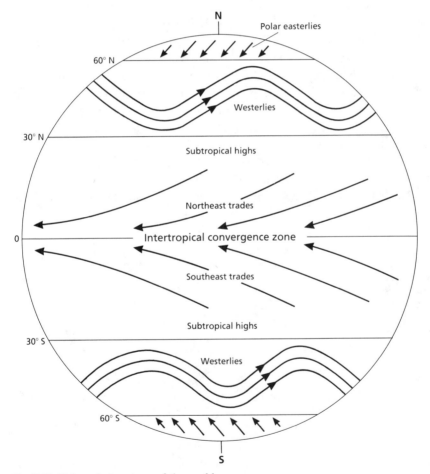

Fig. 8.01 Major wind systems of the world.

northern and southern hemisphere. The Coriolis force deflects the equatorward flows to the west giving rise to the northeast and southeast trades (winds are named after the direction from which they come while ocean currents are named after the direction to which they are going). The air in the trade winds comes from the air which rises above the equator and which flows to higher latitudes then descends back to the earth's surface in the subtropical highs. These circular convection cells are known as the Hadley cells and the zone near the equator where the surface winds converge is called the intertropical convergence zone (ITCZ). The ITCZ moves north and south with the seasons, being furthest north in August and furthest south in February, but is always north of the equator (Gill 1982).

The Hadley cells tend to mix the air between the equator and 30° latitude resulting in low horizontal temperature gradients between 30° S and 30° N throughout the lower 10–20 km of the atmosphere (the troposphere). Poleward of the Hadley cells the temperature at all levels in the troposphere decreases rapidly. This gives rise to a strong horizontal pressure gradient between $\simeq 30°$ latitude and the pole and strong westerly winds. These winds are perpendicular to the pressure gradient because of the Coriolis effect and are in near geostrophic balance. The westerlies are continuous bands of flow right around the earth and are sometimes called circumpolar vortices. The speed of the flow is not constant at all levels but increases with height and reaches a maximum in the so-called jet stream at $\simeq 12$ km over $\simeq 30°$ latitude as illustrated in Fig. 8.02. This figure emphasizes the dominance of the westerly air

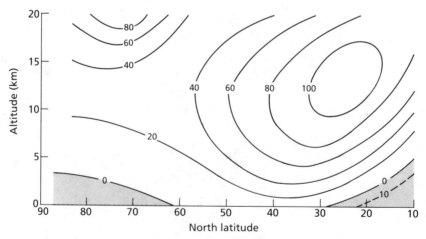

Fig. 8.02 The distribution with altitude and north latitude of the average winter zonal wind speed in km h^{-1}. Winds are towards the east except where hatched. Adapted from Pettersen (1969), with permission of McGraw-Hill, Inc.

flow compared to the small easterly flows of the trade winds near the equator and the polar easterlies. In the summer the jet stream is weaker and moves poleward about 10° in latitude.

The westerlies do not stay at a constant latitude right around the globe but meander north and south in waves that are about 10,000 km long. At any one time there are three to five complete waves around the globe but averaged over a month only three are usually visible. These waves are called planetary or Rossby waves and cause the westerlies at a given longitude to move north and south over periods of a month or so. Since the movement of the weather systems is guided by the westerlies, the shifts in position also alter the paths of the weather systems, and is the process which causes the weather in the mid-latitudes to be different from the average for weeks at a time. As we shall discuss in the next chapter, the position of the westerlies can be shifted in special locations such as the North Atlantic Ocean for much longer periods than Rossby wave fluctuations. The longer term changes can last for years and are known as oscillations.

The winds shown in Fig. 8.01 drive the ocean currents shown in Fig. 8.03. However, the Coriolis force and its variation with latitude make the link between the wind stress and the water motion less than straightforward. At the equator where the Coriolis force is zero the situation is simple and the water moves in the same direction as the wind. The equatorial currents in the Atlantic and Pacific, for example, are mainly parallel to the equator. At other latitudes the Coriolis force is not zero and causes the moving water to be deflected to the right of the wind in the northern hemisphere and to the left in the southern hemisphere. The net direct effect of the wind is a flow perpendicular to the wind in the upper layer of the ocean. This is the Ekman transport which begins the series of processes that generate the currents seen in the upper layers of subtropical and subpolar oceans.

As is evident in Fig. 8.03 the currents in the subtropics, between latitudes 15 and 45°, form large gyres in which the poleward flowing currents on the western sides of the oceans are strong and narrow while the currents throughout the remainder of the oceans are broad and slow. In the North Atlantic Ocean, for example, the subtropical gyre is made up of the broad slow North Atlantic, Canary, and North Equatorial currents flowing east, south and west, respectively. The northward flowing current that completes the gyre is the Gulf Stream which is narrow and rapid. Similar subtropical gyres of broad slow currents with strong narrow poleward flows like the Gulf Stream are found in all the other major oceans except the South Pacific. In the subpolar regions of the North Atlantic and North Pacific, gyres are observed in which the sense of the flow is opposite to that in the subtropical gyres, that is, anti-clockwise. These

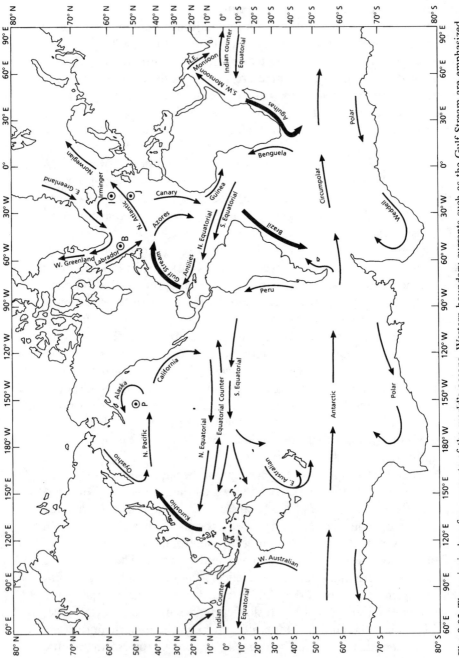

Fig. 8.03 The principal surface currents of the world's oceans. Western boundary currents such as the Gulf Stream are emphasized. Positions of ocean weather ships B(Bravo), 56.5 ° N, 51.0 ° W; I(India), 59.0 ° N, 19 ° W; J(Juliett), 52.5 ° N, 20 ° W and P(Papa), 50 ° N, 145 ° W are indicated by circled dots and letters.

gyres, like the subtropical ones, contain broad slow flows throughout most of the ocean with a strong narrow flow on the western side. The strong narrow flows, the Labrador and Oyashio currents, however, flow towards the equator instead of towards the poles. In the higher latitudes of the southern hemisphere where there are no obstructing continents to force the formation of gyres the main current is the Antarctic circumpolar current which circles the globe at about 50° S.

8.2.1 Theory of the wind-driven circulation

The asymmetry of the currents in the large ocean gyres has been known for centuries but the reason for it remained obscure until Stommel (1948) showed mathematically that the westward intensification of the currents is due to the change in the Coriolis force with latitude. In the following short description we try to outline the main physical processes creating the currents but it is clear from the reviews of Fofonoff (1981), Rhines (1986), Stommel (1965) and Pedlosky (1990) that the subject is vast, complicated and dynamic and we can only 'scratch the surface'. Our account concentrates on the North Atlantic Ocean and begins with the large-scale wind patterns and the Ekman transport.

The important parameter associated with the wind is the force per unit area or stress of the wind drag on the water surface. For the North Atlantic Ocean the yearly average of the strength and direction of the wind stress is shown by Isemer and Hasse (1987) to be similar to the distribution illustrated in Fig. 8.01. There is a maximum eastward stress between 40 and 50° N associated with the Westerlies, a region of relative calm at 30° N associated with the subtropical highs and the maximum westward stress associated with the northeast trade winds at about 15° N. Analyses by Hellerman and Rosenstein (1983) demonstrate that this north–south variation in wind stress is symmetrical across the equator.

To simplify our discussion we reduce the ocean to an idealized rectangle with idealized east–west (zonal) wind stress (Fig. 8.04). Our model ocean extends from 20 to 40° N and from 80 to 20° W with a wind stress which varies sinusoidally from a maximum of 0.1 N m^{-2} towards the east at 40° N to 0.0 N m^{-2} at 30° N then increases again to 0.1 N m^{-2} to the west at 20° N. This idealized wind ignores any mean winds in the north–south direction (meridional winds) but these are small relative to the zonal winds. Mathematically the idealized wind stress can be written

$$\tau_x = 0.1 \sin\frac{\pi}{2L}y, \tag{8.01}$$

where τ_x represents the component of the wind stress in the x direction in units of N m^{-2} and y represents the distance north and south of 30° N which varies between $+L$ and $-L$.

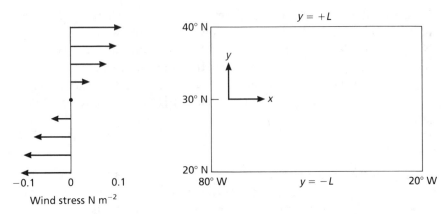

Wind stress N m^{-2}

Fig. 8.04 An idealized rectangular ocean (on the right) extending from 20° N to 40° N and from 20° W to 80° W. The coordinate origin is at 30° N, 80° W and y is positive to the north; x is positive to the east. The wind stress (on the left) is purely zonal and decreases sinusoidally from 0.1 N m^{-2} at 40° N to −0.1 N m^{-2} at 20° N.

Referring back to section 5.2.2 we recall that the net effect of a wind stress on the water surface is to produce a transport in the upper layer of the ocean to the right of the wind known as the Ekman drift. The mass transport, M, of this flow is calculated by dividing the wind stress τ by the Coriolis parameter f as in Eq. 5.06, i.e.

$$M = -\tau/f. \tag{8.02}$$

If we represent the wind stress with Eq. 8.01 the Ekman transport is found by substitution of Eq. 8.01 into Eq. 8.02 to give

$$M_{yE} = -\frac{0.1}{f}\sin(\pi y/2L), \tag{8.03}$$

where M_{yE} is the meridional Ekman mass transport through a one metre wide strip of the ocean in units of kg m^{-1} s^{-1}. This transport is illustrated at the top of Fig. 8.05 which shows a one metre wide slice of the ocean between 20 and 40° N from the surface to the main pycnocline at 1000 m. It is a perspective view looking towards the northeast. At the very top, above the surface of the water is a representation of the wind field with a maximum stress to the east at 40° N and a maximum towards the west at 20° N. The Ekman layer takes up the top 100 m. North of 30° N the westward wind creates the southward Ekman transport shown by the shaded blocks of water with the arrows to indicate movement to the south. South of 30° N the east wind creates a northward transport in the Ekman layer. These oppositely-directed transports therefore create a convergence in the Ekman layer which increases from zero at the extremities to a maximum in the centre of the ocean at

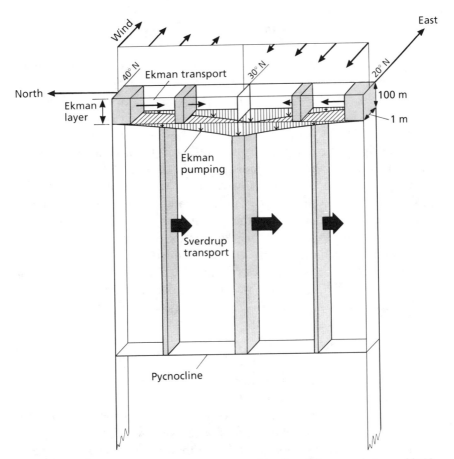

Fig. 8.05 A perspective drawing of a one metre wide slice of the ocean between 20° N and 40° N and from the ocean's surface to the pycnocline at 1000 m depth. The wind stress, changing from eastward at the north end to westward at the south end, is shown at the top of the picture. The top 100 m of the ocean is the Ekman layer which converges towards the centre (30° N) under the influence of the wind. Ekman pumping is indicated at the base of the Ekman layer and the southward Sverdrup transport is indicated in the layer above the pycnocline.

30° N. Because the sea level cannot rise enough to accommodate the convergence it generates a downward velocity in the upper layer which is indicated in the figure by the small arrows at the base of the Ekman layer. The downward flow is referred to as Ekman pumping. (If the velocity at the base of the Ekman layer is upward because of a divergence in the Ekman layer it is called Ekman suction.)

An expression for the vertical velocity associated with the Ekman pumping can be derived from Eq. 8.03 by calculating the derivative of the transport with respect to the horizontal coordinate which in this case is y. The result,

$$dM_{yE}/dy = -\frac{0.1\,\pi}{2Lf}\cos(\pi y/2L),\tag{8.04}$$

gives the amount of water which is piling up at latitude y. The maximum of this function is found at 30° N where the cosine is 1.0. At this latitude with $f = 0.73 \times 10^{-4}\,s^{-1}$ and with $L = 1.1 \times 10^6$ m (10° of latitude) the convergence works out to 2.0×10^{-3} kg m^{-2}s^{-1} or 169 kg m^{-2} d^{-1} which converts to a downward velocity at the base of the Ekman layer of $\simeq 0.17$ m d^{-1}. This decreases to the north and south along with the cosine function and reaches 0.0 m d^{-1} at 20° N and 40° N.

The ocean beneath the wind-driven layer responds to the downward push of the Ekman pumping with a horizontal flow towards the equator that is many times larger than the Ekman transport. This derived flow, called the Sverdrup transport, is dictated by the fact that the Coriolis force varies with latitude. In order to understand this response we briefly review the consequences of the law of conservation of angular momentum or vorticity which we first introduced in Box 5.02.

In oceanography the law is usually expressed by Eq. 5.09, i.e.

$$\frac{f + \zeta}{H} = \text{constant},\tag{8.05}$$

where f is the Coriolis parameter or twice the rate at which the earth rotates about the local vertical relative to the stars, ζ is twice the rate that the water is rotating relative to the earth about the local vertical, and H is the height of the column of water. In discussions of vorticity f is called the planetary vorticity meaning that it is the vorticity given to the water by virtue of it being on the rotating planet at the latitude ϕ. The term ζ is called the relative vorticity representing the vorticity of the water relative to the earth.

The two parts (f and ζ) are illustrated in Fig. 8.06. The cylinder of water on the surface of the earth at latitude ϕ moves with the earth as it rotates about its axis at the angular velocity Ω. The cylinder therefore rotates in an anti-clockwise sense relative to the stars at the rate $\Omega \sin \phi$ as was shown in section 4.2. The planetary vorticity f is twice this rate of rotation or $f = 2\Omega \sin \phi$, which is counted positive in the northern hemisphere. If the cylinder in the figure is given a clockwise rotation relative to the earth, we call this its relative vorticity ζ and clockwise is taken as negative.

The consequences of the conservation law are illustrated in Fig. 8.07. Figure 8.07(a) shows the cylinder of water illustrated in Fig. 8.06 in cross-section. We assume that the cylinder is not rotating relative to the earth ($\zeta = 0$) and possesses only planetary vorticity f which stays constant. If the cylinder is squashed so that H becomes less the water

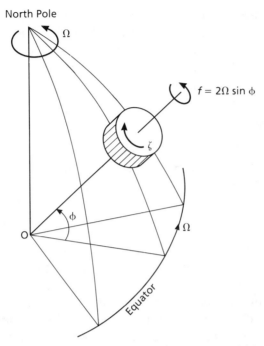

Fig. 8.06 A cylinder of water sits on the surface of the earth at latitude ϕ north of the equator. The earth rotates at an angular velocity Ω causing the cylinder to rotate about its local vertical relative to the stars at $\Omega \sin \phi$ giving the cylinder a positive planetary vorticity $2\Omega \sin \phi$. The cylinder is also shown to be rotating relative to the earth to give it a negative relative vorticity ζ.

spreads out as shown in Fig. 8.07(b). The outward horizontal flow generates a Coriolis force to the right and gives the water enough clockwise rotation or negative relative vorticity to keep the ratio of Eq. 8.05 constant. If the column of water is stretched instead of squashed depth H increases and the water in the cylinder flows inward. The Coriolis force generates an anticlockwise rotation or positive vorticity to the water as shown in Fig. 8.07(c) which again maintains the constancy of the ratio in Eq. 8.05.

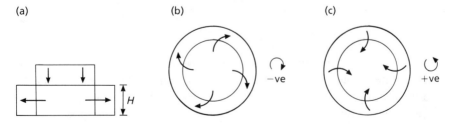

Fig. 8.07 (a) A cross-section of the cylinder in Fig. 8.06 before and after squashing. (b) Overview of the cylinder spreading out as the height diminishes. (c) The same as (b) for a cylinder increasing in height and diminishing in radius.

In the open ocean away from the coasts and intense currents, squashing or stretching water columns does not tend to produce relative vorticity. The water instead moves north or south (the Sverdrup transport) to seek a value of f that keeps Eq. 8.05 true without generating relative vorticity. For this part of the ocean, where ζ is always close to zero, we can rewrite Eq. 8.05 as

$$f/H = \text{constant} . \tag{8.06}$$

Since we wish to know the consequences of a change in the height (dH) of the column of water we calculate the derivative of Eq. 8.06 with respect to the north–south coordinate y, to get

$$\frac{df}{dy} = \frac{f}{H}\frac{dH}{dy} . \tag{8.07}$$

The term df/dy, the rate of change of the Coriolis parameter with latitude is commonly called β. It only exists because there is a change in the vertical component of the earth's rotation with latitude and the effects that depend on its existence are referred to as 'β effects'. If the Coriolis force did not change with latitude there would be no tendency for the water to change latitudes to satisfy Eq. 8.07 and Sverdrup transport would not exist. Since $f = 2\Omega \sin\phi$ we can calculate β as a function of latitude, i.e. $\beta = df/dy = 2\Omega \cos\phi \, d\phi/dy$. The derivative dy can be written as $Rd\phi$, where R is the radius of the earth $(6.4 \times 10^6$ m), which results in the formula

$$\beta = 2\Omega \cos\phi/R . \tag{8.08}$$

The angular velocity of the earth, Ω, is 2π rad d^{-1} or 7.27×10^{-5}rad s^{-1}, therefore at 30° N, where $\cos\phi$ is 0.87, β is $\simeq 2.0 \times 10^{-11}$ m^{-1} s^{-1}. Using β in place of df/dy we rewrite Eq. 8.07 as

$$dy = \frac{f\,dH}{\beta H} , \tag{8.09}$$

and by dividing both sides by the time interval dt the ratios dy/dt and dH/dt are velocities and the equation becomes

$$V_S = \frac{f}{\beta H} W_E , \tag{8.10}$$

where V_S represents the Sverdrup velocity and W_E represents the velocity of the Ekman pumping. This equation can be used to calculate the north–south movement of the squashed cylinders of water at the base of the Ekman layer.

The downward push in the Ekman layer in our idealized model ocean is $\simeq 0.17$ m d^{-1} which is W_E in Eq. 8.10. If $f = 0.73 \times 10^{-4}$ s^{-1},

$\beta = 2.0 \times 10^{-11}$ and $H = 1000$ m, Eq. 8.10 gives a value for V_S of 620 m d^{-1}. This value is based on the maximum value of W_E and is the maximum Sverdrup flow. Since the column is about 1000 m deep the flow represents a transport across $30°$ N of 6.2×10^5 m^3 d^{-1} or 7.2 m^3 s^{-1} through each one metre wide meridional strip across the ocean. If the ocean is 5000 km wide the total Sverdrup transport southward across $30°$ N is $\simeq 36 \times 10^6$ m^3 s^{-1} or 36 Sv, where 1 Sv or 1 Sverdrup equals 10^6 m^3 s^{-1} in honour of H.U. Sverdrup, the man who contributed so much to understanding the oceans.

This flow is illustrated in Fig. 8.05 by the shaded blocks of water between the Ekman layer and the pycnocline that are moving towards the equator. The volume of the flow varies from a maximum at $30°$ N to zero at the northern and southern extremes. The diagram also illustrates the fact that the Sverdrup flow in this example is always towards the south even though the winds and the Ekman drifts are in opposite directions in the northern and southern halves of the ocean. The important variable is of course the vertical velocity in the Ekman layer which is downward at all positions across the ocean.

It is a curious fact, evident in Fig. 8.05, that the maximum Sverdrup transport at these latitudes is much larger than the maximum Ekman transport. Intuitively one might expect that the convergence in the Ekman layer would create an equal and opposite compensating divergent flow to keep the sea level and the depth of the Ekman layer constant in time. The Sverdrup flow, however, does nothing to compensate for the Ekman convergence but creates a much larger convergence and divergence in the water columns in the northern and southern halves of the idealized ocean as shown in Fig. 8.08. The Ekman convergence is a maximum at $30°$ N, while the divergence created by the Sverdrup flow

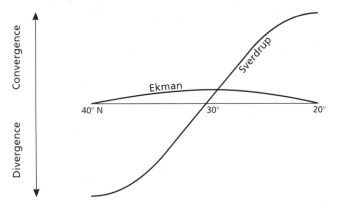

Fig. 8.08 The convergences and divergences due to the Ekman and Sverdrup transports between $20°$ N and $40°$ N.

is a maximum along the northern edge of the basin, and the convergence of the Sverdrup flow is a maximum along the southern edge. The Sverdrup flow is clearly not providing a compensation for the converging Ekman layer. If the Sverdrup flow was able to operate alone, the sea level at 40° N would decrease about 1 metre per day while the sea level at 20° N would increase by the same amount. Such a change in sea level does not occur because large compensating zonal flows are generated which provide the water for the regions of divergence and remove the water from the regions of convergence.

Because a flow towards the east or west does not involve a change in the Coriolis force and hence a change in the planetary vorticity, it is easier for water to move in these directions than north or south. This is the reason that the compensating flows are east and west. Rhines (1986), when discussing this point, says the ocean is 'more rigid' to meridional flows than to zonal ones. An illustration of the zonal flows and the Sverdrup transport are given in Fig. 8.09. The arrows to the left represent the zonal flows that compensate for the convergence and divergence in the Sverdrup flow. The flow from north to south is shown on the right. The thickness of each tube is proportional to the volume of the water being transported through a 1 metre wide slice in the ocean and the small darker arrows indicate the direction of the flow.

Each one metre wide north–south slice across the ocean requires the same zonal flows to compensate for the convergences and divergences in the Sverdrup flow. By placing a series of diagrams like the one in Fig. 8.09 side by side to represent successive north–south slices it is clear that the zonal compensating flows must increase towards the west. This increase is illustrated in Fig. 8.10 where each thin line represents the flow of water crossing 30° N through a 1 metre wide section and which comes from the west along latitude $+y$ and returns to the west along latitude $-y$. The broad arrows on the left indicate the zonal flows at latitudes $+y$ and $-y$.

The diagram in Fig. 8.10 represents the flow delivered from one parallel of latitude north of the centre line to a symmetric parallel to the south of the centre line. A similar picture, from Stommel (1948), is shown on the right in Fig. 8.11 except that the contour lines represent the flow in the whole of the idealized ocean except near the western boundary of the ocean. The contour lines define the paths of the water particles and between two contours the volume transport stays constant. As the lines get closer together the speed of the flow increases. The diagram incorporates both the meridional Sverdrup flow which is constant across each parallel of latitude and the compensating zonal flows which increase towards the west. The pattern of currents illustrated in

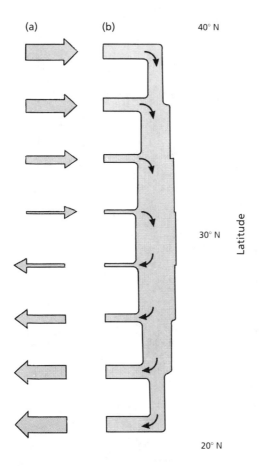

Fig. 8.09 (a) The zonal flows required to compensate for the convergences and divergences in the meridional Sverdrup transport. (b) The Sverdrup flow through a one metre wide north–south slice in the ocean with compensating zonal flows. The width of the tubes is proportional to the volume transport.

this part of the diagram will continue towards the west for as long as the ocean continues in that direction. There is no mechanism in what we have discussed so far to close the gyre with a flow towards the north. The reason for this is again the fact that water is much easier to move east or west than it is to move north or south.

At the western side of the ocean some force must be generated to overcome the resistance of the water to go north to complete the gyre. The required force was shown by Charney (1955) to derive from the inertia of the westward flowing water which must slow down as it approaches the coast. And any acceleration or deceleration of a mass requires a force. The magnitude of such inertial forces are usually insignificant in the ocean, where normally the only forces of consequence are

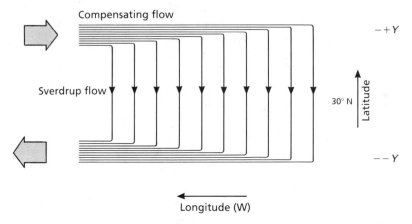

Fig. 8.10 The north–south Sverdrup flow across 30° N is constant at each meridian but the compensating east–west flows at each latitude ($+y$ and $-y$) must increase to the west.

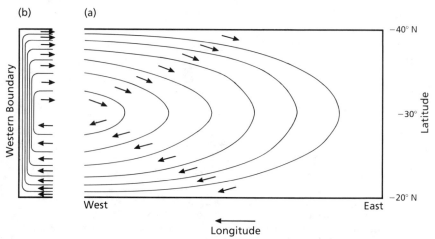

Fig. 8.11 (a) The total wind-driven circulation in the idealized ocean without a western boundary including the Sverdrup flow and the compensating zonal flows. The transport between contour lines is constant. (b) The presence of the western boundary forces the westward flowing water of the gyre into a narrow northward current.

the Coriolis and pressure gradient forces involved in the geostrophic balance. The inertial force is dominant in only a narrow region near the coast but its generation causes the deflection of the westward flowing current towards the north and causes the flow to be squeezed into a narrow rapid flow as shown on the left in Fig. 8.11.

The narrow current forces the density distribution in the ocean to change to satisfy the geostrophic balance discussed in section 4.3. The result is that the pycnocline along with the thermocline and the halocline rise up across the current. This is illustrated in Fig. 8.12 which

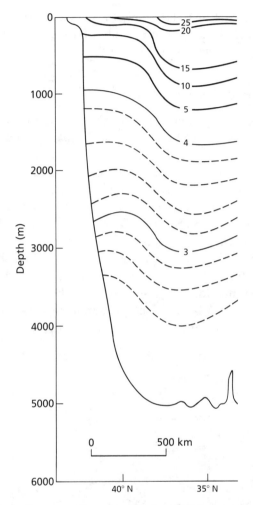

Fig. 8.12 A north–south cross-section of temperature (°C) between Bermuda and Nova Scotia showing the rapid depth change of the isotherms across the Gulf Stream. Adapted from Fuglister (1960).

shows the temperature distribution between Nova Scotia and Bermuda with the Gulf Stream between 38 and 39° N evident by the sudden decrease in the depth of the isotherms between the Sargasso Sea and the north side of the stream. The 10 °C isotherm, for instance, rises from $\simeq 800\,\text{m}$ to $\simeq 200\,\text{m}$ and this sudden rise is often used to map the position of the stream. The rapid rise in the isotherms means that on a level surface there is a rapid drop in temperature when crossing the stream; often called the 'cold wall'.

Because both a western boundary and inertial forces are instrumental in creating the Gulf Stream it has been called a western boundary current or an inertial boundary flow. In our idealized ocean this northward flow completes the gyre of the subtropical wind-driven circulation but how does the model compare with the observed circulation?

8.2.2 The observed circulation

The near-surface circulation of the North Atlantic subtropical gyre presented in Fig. 8.13 is based on the works of many authors. The general path of the Gulf Stream is similar to the original diagram due to Sverdrup *et al.* (1942) (page 684) but the northern and southern recirculation gyres are shown with greater intensity according to Hogg *et al.* (1986) and Worthington (1976), respectively. The maximum Gulf Stream transport of $\simeq 150$ Sv occurs at about 65° W. The branching and path of the stream near the southeast Newfoundland ridge is according to Mann (1967) and Clarke *et al.* (1980), while the branches at $\simeq 45°$ N and $\simeq 51°$ N to the east of Newfoundland are discussed by Krauss (1986). The continuation of the stream's branches in the North Atlantic current to the mid-Atlantic ridge is detailed by Harvey and Arhan (1988) and Sy (1988) and the circulation to the east of the Ridge is

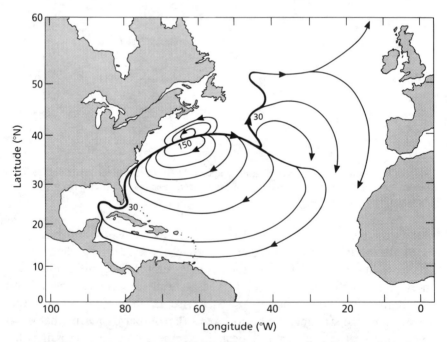

Fig. 8.13 The wind-driven circulation in the North Atlantic subtropical gyre including the Gulf Stream and its branches with some estimates of volume transport in Sverdrups. The rapid narrow currents such as the Gulf Stream are indicated by heavy lines. The slow broad flows are shown by light lines.

adapted from Saunders (1982) and Klein and Siedler (1989). The circulation east of the Bahamas is based on Olson *et al.* (1984) and Stommel *et al.* (1978). Some of the flowlines in the eastern basin are not continued south of 35° N to indicate that much of this flow is thought to descend below the surface layers of the southern part of the gyre.

It is clear from Fig. 8.13 that the circulation pattern of our idealized ocean does not reproduce all the features of the real ocean. For example the location of the maximum flow in the Gulf Stream is not where the model predicts it to be and the maximum flow of the stream is almost four times the Sverdrup transport which we estimated in the previous section to be $\simeq 36 \times 10^6 \, \mathrm{m^3 \, s^{-1}}$. In the model the maximum flow occurs at the latitude where the wind stress is zero and changing from eastward to westward. This point actually occurs at about Cape Hatteras; however, the Gulf Stream continues to increase in intensity north of this latitude. The increase in the flow above the predicted amount is thought to have nothing to do with the Sverdrup transport but is associated with the recirculation gyre between Cape Hatteras and the Grand Banks which is not well understood (see Fofonoff 1981).

One other feature of the Gulf Stream and its continuation, the North Atlantic current, that is not in the model is the fact that the strong narrow current continues towards the east well past the region where it would disperse in the model. Also the branching of the flow which occurs to the east of the Grand Banks of Newfoundland is not part of the model. That the stream does split into branches was first postulated by Iselin (1936) and has been confirmed by various authors since, as mentioned above, but the reason for the splitting is not yet understood. It is obviously associated with the decay mechanism of the current which cannot spread out evenly but only in smaller and smaller well-defined narrow streams.

Another of the curious features of the western boundary currents that does not seem to have a satisfactory explanation and which is not part of our model is the fact that they turn away from the western boundary and flow out into deep water. The Gulf Stream for example leaves the continental shelf at Cape Hatteras and travels northeast over deep water rather than continue over the continental shelf. This behaviour is also seen in the other western boundary currents in Fig. 8.03. The Agulhas current is the most impressive case for when it leaves the coast it turns almost completely around to flow the opposite way. It is said to retro-reflect. The fact that the current is travelling along a western boundary that is trending towards the west as it moves poleward rather than towards the east as does the coast of North America, emphasizes the turning of the current. A number of suggestions to explain this phenomenon have been reviewed by Fofonoff (1981) but he concludes that no satisfactory answer has yet been found.

8.2.3 Meanders, rings, eddies and gyres

(a) Meanders and rings

After the Gulf Stream leaves the coast it continues eastward as a strong narrow stream but at about 65° W it becomes unstable and begins to develop large north–south oscillations or meanders (Fig. 8.14). These meanders cause the cooler water found to the north of the Gulf Stream to be brought further south than usual and the warmer Sargasso Sea water to be transported further north. The meanders do not change the properties of the water significantly and thus the meanders are not biologically unique. Often however the meanders grow too large and the ends separate into isolated rings of water as illustrated. The rings to the south of the stream rotate anticlockwise, enclose colder water from the

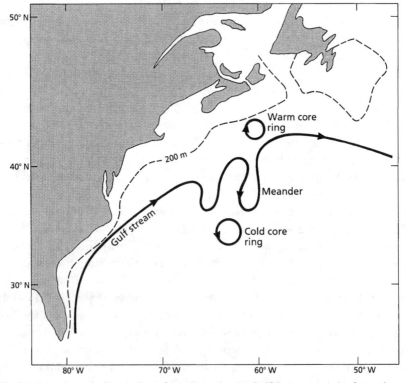

Fig. 8.14 A schematic illustration of meanders in the Gulf Stream and the formation of rings based on Richardson *et al.* (1978) and Parker (1971).

north side of the current and are known as cold-core rings. The clock-wise rotating rings to the north of the stream are the warm-core rings because they contain the warmer water from the Sargasso Sea and are surrounded by the cooler waters found north of the stream. Both kinds of rings are baroclinic phenomena, that is, confined to the thermocline or the upper $\simeq 1000$ m.

Each year there are about 10 cold-core rings formed and each exists for about a year before it is either re-absorbed into the Gulf Stream or loses its unique properties to diffusion (the Ring Group 1981). At any one time the cold-core rings represent about 10% of the surface area of the Sargasso Sea. According to Richardson (1983) the number of warm-core rings formed each year is about half that of cold-core rings. The production rate however is quite variable and sometimes none can be identified in the region north of the Gulf Stream while at other times the region is almost filled with rings. The warm-core rings also tend to be smaller than the cold-core rings with diameters of $\simeq 100$ km as opposed to 100–300 km for the warm-core ones.

The rings appear as pools of anomalous conditions because they contain water from the opposite side of the stream having the other side's physical, chemical and biological properties. As the anomalous conditions change over the $\simeq 1$ y ring's life, through mixing with the ambient water, the biology changes in ways that are unique to the rings as we discuss in sections 8.4.3 and 8.4.4.

(b) Eddies

Rings are a form of eddy but the term eddy is usually reserved for the lower amplitude current variations that are found throughout the ocean. These oscillations, described by Robinson (1983), tend to be barotropic (constant with depth) as opposed to baroclinic and because the ocean is full of them they interact with each other in a way reminiscent of turbulent eddies. For this reason the properties in the eddies change much more rapidly than the properties within the rings so the biology of an eddy is less individualistic than that of a ring.

The time scale of one rotation of an eddy is usually 10–30 days, and the horizontal scale between 10 and 100 km, close to the internal Rossby radius (section 5.2.3). Two terms often used to describe eddies are 'synoptic' and 'meso-scale'. Synoptic is borrowed from meteorology, where it is used to indicate the eddies in mid-latitudes associated with cyclonic storms. It implies a horizontal scale that can be sampled more or less simultaneously. Meso-scale is an ill-defined term indicating an eddy smaller than ocean basin scale, of the order of 100 km rather than 1000 km.

The energy level of eddies tends to be highest in those close to narrow boundary flows like the Gulf Stream. Currents in these eddies may be about 1 m s^{-1} while currents in eddies remote from major currents (e.g. in the southeast Pacific) are nearer 0.01 m s^{-1}. It has been suggested (Robinson 1983) that most open-ocean eddies are formed indirectly from strong currents, with a transfer of energy through the eddy field in a way resembling the cascade process described in section 2.2.3. However, neither the location nor the mechanisms of such processes are yet known.

Two other types of eddy are recognized. One type is clearly formed by the interaction between flows and bathymetric irregularities. A second kind is the result of wind stress on the surface.

As pointed out by Swallow (1976), the variability associated with ocean eddies had been observed for many years before their full importance was discovered but they were thought to be transient features associated with the wind or the internal tides. Swallow and Hamon (1960) first observed the currents in the deep open ocean with neutrally buoyant floats and discovered that the currents varied by 0.5 m s^{-1} rather than the expected 0.01 m s^{-1}. Further work revealed that, in total, the eddies in the ocean actually contain more kinetic energy than the mean flows.

(c) Gyres

The word gyre is usually used to describe a circular current that is confined by or associated with bathymetric features. The term covers a wide range of spatial scales. Thus the currents of the Gulf Stream, the North Atlantic current, the Canary current and the North Equatorial current make up the North Atlantic subtropical gyre at the ocean basin scale. Over shallow banks such as Georges Bank in the Gulf of Maine there is a continuous current flowing around the bank located over the edge. This circulation is the Georges Bank gyre.

8.2.4 Rossby waves

Rossby waves are large slow horizontal oscillations in the open ocean. Some of these have recently been associated with the low frequency variations in the equatorial Pacific which lead to El Niño (Chapter 9). The waves can have amplitudes of hundreds of kilometres, wavelengths of hundreds to thousands of kilometres and periods of tens of days. Unlike the surface wind-driven waves and the internal waves which arise because of the force of gravity, Rossby waves exist because of the north–south variation in the vertical component of the earth's rotation. This is the β effect which we showed in section 8.2.1 to be responsible for the westward intensification of the wind-driven circulation.

To help explain how this factor can generate a wave motion, consider the eastward flowing current shown in Fig. 8.15. The water at A besides moving east is rotating around to the left with the rotation of earth at a rate of $\Omega \sin \phi_A$ (as was pointed out in the discussion of the Coriolis effect in section 4.2 and Fig. 4.01). Let us assume that the water is deflected northward towards B. The water carries with it the rotation $\Omega \sin \phi_A$ but it is moving north where the vertical component of the rotation is $\Omega \sin \phi_B$ which is greater than $\Omega \sin \phi_A$ because $\phi_B > \phi_A$. So the water in the current at B is surrounded by water that is turning faster around to the left. But at B the reference frame is the earth and the water imported from the south is rotating, relative to the local conditions, around to the right or back towards the latitude where it began.

If the water now heads south and passes the original latitude and heads for C it will arrive in a region where the rate of rotation $\Omega \sin \phi_C$ is less than the water in the current. The water in the current then appears to turn to the left relative to the local vertical and again head back towards the original latitude. The wave motion is created because the poleward increase in the earth's rate of rotation provides a restoring force to north–south displacements of eastward currents. The same argument presented in terms of potential vorticity is given in Box 8.01.

That Rossby waves cannot be formed on a westward flowing current can be easily seen by thinking of the current in Fig. 8.15 going in the

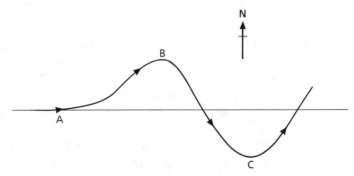

Fig. 8.15 The Rossby wave; displacements to the north or south in an eastward flow are turned back towards the original latitude by the generation of relative vorticity.

Box 8.01 Conservation of vorticity in Rossby waves

The conservation of potential vorticity is expressed by Eq. 8.05. If we assume the thickness H stays constant the ratio in Eq. 8.05 must remain constant for a water parcel as it moves, thus we can write

$$f_A + \zeta_A = f_B + \zeta_B, \tag{8.11}$$

where the subscripts A and B refer to different latitudes. At the beginning of the path shown in Fig. 8.15 we assume that ζ_A is zero. Thus at the latitude B the relative vorticity must be equal to the difference in the planetary vorticities of the two positions, i.e.

$$\zeta_B = f_A - f_B. \tag{8.12}$$

Since in our example $f_B > f_A$ the relative vorticity will be negative or clockwise. Its magnitude can be easily calculated. Assume, for example, that the latitudes at A and B are 30° N and 40° N, respectively, where the values of f are $0.73 \times 10^{-4}\,s^{-1}$ and $0.93 \times 10^{-4}\,s^{-1}$. The difference is the value of ζ_B and is $-0.2 \times 10^{-4}\,s^{-1}$.

The simplest form of the Rossby wave is the one which is barotropic or constant throughout the depth. Following von Schwind (1980) these waves have a phase speed of

$$C = -\beta \lambda^2 / 4\pi^2, \tag{8.13}$$

where λ is the wave length and β is the north–south gradient in the planetary vorticity given by Eq. 8.08. For $\beta = 2 \times 10^{-11}$ and $\lambda = 1000$ km the phase speed is $0.5\,m\,s^{-1}$ towards the west. In fact the phase speed of all Rossby waves is towards the west. The Rossby waves existing in the upper layer of a two layer ocean and those existing on currents travelling in directions other than east are more complicated than the simple ones discussed here and are explored in Apel (1987) and Gill (1982).

opposite direction. A northward displacement generates clockwise rotation as in the eastward current but in this case it tends to push the water farther north rather than back towards the original latitude. Similarly a southward displacement of a westward flowing current generates anticlockwise rotation which also tends to push the water further south. Thus, for a westward flowing current displacements to the north or south do not generate a restoring force to push the water back to the original position and Rossby waves on such a current are not possible.

Rossby waves can also be formed when the potential vorticity is changed because of a change in depth. These are topographic Rossby waves and operate the same way as the ones which depend on the β effect except that we assume f to be constant and that H varies in Eq. 8.05. If we assume in Fig. 8.15 that the depth of the bottom is less in the northern half of the diagram than in the southern half, then the topographic effect will be capable of generating a wave. Consider the flow of water towards the east as before. A northward displacement causes H to decrease. For the ratio of Eq. 8.05 to remain constant the numerator must also decrease which means a decrease in ζ since f is assumed constant. The decrease in ζ is a clockwise rotation tending to restore the water back to the original position. A displacement to the right leads to an increase in H and an increase in ζ. The anti-clockwise rotation pushes the water back towards the original position and a wave is formed. As in the β effect waves the topographic Rossby waves can only exist for certain directions of flows. A flow along an isobath with shallow water on the right rather than the left cannot support a barotropic Rossby wave. In some cases the topographic and β effects are both present, such as in the eastward current over a bottom sloping up to the north. The behaviour of the Rossby wave will depend on the magnitude of the two effects.

8.3 DISTRIBUTION OF BIOLOGICAL PRODUCTION IN OCEAN BASINS

We have now examined the consequences of differential solar heating at low and high latitudes which, in conjunction with the rotation of the earth, leads to westerly and trade winds. These, in turn drive major anticyclonic subtropical gyres in each of the major ocean basins. Associated with them are powerful western boundary currents that transport large quantities of heat away from equatorial regions. These currents meander and cut off large gyrating bodies of water, the warm- and cold-core rings, which lead independent existences for several months.

We now examine the ways in which organisms have adapted to the various physical regimes. Some use the boundary currents for long range transport between breeding grounds and feeding areas. Others make several circuits of an ocean gyre while growing to maturity. We shall see that the standard pattern of primary production for open-ocean situations (Chapter 3) is much modified by the presence of rings and gyres, with the result that primary and secondary production in the central basins is considerably higher than we thought a decade or so ago.

The biological phenomena associated with the major eastern boundary currents were dealt with in Chapter 5. These are the regions of major coastal upwelling and home for major stocks of sardines and anchovies. When we turn our attention to the western boundary currents and to the whole gyral pattern of circulation we find that well-known species like squid, eels and salmon make migrations of thousands of kilometres by travelling with these currents. We also find that the larvae of some kinds of coastal invertebrates live long enough in the plankton to make, for example, the crossing of the North Atlantic before settling on the opposite side from where they were hatched. This leads to changes in our traditional ideas about gene flow between coastal populations on opposite sides of ocean basins.

8.3.1 Squid and the western boundary currents

Illex illecebrosus is caught in great quantities off the east coast of North America from Georgia to Newfoundland (Coelho 1985). From January to early March larvae occur in slope water along the northern edge of the Gulf Stream, roughly from Cape Hatteras to the latitude of Cape Cod. In May juveniles are found in warm waters along the southern slope of the Grand Banks, and from there they migrate to Canadian inshore waters in late June and July. They grow about 1.5 mm d^{-1} during the summer, and by November many of the males are mature, but the females are less advanced. There appears to be only one year-class and it is thought that the adults return south to spawn on the northern edge of the Gulf Stream, thus completing their life cycle in 1 year. The recapture of tagged squid confirms that this migration can occur.

Since squid are also caught inshore from Cape Hatteras to Cape Cod, it is inferred that part of the stock leaves the Gulf Stream at the latitude of Cape Cod and moves into the southward-flowing coastal currents. Warm-core rings (see section 8.4 below) may play an important part in this shoreward movement.

Trites (1983) used oceanographic data to infer the probable spawning grounds. Physiological studies suggested that the squid need water above 13 °C for successful breeding, and in winter this occurs near the bottom only in water from Cape Hatteras south. Consideration of the known distribution of larvae suggest that the spawning grounds are close to Cape Hatteras and that after hatching the larvae are entrained in the Gulf Stream and travel northward with it. Here again, it is possible that the formation of warm-core rings and their subsequent drift westwards is a mechanism for transporting the juveniles from the Gulf Stream south of the Grand Banks to the continental shelf of Canada.

There is an analogous association between the Japanese squid *Todarodes pacificus* and the Kuroshio current (see map, Fig. 8.03). In this case there are three stocks breeding at different seasons: winter, summer and autumn. It is the winter-spawning group that is associated with the Kuroshio current. After spawning in the period January–April in the East China Sea, the larvae and juveniles travel north with the Kuroshio current, then turn inshore and are caught between the islands of Honshu and Hokkaido in summer.

The summer spawning is in another part of the East China Sea from which the larvae are entrained into the Tsushima current which flows north between the islands of Japan and the mainland. This current meets a southward-flowing cold coastal current, the Liman current, and the summer-spawned squid are fished along the boundary between the two. This is a clear example of the use of these western boundary jet currents as 'rapid transport', enabling the eggs and larvae to develop in winter in warm water while the adults travel north with minimum energy expenditure to exploit the rich feeding grounds further north.

8.3.2 Eels and the North Atlantic gyre

The classic work of Schmidt (1922) on the life history and migrations of eels is well known. As a result of extensive collection of the translucent, leaf-like larvae (known as leptocephali) he concluded that there were two species in the North Atlantic, the European eel, *Anguilla anguilla,* and the American eel, *Anguilla rostrata,* and that the smallest larvae of both species were found in the region of the Sargasso Sea, with clear evidence of increasing mean size of European eel larvae as one moved closer to Europe, and increasing size of American eel larvae as one moved north from the Sargasso Sea towards Newfoundland. He therefore proposed the startling theory that all European eels are derived from a stock that breeds in the Sargasso Sea in winter, that the leptocephali take over two years to reach the coasts of Europe, where they metamorphose into the adult form and migrate up the rivers and lakes to places far inland. No less startling was the view that this stock is maintained by eels that make the incredibly long breeding migration down the rivers of Europe and thousands of kilometres westward across the Atlantic Ocean to breed in the Sargasso Sea. One wonders whether eels have used the same breeding ground since the two sides of the Atlantic were much closer together, and have continued to colonize all the rivers available in Europe and North America, even though the continents have drifted apart.

In general, the case for the existence of two distinct species has been substantiated. Counts of muscle blocks (myomeres) in the larvae yield a mean of about 114 for European eels and about 106 for American eels. Counts of vertebrae in the adults yield similar differences in the mean values. Kleckner and McCleave (1985) made an intensive study of the distribution of American eel larvae in relation to major currents, and in the process obtained some new evidence about European eels. They concluded, from the distribution of the smallest larvae, that the spawning ground of the American eel was east of the Bahamas and north of Hispaniola. This is the location of a warm, high-salinity water mass known as the 'subtropical underwater'. Spawning was thought to occur in February and the larvae were found to grow rapidly at a rate of about 0.24 mm d^{-1} from February to October. A large fraction of the population was probably transported northwest in the Antilles current and entered the Gulf Stream north of the Straits of Florida by April. By May American eel leptocephali were abundant in the Gulf Stream opposite Cape Hatteras and beyond, and during July and August were common between Cape Hatteras and the southeast Newfoundland rise. From August to November the larvae were common in collections from slope water, inshore of the Gulf Stream, south to Cape Hatteras.

Kleckner and McCleave (1985) considered possible mechanisms by which the larvae escaped from the Gulf Stream and entered the slope water, before metamorphosing and migrating up the estuaries of eastern USA and Canada. They considered the size and frequency of warm-core rings (see section 8.4 following) and concluded that too few rings would form during the 2 or 3 months of summer to account for the passive transport of the leptocephali from the Gulf Stream to the slope water. They concluded that the larvae must actively migrate across the frontal region of the Gulf Stream in the area between the New England Seamounts (south of Cape Cod) and the southeast Newfoundland rise, where there are many meanders and the speed of the current is somewhat reduced. They would then be carried in a southwesterly direction along with the mean flow of the slope water.

Eels which did not escape from the Gulf Stream would be carried in one of the branches of the Gulf Stream found to the east of Newfoundland. A good proportion of these would join return flows in the mid-Atlantic (see Fig. 8.13), thus accounting for the significant numbers of larvae taken in this region during the summer. An interesting difference in the distribution of the larvae of the two species of eel is that European leptocephali were abundant in the Gulf Stream water but virtually absent from the slope water. This reinforces the idea that active swimming is involved in the movement of American eel larvae out of the Gulf Stream and into the slope water.

In the samples analysed by Kleckner and McCleave (1985) the main concentrations of European eel larvae appeared to originate from a water mass about 10° eastward of the American eel larvae. Nevertheless, many of them appeared to reach the Gulf Stream, for they were commonly found along with larvae of the American eel. There has been much controversy about the length of time it takes European eel larvae to drift across the Atlantic and reach the coastal waters of Europe. Schmidt (1922) suggested about 2.5 years, but Boëtius and Harding (1985) re-examined Schmidt's original data and concluded that it may be possible for European eel larvae (as it is for American eel larvae) to grow to a size appropriate to metamorphosis in a little over a year. The source of Schmidt's confusion appears to be a group of larvae of intermediate size that can be found in mid-Atlantic waters at any time of the year. A possible explanation is that larvae that travel with the Gulf Stream but join the return currents that bring them into the mid-Atlantic gyre may find themselves in poor conditions for feeding and have their growth sharply arrested. It is possible that the larvae of the European eel that succeed in reaching the European rivers are those that travel with the Gulf Stream and stay in those branches that carry them most directly towards the coast of Europe, while those that either never enter the Gulf Stream, or if they do are carried in a return flow to the mid-Atlantic gyre, do not recruit to the stock of breeding adults. In the terminology of Sinclair (1988) these might be regarded as 'vagrants'.

To complete our mental picture of the way in which eels make use of the major currents of the North Atlantic, we may suppose that the returning adults would travel west by a subtropical route, making maximum use of the Canary and north equatorial currents. In this way, their life history would be elegantly adapted to the North Atlantic gyre, with the successful group of larvae riding the western boundary and North Atlantic currents, while the breeding migrations of the adults follow the return flows. This story is a clear indication of how recent improvements of our understanding of the physical oceanography of the North Atlantic have thrown new light on the biology of these remarkable fish.

8.3.3 Salmon and the Alaskan gyre

The life history of eels involves breeding at sea and ascending the rivers to feed and grow. The migrations of salmon are in the reverse sense: they breed in the rivers but migrate out to sea to feed and grow. For many years the whereabouts of Pacific salmon, when they went to sea, were very poorly known, but in the decade of the 1960s an effort was made to obtain the missing information. A great deal of research was summarized by Royce *et al.* (1968). There are five kinds of salmon breeding along the Pacific coast of North America. All belong to the genus *Onchorhynchus* and they are pink salmon, *O. gorbuscha*; coho salmon, *O. kisutch*; chinook salmon, *O. tsawytscha*; chum salmon, *O. keta*; and sockeye salmon, *O. nerka*. Of these, the migration of the pink salmon is the shortest and easiest to understand. Spawning takes place in the rivers from mid-July to mid-October. Fry hatch and emerge from the gravel in the following spring and immediately migrate down to the estuaries and coastal waters. They migrate out into the ocean proper between July and September at about 1 year of age. It is estimated that during the next year they swim between 5500 and 7500 km. They do so by swimming actively downstream with the Alaskan gyre. During the summer they move rapidly north and west, travelling parallel with the coast and feeding voraciously. In winter they move about $10°$ south into the west wind drift and in the following spring they swim rapidly towards their parent river, to spawn at 2 years of age. It is estimated that their average rate of travel is about 18.5 km d^{-1}, but they are substantially helped in this by the Alaskan gyre (see map, Fig. 8.03). During their travels they have exploited the marine plankton, grown to breeding size and matured large gonads.

Further to the west, the pink salmon stocks originating on the coast of Asia in East Kamchatka are believed to follow the East Kamchatka current south until they join the eastward-flowing Kuroshio extension and North Pacific current with which they travel through the winter. In the following spring they migrate back to their rivers of origin. In both these cases, the fish travel great distances with the help of the prevailing currents.

Sockeye salmon take longer to complete their life history, spending one or two winters in fresh water before running down to the sea. Those from southern British Columbia are thought to spend 2 years at sea, during which they make approximately two circuits of the Alaskan gyre, but those from rivers on the shores of the Bering Sea stay at sea for 3 years and appear to make two or three circuits of an extended loop that takes in both the Alaskan gyre and the Bering Sea gyre.

8.3.4 Transport of invertebrate larvae across ocean basins

Before we leave the question of the biological importance of the major ocean currents, it is interesting to note that Scheltema (1966) obtained the first good evidence for the transatlantic transport of a larva of a benthic invertebrate. A few years before, the eminent marine benthic ecologist Thorson had concluded that very few invertebrate larvae had even the slightest chance to cross the larger ocean basins. The implication was that invertebrate populations in shallow water on the east and west coasts of the Atlantic were genetically quite distinct.

Scheltema (1966) found veliger larvae, probably of the gastropod *Cymatium parthenopeum* throughout the Gulf Stream and North Atlantic drift. He calculated from drift-bottle experiments that the average rate of transport of water from the Bahamas to the Azores is 0.4 knots and he succeeded in keeping the larvae alive in the laboratory for a time longer than that necessary for the transatlantic crossing. He pointed out that *C. parthenopeum* is found on the eastern side of the Atlantic from the Azores to South Africa and on the western side from Bermuda to Brazil. There is no need to postulate an introduction by man: the survival of larvae throughout a transatlantic crossing may not be so very unusual.

8.4 BIOLOGY OF EDDIES AND RINGS ASSOCIATED WITH MAJOR CURRENTS

As we have seen, the western boundary currents in both northern and southern hemispheres tend to be fast, deep and relatively narrow, so that a great deal of energy is concentrated in a relatively small cross-section. These currents have a tendency to meander, and in doing so they form eddies and gyres of various kinds. One kind remains attached to the boundary of the current, and is known as a frontal eddy. Another kind breaks off as a distinct ring with an independent existence. For example, the Gulf Stream rings discussed in section 8.2.3 may persist as distinct entities for many months, and are prime targets for investigation of their physical and biological properties. Studies have been made of rings associated with the Gulf Stream in the western North Atlantic, the Kuroshio current in the western North Pacific, and the east Australia current in the western South Pacific.

8.4.1 Gulf Stream frontal eddies

The Gulf Stream (see Figs 8.03 and 8.13) as it flows in a northerly direction offshore of Miami, Florida, is about 30 km wide, 300 m deep and has a flow in excess of 25 Sv. It flows close to the edge of the continental shelf until it reaches Cape Hatteras by which time it has approximately doubled its flow. It is now approximately 50 km wide and about 1000 m deep. Surface velocity is in the range 2–5 knots and the temperature 25–28 °C. It is bounded on the west side by waters of the continental slope and continental shelf that are much cooler than the Gulf Stream, and on the east side by slightly cooler waters of the Sargasso Sea with a temperature range of 20–24 °C. As discussed in section 6.2.1 there is a strong front on the westward edge, sometimes known as the 'cold wall' of the Gulf Stream (Fig. 8.12).

A consequence of the meandering of the Gulf Stream is that its distance from a particular point on the continental shelf is constantly changing. At a time when this distance is greatest, it is common to find a frontal eddy between the Gulf Stream and the coast (Yoder *et al.* 1981). This often takes the form of a finger-like extension of the Gulf Stream protruding into the shelf water and folding back to enclose a core of cold water (Fig. 8.16). This cold core differs from that contained in cold-core rings formed further north (see below) in being formed by upwelling of North Atlantic central water from deep in the Gulf Stream. A simplified explanation is that wherever the Gulf Stream in its meandering moves away from the coast, water from deep in the Gulf Stream upwells in the space created. Formation of these structures is a common event south of Cape Hatteras, a new one being formed on average once every fortnight. The upwelled water is rich in nitrate and Lee *et al.* (1981) estimated that this mechanism introduced about 55,000 tonnes of nitrogen annually to the outer shelf.

Yoder *et al.* (1981) made a study in April 1979 that showed clearly that a vigorous diatom bloom covering an area of more than 1000 km^2

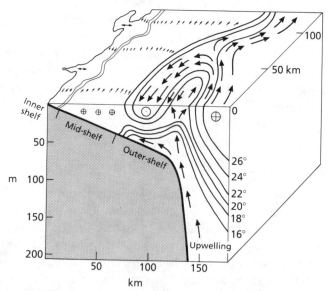

Fig. 8.16 Schematic diagram of a Gulf Stream frontal eddy on the Georgia shelf. Reproduced with permission from Lee *et al.* (1981), Pergamon Press.

occupied the cold core of upwelled water, in the region of the 200 m isobath. They speculated that this production helped explain why the shelf break south of Cape Hatteras is a breeding centre for Atlantic menhaden (*Brevoortia tyrannus*) and for bluefish (*Pomatomus saltatrix*). Atkinson and Targett (1983), using acoustic methods of fish biomass assessment, found that larger fish biomasses were associated with areas of upwelling. The distribution of sea birds is also influenced by these frontal eddies. Haney (1986) found that the shearwaters *Puffinus gravis* and *Calonectris diomedea* and the storm-petrels *Oceanites oceanicus* and *Oceanodroma castro* were most abundant in the upwelled cold core of the eddy, while the black-capped petrel *Pterodroma hasiata* and the bridled tern *Sterna anaethetus* were most abundant in the oligotrophic warm filament inshore of the cold core.

More detailed follow-up studies (Yoder *et al.* 1983) showed that in winter and early spring when the waters were not stratified the upwelling events led to primary production of about $2 \text{ g C m}^{-2} \text{d}^{-1}$ (for a total of about 175 g C m^{-2} during the 6 months of winter and spring). In summer and autumn, when the shelf water was stratified, upwelled water penetrated onto the shelf as a subsurface intrusion. Phytoplankton in this intrusion used the nitrate in about 10 days, and likewise fixed carbon at the rate of about $2 \text{ g C m}^{-2} \text{d}^{-1}$. Strong southwesterly winds were found to cause Ekman offshore transport which both favoured the formation of an offshore meander of the Gulf Stream and favoured the inshore movement of upwelled water at intermediate depths, to compensate for the offshore transport of the surface layer (Atkinson *et al.* 1984). It is now thought that this upwelling caused by the meandering offshore of the Gulf Stream is the primary mechanism for providing nutrients and stimulating production on the outer continental shelf of southeastern USA. This effect operates south of Cape Hatteras, where the Gulf Stream runs relatively close to the coast. North of Cape Hatteras the Gulf Stream moves away from the coast and instead of cold-core frontal eddies we find warm-core ring formation on the shoreward side of the current (see section 8.2.3).

Tranter *et al.* (1986) showed that slope water intrusions occurred near Sydney, Australia, and that they seemed to occur when eddies or meanders of the East Australian current came within 90 km of the shelf break. They made comparisons with the upwelling events associated with the western boundary of the Gulf Stream, but pointed out that the mechanism producing the intrusions must be different, because the continental shelf of that part of Australia is much narrower and does not permit the formation of cold-core frontal eddies of the type found off Florida and Georgia.

8.4.2 Formation of Gulf Stream rings

North of Cape Hatteras the Gulf Stream leaves the edge of the continental shelf and begins to flow in a more or less easterly direction. Every month or two a meander cuts off distinct rings, as in Fig. 8.14. As we saw in section 8.2.3, cold-core rings form on the south side of the Gulf Stream, and warm-core rings on the north. Once free of the Gulf Stream they tend to move in a southwesterly direction, more or less parallel with the Gulf Stream, at a speed of about 3–5 km d^{-1}. Eventually, most make contact with the Gulf Stream and are reabsorbed into it, although some lose their identity by diffusion into the surrounding water.

8.4.3 Ecology of cold-core rings

From first principles it would seem that cold-core rings have interesting potential to stimulate biological production. The water of the core is derived from the slope of the continental shelf and typically has a much larger concentration of plants, animals and nutrients than the Sargasso Sea into which it is propelled. In the period 1976–1977 a number of scientists who called themselves the Ring Group conducted four multi-disciplinary research cruises on a total of six rings, but the most intensively studied was ring Bob which was formed in February and March 1977, interacted with the Gulf Stream in April and May, and finally coalesced with it off Cape Hatteras in September. Their results to 1981 were summarized in an article in *Science* (the Ring Group, 1981).

A cross-section of the ring soon after formation (Fig. 8.17) showed bell-shaped isotherms, with 10–16 °C isotherms located about 600 m higher in the water column than their normal Sargasso Sea depth. One important effect was to bring nutrient-enriched water into the euphotic zone.

The biological characteristics of the ring changed more rapidly than the physical. In April the surface concentration of chlorophyll at the centre was about 4 $\mu g\,l^{-1}$ compared with less than 0.1 $\mu g\,l^{-1}$ in the Sargasso Sea water, but the concentration declined by a factor of eight between April and August. As the surface waters were heated by the sun there was a shift to smaller species and a greater species diversity, so that the phytoplankton community came to resemble more closely that of the Sargasso Sea. By August the chlorophyll maximum had moved

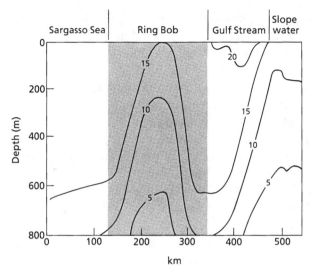

Fig. 8.17 Diagram of a vertical section through warm-core ring Bob to show the elevation of the 5, 10 and 15 °C isotherms. After the Ring Group (1981).

from the upper 20 m to a mean depth of 80 m and its concentration was less than 0.6 $\mu g\,l^{-1}$.

Overall, it was estimated that primary production in the rings was about 50% higher than that in the Sargasso Sea. In general cold-core rings were found to have a zooplankton biomass 1.3–1.8 times higher than the surrounding Sargasso Sea and as might be expected, the species composition was that characteristic of slope waters. While in the Sargasso Sea itself the greater part of the zooplankton biomass was concentrated in the uppermost 100 m, in the cold core there was a high zooplankton biomass down to about 800 m. This was thought to be the result of downward migration. As the surface waters warmed, species from the cooler slope water migrated downward in an attempt to maintain themselves in their preferred environment.

The euphausid crustacean *Nematoscelis megalopsis* is characteristic of slope water, but is often carried into the Sargasso Sea in cold-core rings. In the slope water most individuals make limited vertical migrations, staying above 300 m both night and day. A similar vertical distribution was observed in newly formed rings, but in older rings the population was found mostly between 300 and 800 m. Physiological studies of animals from these depths showed that their respiration rate was depressed 5–20%, production of eggs ceased and fewer and fewer males were found. It was concluded that they had found insufficient food to maintain growth and reproduction. Their numbers gradually decreased and in a 17-month-old ring no *Nematoscelis* were present.

Similar patterns, showing export of organisms from the slope water in the cold core and gradual decline in numbers as the ring gets older have been reported for the copepod *Paraeuchaeta norvegica,* the pteropod *Limacina retroversa* and the myctophid fish *Benthosema glaciale.* On the other hand, there are some species that become more abundant in the rings than in adjacent waters, apparently benefiting from the changing conditions. These include the pteropod *Limacina inflata* and the myctophid fishes *Hygophum benoiti* and *Lympanyctus pusillus.* It is not clear whether these species are present in the rings when they form or enter them from the Sargasso Sea.

The traditional view of the Sargasso Sea as of uniform low productivity has now to be modified by this information about the formation and life of the cold-core rings. About eight rings are formed per year and at any given time about 10% of the area of the Sargasso Sea is occupied by these rings. Phytoplankton and zooplankton species normally confined to the cooler slope waters are carried into the Sargasso Sea, and for a time appear to thrive and multiply, before being driven downwards by the heating of surface waters. For certain organisms the effect is dramatic. Foraminifera have a standing stock in rings up to 18 times as large as in the equivalent area of Sargasso Sea water.

Mesopelagic fishes are small species which live in considerable numbers deep in the water column, often forming distinct layers that can be detected by acoustic methods, and making marked diurnal migrations. Backus and Craddock (1982) reported on quantitative samples from slope water, cold-core rings and the Sargasso Sea. They used the depth of the 15 °C isotherm as an indicator of the nature of the habitat. This isotherm at or near the surface in slope water is at 600–900 m in the Sargasso Sea but, as we have discussed, is greatly elevated in a newly formed cold-core ring, sinking slowly as the ring ages and dissipates energy. They found that the abundance of mesopelagic fishes was high in the slope water, low in Sargasso water, and decreased with age in the cold-core rings. The slope water species tended to become more scarce as the rings aged, a few species were more abundant in rings than elsewhere, and Sargasso Sea species tended to invade from the top as the water warmed. Undoubtedly, cold-core rings have the effect of raising the average productivity of the western Sargasso Sea, and of introducing a range of species not otherwise to be found there.

8.4.4 Ecology of warm-core rings

The internal structure of a warm-core ring is in many respects the opposite of that of a cold-core ring (Fig. 8.18). Cold, nutrient-depleted

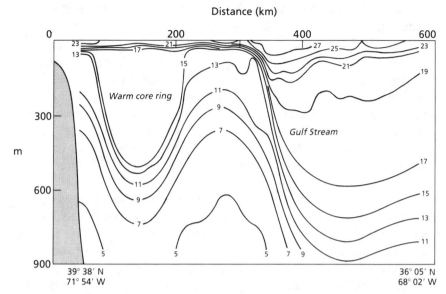

Fig. 8.18 Cross-section of a warm-core ring lying between the Gulf Stream and the continental slope, showing isotherms (°C). After Csanady (1979).

waters are found at greater depth in the core than elsewhere, so isotherms and nutriclines are depressed at the centre. Yentsch and Phinney (1985) discussed this structure in an interesting way (Fig. 8.19), pointing out that the outer part of the ring, the high velocity region, is formed from the 'cold wall' of the Gulf Stream, which has the usual frontal structure of sloping isopycnals (see Chapter 6.). It is to be expected that there will be transport of water along lines of equal density

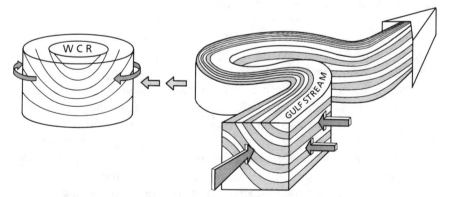

Fig. 8.19 Diagram showing how the Gulf Stream with its isopycnals sloping upwards to the left, gives rise to a warm-core ring with its isopycnals sloping upwards towards the perimeter. Arrows show flow of the Gulf Stream, isopycnal flow and rotation of the warm-core ring. After Yentsch and Phinney (1985).

(isopycnal mixing, Woods 1977) and that this will lead to the upwelling of nutrients in the high velocity peripheral region.

By analogy with what is known about cold-core rings, it might be supposed that warm-core rings, which have at their centre nutrient-depleted Sargasso Sea water, would at first be biologically unproductive and would slowly evolve to take on more and more the character of slope water in which they are immersed. One would not suppose that warm-core rings would appear in satellite images as regions of higher chlorophyll-a biomass than the surrounding water. Yet this is what happens from time to time (McCarthy and Nevins 1986; Tranter *et al.* 1980).

Yentsch and Phinney (1985), synthesizing the information available, have put forward the hypothesis that two distinct physical mechanisms regulate primary production in warm-core rings. The first has already been mentioned: isopycnal mixing brings deep, nutrient-rich water to the surface in the periphery of the ring. The authors show satellite images in which one can distinguish the low-chlorophyll core of Sargasso Sea water and the outer ring of Gulf Stream water with a higher chlorophyll content.

In a modelling study, Franks *et al.* (1986) invoked isopycnal flow in a rather different way. They proposed that as the ring ages, energy is lost due to friction and the permanent pycnocline, much depressed in the newly formed ring, gradually rises towards the level of the permanent pycnocline in the surrounding waters. As this occurs, the volume of the ring decreases and water is forced out along isopycnals. When this physical process was modelled along with the biological dynamics, a lens of enhanced phytoplankton biomass was formed at a depth of about 20 m in the ring centre. This was attributed to upward movement of nutrient-rich water at the ring centre.

The second mechanism proposed by Yentsch and Phinney (1985) as influencing primary production in warm-core rings is convectional mixing followed by stabilization. Since the warm-core rings, during their lifetime of several months are lying north of the Gulf Stream it is inevitable that surface waters will become cooler than those at depth, derived from the Sargasso Sea, stimulating convective mixing and formation of a deeper well-mixed layer. The latter, having discrete boundaries and a uniform temperature, is referred to as the thermostad. Since isotherms are depressed in the centre, convective mixing in the core may carry phytoplankton to greater depths than in the outer ring, reducing photosynthetic production. Conversely, convective mixing in the outer ring may bring to the surface water that has been enriched by the isopycnal flow of nutrient-rich water, thus stimulating phytoplankton production.

In addition to the changes resulting from the northward movement of Sargasso Sea water in a warm-core ring, seasonal changes have been observed in the rings related to the events typical of slope and shelf waters. Because of the distinctive features of the warm-core ring, the Gran effect (the formation in spring of a shallow mixed layer and its break-up in the autumn), occurs out of phase with the surrounding water. As a result, there may be times when a warm-core ring is undergoing its 'spring bloom' while the surrounding shelf waters are not. Events are complex and incompletely understood, but empirical studies show that warm-core rings may undergo bursts of relatively high production, associated with the vertical transport of considerable quantities of nitrate (Tranter *et al.* 1980; McCarthy and Nevins 1986; Hitchcock *et al.* 1987). Towards the end of its life, the thermostad of a warm-core ring may be completely covered by waters flooding in from the surrounding waters, thus making the mixed layer indistinguishable from surrounding slope waters. Over the lifespan of a warm-core ring, from its formation from Gulf-Stream meander to its eventual rejoining the Gulf Stream further south and west many months later, its primary productivity seems to be not very different from that of the slope water surrounding it (Hitchcock *et al.* 1985). This conclusion is quite different from earlier suggestions that warm-core rings, being composed of nutrient-depleted Sargasso Sea water, are islands of low productivity surrounded by much more productive slope water.

Zooplankton in warm-core rings

Species identification reveals that many zooplankton are advected into warm-core rings from the surrounding slope water (Roman *et al.* 1985), and physical oceanographic studies using salinity as a conservative property confirm that there is significant lateral exchange of water masses (Olson 1986). In general, a newly formed ring is found to have a low zooplankton biomass in the core, consistent with its origin from the Sargasso Sea, but within a few months there is often a biomass of zooplankton as high or higher than in the slope water. In general this is thought to be the result of *in situ* growth, resulting from the enhanced production of phytoplankton discussed earlier, and its consumption by zooplankton. A concrete example of such production was obtained by Cowles *et al.* (1987). After gale-force winds had deepened the mixed layer and introduced nutrients to the surface waters of a warm-core ring in 1981, micro-zooplankton, mainly young cyclopoid copepods, doubled in biomass. Adult copepods showed high grazing rates and strong egg production. The response could be detected within 10 days of the storm event.

In contrast to all these changes, mesopelagic fish and siphonophores that characteristically form a non-migratory acoustic deep scattering layer in the Sargasso Sea were found to be captured in the thermostad of a warm-core ring and could be traced acoustically throughout the life of that ring (Conte *et al.* 1986). There was no evidence of a significant change in abundance and their vertical distribution remained approximately constant for several months in spite of changes in temperature and salinity.

In the next section we shall consider the biological consequences of a warm-core ring coming into contact with the edge of the continental shelf.

Interaction of warm-core rings with the continental shelf

Satellite imagery shows that it is not uncommon for warm-core rings to make contact with the continental shelf and for large volumes of cold water to be entrained and dragged off the shelf by the rotational movement. It also seems likely that on the western or southern side of the rings considerable quantities of warm water are injected onto the shelf. It was natural, therefore, that when Markle *et al.* (1980) discovered fish larvae of tropical and subtropical origin on the Scotian Shelf they should attribute their presence to the action of warm-core rings. Similar observations have been made off southern New England.

Friedlander and Smith (1983) identified water advected offshore by a warm-core ring and contained within it were abundant sand lance larvae (*Ammodytes* sp.) that had been carried off the shelf along with the copepods that form their food. The idea then developed that a strong advection of a shelf water mass at a time when the larvae of commercially important species of fish are present could have harmful effects on the year-class strength of those species.

Wroblewski and Cheney (1984) explored a warm-core ring off the coast of Nova Scotia and found that larval and juvenile white hake were being carried as much as 140 km seaward of the shelf break. They were in much poorer condition than the same species on the continental shelf and the authors thought that they represented a loss from their source populations. The ring being studied also contained many fish larvae of tropical and subtropical origin belonging to the same species reported as immigrants to the Scotian shelf by Markle *et al.* (1980).

A modelling study (Flierl and Wroblewski 1985) showed that at a time when commercially important fish larvae are being carried by the residual current southwest along the Scotian shelf, the advection caused by a stationary warm-core ring could cause a 35–50% drop in abundance, depending on the ring size, and a slowly moving ring could, in certain circumstances, remove most of a year-class. This initiative was followed up by Myers and Drinkwater (1989), who analyzed weekly satellite images for 1973–1986 to determine positions and numbers of warm-core rings, and location of the Scotian shelf-slope front during those years. These data were combined with estimates of the timing and duration of spawning and the duration of larval stages, to provide 'susceptibility functions' for 17 groundfish stocks. They found evidence that increased warm-core ring activity was associated with reduced recruitment in 15 of the 17 groundfish stocks. A similar analysis of seven pelagic stocks and one shellfish stock showed no consistent evidence of the same effect, but it was pointed out that recruitment data for the pelagic stocks are less reliable than for the groundfish stocks. As the authors said, it is possible that the true mechanism causing reduced recruitment may be not the warm-core ring itself but some hydrographic feature that is correlated with warm-core ring activity. Nevertheless, off-shelf transport due to entrainment in a warm-core ring is the best explanation available at present.

To summarize to this point, it appears that cold-core rings are responsible for introducing slope-water species into the Sargasso Sea, and that warm-core rings introduce tropical and subtropical species not only into slope waters but even onto the continental shelf. Cold-core rings, when newly formed, have a productivity much higher than the Sargasso Sea, but as their surface layer warms and their cold-water fauna is driven deeper and deeper, their productivity falls to something close to the productivity of the Sargasso Sea. Warm-core rings, by analogy, were expected to begin life with a low productivity characteristic of Sargasso water, and to slowly come to resemble the more productive slope water. In practice it seems that warm-core rings have mechanisms for generating upwelling of nutrient-rich water especially near the perimeter in the high velocity zone. Stratification as a result of spring warming often leads to levels of primary production in the rings greater than that in the surrounding slope water and to subsequent rapid growth of zooplankton populations. Over the life of a warm-core ring its productivity appears to be about the same as that of the slope water.

Perhaps the most interesting aspect of the whole story is the sequence of events that takes place when a warm-core ring comes in contact with the continental shelf. Cold water is advected off the shelf in large amounts, and at times when planktonic fish larvae are present in the water, this leads to reduced larval survival and subsequently to reduced recruitment in the parent stock. On the other side of the ring, warm water is probably being injected onto the shelf and is the means of introducing adults and larvae of tropical and subtropical origin into temperate shelf waters.

8.5 ECOLOGY OF THE CENTRAL GYRES

In section 8.3 we discussed the major currents associated with the periphery of the huge subtropical gyres. We shall now look at the relatively quiet water that occupies the central gyral regions. The anti-cyclonic gyre of the North Pacific has been the subject of particularly intensive study and the focus of a major debate about the general level of biological productivity of the subtropical gyres. There is also a long history of observations in the Sargasso Sea.

8.5.1 Primary production in the subtropical gyres

Table 8.1 from Blackburn (1981) summarizes the data from more than a decade of studies that indicated that subtropical gyres are the least productive parts of the ocean. Primary production, measured by ^{14}C uptake, was believed to be less than $0.1 \text{ g C m}^{-2} \text{ d}^{-1}$, with the result that biomass of both phytoplankton and zooplankton is low, the water

Table 8.1 Approximate values for biological and related measurements in three kinds of ocean environments. From Blackburn (1981).

Measurement *	Subpolar gyres	Equatorial divergences	Subtropical gyres
Mixed layer depth, summer (m)	15–30	25–65	15–20
Mixed layer depth, winter (m)	>120	25–65	65–120
Euphotic layer depth (m)	20–50	45–85	75–150
NO_3-N, 0 m (μg-at 1^{-1})	5–25	5–10	0–1
NO_3-N, 100 m (μg-at 1^{-1})	10–25	10–25	0–5
Primary production (g C $m^{-2} d^{-1}$)	0.1–0.5	0.1–0.5	≤0.1
Chlorophyll (mg m^{-2})	15–150	15–30	5–25
Herbivore mean weight (μg C)	40–400	4–40	4–40
Zooplankton, 0–200 m (g m^{-2})	30–50	7–25	<7
Zooplankton, 0–1000 m (g m^{-2})	150	8–13	9
Micro-nekton, 0–200 m (g m^{-2})	0.6	0.6–0.8	0.3
Micro-nekton, 0–1000 m (g m^{-2})	2	1–3	1
Benthos (g m^{-2})	0.1–1.0	0.1–1.0	<0.05

* Primary production and chlorophyll integrated for the euphotic layer. For zooplankton and micro-nekton, 1 ml displacement volume assumed to equal 1 g.

is very clear with the euphotic zone extending to 75–150 m depth, and nitrate levels at the surface are in the range 0–1 μg-at l^{-1}. This low productivity extends to the fish in the water column and the invertebrates on the bottom. Consideration of these facts led to the description of the subtropical gyres as 'biological deserts'.

In section 3.2.5 we considered the mechanisms controlling productivity in such a situation, approaching them from the point of view of vertical structure of the water column. Stratification is a permanent feature, with the thermocline separating a lower, cooler layer having high concentrations of nutrients from an upper, warmer mixed layer having an extremely low level of nutrients. In the upper parts of the mixed layer phytoplankton production is approximately balanced by zooplankton consumption, and the ammonia excreted by the zooplankton provides a nitrogen source for a continued low level of phytoplankton production. Nitrogen and other nutrients are therefore in some approximate steady state of recycling and the primary production based on this cycle is referred to as regenerated production.

At the same time, a certain amount of nitrate is transported upwards through the nutricline by turbulent diffusion. This leads to more rapid growth of the phytoplankton population and the formation of a zone of maximum phytoplankton biomass, the 'chlorophyll maximum', just above the nutricline. The upper boundary of this zone is set by the supply of nutrients from below, and the lower boundary is set by the availability of light from above. This primary production based on nitrate-nitrogen is referred to as 'new production' and the ratio of new production to total production is called the 'f' ratio. Measurements of the f ratio for the North Pacific subtropical gyre suggested that its value lay in the range 5–10%.

This view of the gyre as being in a relatively steady state with a low level of primary production and low f ratio has been challenged by a number of different measurements summarized in Eppley (1980). Use of a particle-counter to measure changes in particle volume suggest that phytoplankton in oligotrophic subtropical gyres are completing a new generation about once every 3 h, which would lead to an estimated production rate an order of magnitude greater than previous estimates.

Measurements by means of particle traps of the rate of sinking of particles out of the euphotic zone were interpreted as reflecting the rate of new production, since in steady state the amount sinking out must be balanced by new production, rather than by production that is being recycled. The traps collected an average 68 mg C m^{-2} d^{-1}. If the f ratio is 5% this means a total production of 1.360 g C m^{-2} d^{-1} and if 10%, total production is 0.680 g C m^{-2} d^{-1}. Comparing these figures with the value of $<0.1\,\mathrm{g\ C\ m^{-2}\ d^{-1}}$ cited by Blackburn (1981), Eppley (1980) concluded that there is an order of magnitude uncertainty about the level of primary production in the large ocean gyres.

Shulenberger and Reid (1981) used a much larger-scale approach than anyone preceding them. They pointed out that in the open mid-latitude Pacific in summer there is to be found a subsurface oxygen maximum in which there are values up to 120% of saturation. This maximum lies below or within the pycnocline, and certainly below the warm mixed layer, which they refer to as a density cap. It is usually located above the chlorophyll maximum but below the primary production maximum (see Fig. 3.01). As Fig. 8.20 shows, values in excess of 100% saturation are found throughout the North Pacific and the area of greatest values is centred on 40° N 160° W.

The authors argued that this oxygen supersaturation accumulates in summer as a result of photosynthesis, and is prevented from equilibrating with the atmosphere by the density cap. Equilibration occurs in winter when the mixed layer deepens and the supersaturated water is circulated in the mixed layer. Allowing 120 days for the build up of the summer oxygen excess, they calculated that at a minimum, the rate of photosynthesis must be greater than the rate indicated by ^{14}C incubations by a factor of about four.

Platt (1984) challenged the Shulenberger and Reid conclusions on the grounds that the oxygen concentration at the beginning of the summer season was not known, the time scale of accumulation was not known, and the data used to compare an instantaneous carbon flux with

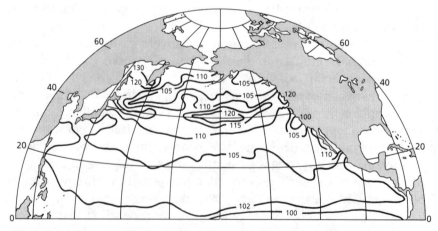

Fig. 8.20 Percentage oxygen saturation at depth of maximum value in the North Pacific. From Shulenberger and Reid (1981).

a time-averaged oxygen flux were inadequate from a number of points of view. Reid and Shulenberger (1986) published a rebuttal and held to their position. Platt and Harrison (1986) then drew attention to the fact that changes in oxygen concentration measure new production, P_{new}, (since regenerated production is approximately in balance with consumption, and net oxygen change resulting from these processes should be zero), whereas ^{14}C measures total production P_T. Scaling up the oxygen changes to estimate P_T would lead to impossibly high estimates of total production.

In a subsequent review paper Platt *et al.* (1989) conceded that the evidence from oxygen accumulation, sediment trap studies and direct determinations of the rate of nitrate uptake by phytoplankton in oligotrophic waters shows that the value of P_{new} in these waters is much higher than previously believed. This is a very important conclusion in relation to the problem of CO_2 accumulation in the atmosphere and the role of the oceans in helping to counteract it. In the long term, organic carbon sinking out of the mixed layer must be balanced by new production, which involves removal of CO_2 from the atmosphere. This topic will be discussed more fully in Chapter 9. The explanation of the higher values of P_{new} lies almost certainly in the occurrence of intermittent episodes of upwelling of nutrients. We have to abandon the view that the central gyres are uniform and quiescent and think of them as filled with eddies of varying energy levels (Kerr 1985). Cyclonic eddies will raise the pycnocline and are capable of bringing about upwelling of nutrients. The problem is to get measurements of these transient episodes of elevated production either directly or by using methods which integrate their effects over time. Satellite monitoring may provide the key to direct measurements while sediment traps and methods involving monitoring oxygen accumulation in the water have the potential to integrate results over time.

Meanwhile, a few people have made direct observations of transient episodes of enhanced primary productivity. McGowan and Hayward (1978) examined biological and physical data from the central Pacific gyre over the period 1964–1974 (10 cruises). In the summer of 1969 the rate of primary production was double what it had been in the previous summer and during four subsequent summers. The zooplankton biomass was also significantly higher. The physical data showed a negative temperature anomaly at the surface and, at a depth of 150–250 m, high variance in temperature with frequent occurrence of temperature inversions. This was taken as evidence of internal wave activity. In all other years of observation, the internal wave activity appeared to have taken place at depths of 300–400 m. The occurrence of these internal waves nearer to the surface were thought to have induced upward mixing of cooler, nutrient-rich water into the euphotic zone.

More recently, Laws *et al.* (1987) measured primary production at a station about 550 km north of the Hawaiian islands. The mean gross primary production measured on three occasions between 21 August and 2 September 1985 was 450 ± 37 mg C m^{-2} d^{-1}, which is about twice the average productivity level recorded at that station over the years 1968–1980. The question of whether this was a short episode of increased productivity, or whether earlier measurements were in error, was left open.

8.5.2 Secondary production in the subtropical gyres

Table 8.01 includes data for the biomass of zooplankton and of micronekton. There are no routine methods for measuring secondary production, although some estimates will be made. However, a good indication of the low levels of secondary production can be seen from the fact that the biomass in the upper 1000 m of the water column is about 9 g wet weight m^{-2}, while in the subpolar gyres it is 150 g. Micro-nekton (animals about 1–10 cm long) and benthos also have biomasses much lower than in the subpolar gyres. Figure 8.21 shows a generalized relationship of zooplankton biomass to latitude in the Pacific. There are very high values at latitudes greater than 40°, extremely low levels in the subtropical gyres, and some increase in the area of the equatorial divergences. The zone of greatest concentration reflects the level of primary productivity. Where this is low, the zooplankton are concentrated high in the water column, because most food is used before it reaches the deeper levels. By contrast, in the subpolar gyres there is enough food to support a dense population of zooplankton to depths of 3000 m.

The structure and dynamics of the fish community in subtropical gyres has been reviewed by Mann (1984). In the euphotic zone (0–100 m) are found fast-swimming predators such as the tunas, bill-fish and sword-fish. Most of them breed in tropical waters but make major excursions into the subtropical gyres. For example, bluefin tuna (*Thunnus thynnus*) breed in the Caribbean area, medium-sized fish migrate north in summer as far as the latitude of Cape Cod, but large, older fish may migrate as far as Newfoundland or cross the Atlantic and spend the summer in Norwegian waters. These large predators feed on a variety of smaller fishes, on squid and on larger crustaceans such as euphausids.

From 200 m to a depth of 1000 m is termed the mesopelagic zone. From acoustic studies we know that there is a 'deep scattering layer' caused by small fish (mostly < 10 cm) and larger invertebrates. Many of

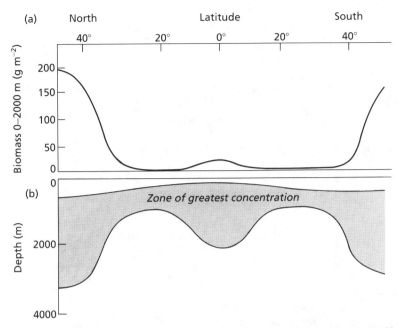

Fig. 8.21 (a) Idealized diagram of distribution of zooplankton biomass along a Pacific transect running north–south across the equator. (b) Vertical section along same transect showing zone of greatest concentration extending more deeply in regions of greater biomass. From Blackburn (1977) using data of Vinogradov.

these migrate to the euphotic zone to feed at night. This layer is more or less continuous all the way across the major ocean basins. Finally, below 1000 m is the bathypelagic zone occupied by fishes with dark colouring, small eyes, weak musculature but large mouths. Some, like the angler-fishes, have elaborate lures to attract prey. They appear to be adapted to life in a food-poor environment, expending little energy on swimming but taking a large meal whenever something comes within range. Figure 8.22 summarizes the biomass and production data (in kcal m^{-2} and kcal $m^{-2} y^{-1}$, respectively) for the different depth zones of a subtropical gyre. Using the approximate conversion 1.0 g wet weight equals 1.0 kcal, we see that the greatest values are in the mesopelagic fishes. The top carnivores, which are the main commercially important fishes, produce on average only 0.02–0.03 g m^{-2} of new tissue per annum. In addition, the smaller fishes on which they feed are thought to produce 0.5–1.3 g $m^{-2} y^{-1}$. This may be compared with 5–10 g m^{-2} in many continental shelf areas and 10–20 g m^{-2} in particularly favourable areas such as George's Bank.

It is not possible at this stage to use the information on fish productivity to throw light on the question of how much primary production

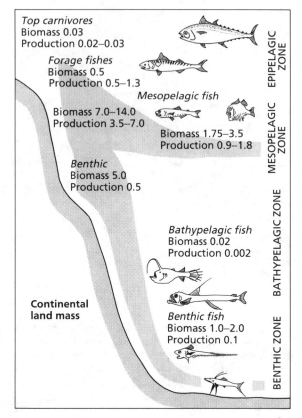

Top carnivores
Biomass 0.03
Production 0.02–0.03

EPIPELAGIC ZONE

Forage fishes
Biomass 0.5
Production 0.5–1.3

Mesopelagic fish

Biomass 7.0–14.0
Production 3.5–7.0

Biomass 1.75–3.5
Production 0.9–1.8

MESOPELAGIC ZONE

Benthic
Biomass 5.0
Production 0.5

Bathypelagic fish
Biomass 0.02
Production 0.002

BATHYPELAGIC ZONE

Continental land mass

Benthic fish
Biomass 1.0–2.0
Production 0.1

BENTHIC ZONE

Fig. 8.22 Diagram summarizing mean values of fish biomass (kcal m^{-2}) and production (kcal m^{-2} y^{-1}) in the open ocean. Shading symbolizes decreasing biomass as one moves from the shelf break towards the centre of the gyre. From Mann (1984).

occurs in the upper layers of the ocean. As we have seen, the major uncertainty is about the level of 'new' production. Fish which live and feed in the mixed layer can be regarded as part of the cycle of regenerated production, while fish that feed in the upper layer but migrate to depth during the day may contribute to the removal of nitrogen from the euphotic zone, and hence form part of the system that is fed by the upwelling of new nitrogen. More work is needed to unravel the complicated network of nutrient transfers.

8.6 SUB-ARCTIC GYRES

North of the main anticyclonic subtropical gyres in both Atlantic and Pacific Oceans are smaller, cyclonic sub-Arctic gyres (Fig. 8.03). In or near to these gyres are ocean weather stations, some of which have been extensively used to obtain time series of oceanographic data. Parsons and Lalli (1988) analysed data from ocean weather stations P (in the Pacific) and B,I,J (in the Atlantic) (see Fig. 8.03), together with data from the Continuous Plankton Recorder Programme (Colebrook 1979) (see Chapter 9) to synthesize a comparative view of plankton ecology in northern parts of the two major ocean basins.

The dynamics of the spring bloom in temperate waters was dealt with in section 3.3.2 using mainly material from the Atlantic Ocean. We shall now see that there seem to be systematic differences in the dynamics of the bloom between Atlantic and Pacific basins. The primary production data indicate that at ocean weather station I (OWSI) in the Atlantic, the arrival of the spring bloom in April is marked by a sharp increase in primary productivity from about 20 to over 300 mg C m^{-2} d^{-1}, while at OWSP in the Pacific there is a slow increase in productivity between January and July, over about the same range of values. Patterns of biomass change are also different in the two basins. In the Atlantic phytoplankton biomass changes by an order of magnitude between winter and summer, while in the Pacific it only doubles during the same period.

Differences between the two basins have their origins in the different winter depths of the mixed layer. In the North Atlantic it is > 200 m, while in the North Pacific it is about 100 m. (These differences are, in turn, explicable in terms of the global hydrological cycle, see Chapter 10.) As a result, the phytoplankton in the Pacific are exposed to a more favourable light regime during winter and are able to maintain a moderate level of biomass and production. About 80% of this biomass is in the form of small (<20 μm) flagellate cells. These are grazed by protozoa, which in turn are preyed upon by two or three common species of copepod, *Neocalanus plumchrus, Neocalanus cristatus* and *Calanus pacificus* which are relatively large and have only one generation per year. Hence, modest levels of both phytoplankton and zooplankton biomass are maintained through the North Pacific winter. When the surface waters warm in spring and the mixed layer becomes shallower, the zooplankton grazing almost keeps pace with the increase in phytoplankton productivity, so that phytoplankton biomass increases only slowly.

In the North Atlantic, by contrast, the mixed layer depth is sufficiently great that phytoplankton production virtually ceases during winter. The dominant copepod, *Calanus finmarchicus* is adapted to this period of famine by having overwintering larvae living in cold deep water >300 m. When the spring bloom of diatoms starts in surface waters there is a lag in the start of zooplankton grazing. Phytoplankton biomass increases to high levels and much of it may sink without being grazed in the plankton. Then, between April and June the *Calanus finmarchicus* larvae rise to the surface layer, complete their development and begin breeding. In most places these calanoids complete several generations in quick succession. Zooplankton production is much more narrowly pulsed in the Atlantic than it is in the Pacific.

In spite of these ecological differences, the two areas seem to have about the same total levels of phytoplankton and macro-zooplankton production, and about the same populations of predators (euphausids, jellyfish, chaetognaths, myctophids, squid, etc.). It is thought that the more constant biomass of plankton in the Pacific system supports a greater amount of pelagic fish production, (salmon, for intance) while the sinking of the diatom bloom in the Atlantic may support a larger benthic fish production.

8.7 CONCLUSIONS

We have seen in this chapter that ocean circulation is closely linked to atmospheric circulation. The difference in solar heating between low and high latitudes, in conjunction with the rotation of the earth, leads to westerly and trade winds, which drive major anti-cyclonic subtropical gyres in each of the ocean basins. Associated with these are powerful western boundary currents that transport large quantities of heat away from equatorial regions. These currents meander and cut off large gyrating bodies of water, the warm-core and cold-core rings that lead independent existences for several months. Organisms have adapted to exploit the properties of the currents and the rings. For example, squid breed in warm subtropical waters, then 'ride' the western boundary currents to exploit the higher levels of productivity in temperate latitudes. Pacific salmon swim thousands of kilometres with the currents of the sub-Arctic gyres and exploit the abundant food supply during their period of rapid growth. Eel larvae rely on the Gulf Stream to carry them from their subtropical breeding grounds to the coasts of Europe.

The anticyclonic gyres are permanently stratified (though with seasonal changes in the thickness of the mixed layer) and surface waters have constant low levels of nutrients. It used to be thought that primary production was uniformly low, relying mostly on the recycling of nutrients between the grazers and the phytoplankton. It is now thought that various mechanisms, at present little understood, lead to transient episodes of upwelling of nutrient-rich water, so that even in these gyres the level of 'new' production is much higher than previously supposed.

The subarctic gyres have alternating periods of summer stratification and winter mixing so that there is a seasonal injection of nutrients into surface waters. In the North Atlantic it appears that a bloom of phytoplankton develops when the zooplankton populations in surface waters are at a low level, so that much of the plankton biomass sinks without being grazed. In the Pacific, on the other hand, the depth of the winter mixing is more restricted, and phytoplankton are able to maintain a modest level of primary production throughout the winter. This supports a population of zooplankton which, when the spring bloom begins, is able to keep pace with the growth of phytoplankton and prevent the accumulation of biomass. Hence, biological production in the North Pacific appears to be less strongly pulsed than in the Atlantic.

While the major ocean gyres and their associated ring formations are well recognized and have been extensively studied, it is now believed that eddies permeate all parts of the oceans. Their biological properties have been little studied, but it is thought that some of them may be responsible for the transient upwelling events that stimulate the primary productivity of surface waters far above their long-term mean values. One of the challenges for oceanography in the coming decade is to develop methods of locating and studying these transient events.

9

Variability in Ocean Circulation: its Biological Consequences

9.1 INTRODUCTION

During the present century our technological advances in fishing have brought us to the point where it is possible to seriously overfish a stock so that it is no longer biologically viable. Fisheries scientists began to try to define the maximum catch that could be taken on a yearly basis while still permitting the stock to remain vigorous; they called it the maximum sustainable yield. At first these calculations were done with the environment of the fish (physical and chemical conditions, food supply, predator pressure, etc.) held constant. Gradually, fisheries scientists became aware that these assumptions were unrealistic. Large fluctuations in abundance occurred that appeared to be unrelated to pressures of the fishery. In the English Channel between 1925 and 1936, the herring stock went into decline and was replaced by pilchards. In various upwelling systems around the world sardines were replaced by anchovies, and vice versa, and cores taken from anaerobic sediment showed that these changes had also occurred before human fishing pressure had become a significant factor.

Perhaps the best known fluctuations of all are those associated with the El Niño phenomenon off the coast of Peru. Changes in oceanographic conditions (see Chapter 5) led on many occasions to drastic declines in

342

numbers of anchoveta, with spectacular consequences in terms of deaths of dependent sea birds and mammals.

Not all changes have been negative. During the period 1940–1970 the cod stocks greatly expanded their range and abundance on the west coast of Greenland, and in 1960–1970 there was a great expansion of the stocks of cod and related species in the North Sea.

Contemporary fisheries management policy therefore seeks to control fishing effort against a background of natural fluctuations in stock sizes. For many years the fluctuations seemed totally unpredictable, but as people began to study oceanographic processes on larger and larger scales, it gradually became apparent that many local changes in fish stocks were related to large-scale processes in the ocean. In this chapter we shall examine two of these processes. In the Pacific Ocean we shall see that coupling between the ocean and the atmosphere causes oceanographic oscillation between Peru and Indonesia — the southern oscillation — which affects the fish stocks of Peru and many parts of the North Pacific. In the Atlantic Ocean we shall see that changes in the pattern of the westerly winds running along the boundary between the arctic air mass and the subtropical air mass gives rise to important shifts in weather patterns. From about 1890 to 1940 conditions in the northeast Atlantic became progressively warmer; southern species of invertebrates and fish extended their range northward, several stocks expanded dramatically, and the balance of species present in the plankton of the English Channel changed radically. Between about 1940 and 1980 there was in the same area an increase in the incidence of cold northerly winds and the biomass of phytoplankton and zooplankton, though variable, showed a steady downward trend in biomass. A number of important fish stocks declined (though at least one showed dramatic improvement) and then towards the end of this period the plankton and fish community of the English Channel reverted to its pre-1930 condition.

In order to try to understand these changes, it is necessary to study the major features of the dynamics of the earth's atmosphere, and the way in which it interacts with the ocean currents.

9.2 PHYSICAL VARIABILITY IN THE PACIFIC AND ATLANTIC OCEANS

9.2.1 El Niño–southern oscillation (ENSO)

The anchovy fishery off Peru and Ecuador used to yield, in a good year, about 10^7 tonnes of fish. The fishery exists because equatorward winds along the coast cause nutrients to be upwelled from below the shallow thermocline as described in section 5.2.2. The catch rate is not constant throughout the year for in November or December warm water from the equatorial region moves south and disrupts the upward flow of nutrients. This change in the oceanographic conditions, because it occurs around Christmas, is called El Niño after the Christ child. In normal years the disruptive phenomenon lasts only a few months and the high nutrient up-welling returns, but every 3–7 years the high nutrient flow is cut off for up to a year. This has a devastating effect on the anchovy fishery which can decline to one-fifth of the peak catches. The term El Niño is now used ex-clusively for this catastrophic phenomenon.

Until about 30 years ago El Niño was believed to be caused by a change in the upwelling-favourable winds and thus an event local to the coasts of Peru and Ecuador. It is now known that El Niño is a small part of an ocean wide oscillation in the atmosphere called the southern oscillation (SO). In this oscillation the usual pressure gradient from the region of high atmospheric pressure in the southeast Pacific Ocean to the region of low pressure in the area of Indonesia becomes higher or lower than average. The combination of El Niño and the southern os-cillation has led to the acronym ENSO.

The southern oscillation is associated with large changes in the cli-mate of the equatorial region and to a lesser extent the subtropical regions. It has been studied intensively over the past 20 years by both sea-going oceanographers and theoreticians and there have been tremen-dous increases in our knowledge and understanding of the processes involved. These are well summarized in the reviews by Enfield (1989), Mysak (1986), Canby (1984) and in a special volume of *Oceanus* (Vol. 27, No. 2, 1984). Our description of ENSO events draws from these articles and begins with a review of the circulation, near the equa-tor, in both the atmosphere and the ocean.

As we saw in section 3.2.6 and Fig. 3.06, the trade winds on either side of the equator in the middle of the Pacific Ocean converge at the intertropical convergence zone (ITCZ) which lies at about 5° N in Feb-ruary and 10° N in August. The South Equatorial current lies from about 20° S to $\simeq$5° N. Between its northern limit and $\simeq$10° N is the eastward Equatorial counter-current and the North Equatorial current flows to the west between 10° N and 20° N (see Fig. 3.06).

The circulation in both the ocean and the atmosphere along the equa-torial Pacific is shown in a vertical section viewed from the south in Fig. 9.01. In the atmosphere the trade winds blow westward over the sea surface towards a low pressure area in the western Pacific located at about 180° W. In the low pressure region the air rises because of the heat gained from the warm ocean. The rising air loses its moisture through

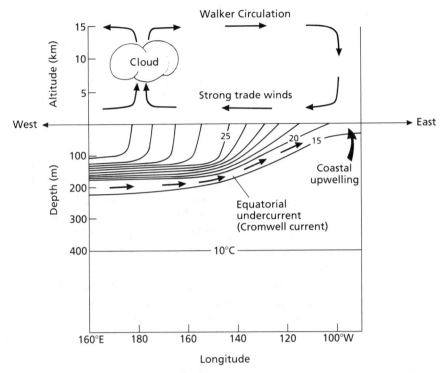

Fig. 9.01 Vertical cross-section of the atmosphere and the ocean viewed from the south showing zonal Walker circulation in the atmosphere and the thermocline which shallows in the east in response to wind stress at the surface. Also noted are the coastal upwelling off the coast of South America and the Equatorial undercurrent flowing east at the base of the thermocline.

rain and eventually circulates back towards the east at upper levels and descends over the eastern Pacific high pressure cell to complete the circuit. This vertical east–west circulation cell is one of a series which circle the equator known as the Walker circulation (Webster 1983).

In the ocean the trade winds push the water towards the west (Coriolis effect is zero at the equator) in the South Equatorial current. The wind stress towards the west is balanced by a pressure gradient in the opposite direction in the form of a slight rise in sea level towards the west. The pressure gradient is not transmitted into the deep ocean because the density structure in the upper layer of the ocean adjusts to cancel the pressure difference. This is accomplished by an increase in the thickness of the low density mixed layer from 30 to 50 m in the east to $\simeq$150 m at the middle of the ocean. In the western half of the ocean the pressure gradient is mainly balanced by the temperature increase towards the west in the mixed layer which stays roughly constant in thickness (see Fig. 9.01).

The large increase in the sea surface temperature from $\simeq 20$ °C in the east to $\simeq 30$ °C in the west is due primarily to upwelling along the equator. The upwelling is created by horizontal Ekman transports away from the equator (Fig. 3.06), created by the easterly trade winds. The upwelling is especially large near the equator because the Coriolis force becomes very small. For example the equation for the Ekman transport (Eq. 5.06),

$$M = -\tau/f, \tag{9.01}$$

indicates that for a given wind stress τ the transport M increases as f, the Coriolis parameter, decreases. When f is zero the equation predicts an infinite transport which is, of course, not realistic but high Ekman transports close to the equator are observed. In the eastern Pacific the thermocline lies at $\simeq 50$ m and upwelling brings water from below the thermocline up into the mixed layer but towards the west where the thermocline gets deeper the upwelling water comes more and more from the warm mixed layer rather than the cool subthermocline waters. This change in the temperature of the upwelled water is what causes the east–west rise in the temperature of the mixed layer and the sea surface.

In the El Niño or warm phase of the ENSO cycle the first indication that the normal situation is changing is a decrease in the trade winds in the central and western Pacific. The resulting decrease in the east–west pressure gradient along the equator allows the thermocline to relax from its normal tilt by rising in the west and descending in the east. This change propagates from the west to the eastern side of the ocean in 2–3 months in the form of equatorially trapped baroclinic Kelvin waves. These are the same waves we examined in section 7.2.3 but these are trapped at the equator (Gill 1982) by the vanishing of the Coriolis force rather than by a coast. Also they move much more slowly because they are baroclinic (only in the upper layer) rather than barotropic. At the eastern side of the ocean these waves split and continue poleward along the west coasts of north and south America.

As the east–west tilt in the thermocline decreases the mixed layer in the east becomes deeper and the water being upwelled along the equator is drawn from warmer sources. This causes the surface temperature along the equator to increase. Figure 9.02 shows the extent and degree of this warming at the 'mature phase' of El Niño as calculated by Rasmussen and Carpenter (1982) who averaged temperature anomalies from the six El Niño episodes between 1951 and 1973. The figure indicates that a significant change in surface temperature occurred over an area roughly 10,000 km long and 4000 km wide. The maximum of the averaged anomaly lies on the equator at $\simeq 130°$ W longitude and is

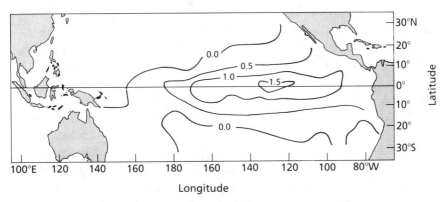

Fig. 9.02 Temperature anomalies in the equatorial Pacific Ocean, averaged from the 'mature phase' of six El Niño episodes between 1951 and 1973. Adapted from Rasmussen and Carpenter (1982).

$\simeq 1.6\ °C$. In the severe El Niño of 1982–1983 the maximum observed anomaly was close to 6 °C.

As the sea surface temperature increases, the region of ascending moist warm air over the western Pacific (Fig. 9.01) expands toward the east. The volume of rising moist air consequently increases. As the air rises it cools and the moisture condenses to form clouds in the upper atmosphere and the heat that caused the water to evaporate in the first place (the latent heat) is released. This creates a large additional source of energy and the convection grows in intensity which in turn increases the already anomalous circulation in the atmosphere. Thus the chain of events which started with an anomaly in the wind speed ends up, through positive feedback, increasing the anomaly in the wind speed.

The coastal upwelling off Peru and Ecuador shown in Fig. 9.01 at the far eastern end of the section continues as the trade winds diminish. But now with the deeper mixed layer the upwelling water at the coast is the warm nutrient-poor water from the mixed layer rather than the cool nutrient-rich water from beneath the thermocline. Thus the warm water and the disappearance of the anchovy fishery called El Niño is seen to be a relatively small part of the ocean-wide changes in both the ocean and the atmosphere rather than a local phenomenon as originally believed.

The end of El Niño, like the beginning, appears to start in the western Pacific. A rise in the thermocline, thought to originate north and south of the equator, emerges in the west and propagates towards the east along the equator. This raises the cooler subthermocline water closer to the surface which combined with the equatorial upwelling leads to a decrease in the sea surface temperature. The lower temperatures slow down the strong convection in the atmosphere and lead subsequently to an increase in the trade winds back to the original situation.

The rise in the thermocline at the western end of the equatorial Pacific triggering the end of El Niño is generated by the warm phase of El Niño. The change in the wind pattern along the equator including the decrease in the easterly trade winds and the eastward movement of the convection region creates regions north and south of the equator where the upper layer thickness is less than normal. These shallower thermoclines move westward as slow-moving Rossby waves until they hit the western boundary where they form baroclinic Kelvin waves. These move the shallow thermocline towards the equator and then along the equator from the west to the east. By this scenario, the return of the cold phase of El Niño is triggered by events in the warm phase. This leads to the idea that the whole phenomenon is an oscillator.

The contention that ENSO may be a natural oscillator, that is, a self-contained closed loop of variations and interactions between the ocean and the atmosphere is made by Graham and White (1988) and Enfield (1989). They point out that an oscillator requires first a mechanism that makes small instabilities grow (positive feedback), and mechanisms that eventually damp out the instability (negative feedback). The timing of the oscillation is controlled by the timing of the feedback mechanisms. The long period of the ENSO oscillator of 3–7 years is thought to be controlled by the slow rate that the shallow thermocline anomalies, generated north and south of the equator during the warm phase, move westward. The changes in the thickness of the upper layer move east along the equator much more rapidly in baroclinic Kelvin waves.

Thinking of ENSO as an oscillator we can summarize the seven stages in its evolution as follows:

1 the decrease in the easterly trade winds in the central Pacific causes the thermocline to relax its east–west tilt by rising in the west and deepening in the east;

2 the increased thermocline depth in the east leads to an increase in the sea surface temperature and an increase in the size and intensity of the atmospheric convection zone which moves eastward;

3 the decrease in the easterly winds (reversal in severe cases) in the western Pacific produces cyclonic circulations (anti-clockwise in the north) in the atmosphere on either side of the equator which produce rises in the thermocline through increased Ekman suction;

4 these regions of shallow thermocline become baroclinic Rossby waves which propagate slowly westward at about 0.5 m s^{-1} at the equator but at $\simeq 0.1 \text{ m s}^{-1}$ at $12°$ latitude (Graham and White 1988).

5 when the Rossby waves get to the western boundary they transfer their energy to Kelvin waves which transport the shallower thermocline towards the equator and then eastward along the equator;

6 the shallower thermocline along the equator allows the equatorial upwelling to reach into the colder water below the thermocline which lowers the sea surface temperature in the east along the equator; and

7 the cooler surface tends to slow down the atmospheric convection which moves back westward and leads to an increase in the easterly trade winds back to the original condition.

If the oscillator theory is right there is no need to find a trigger to start the oscillation for it is an endless series of interconnected events. There are, however, still many gaps in the story especially the details of the processes in the western end of the Pacific. For example, the Rossby and Kelvin waves which are postulated to propagate the shallow thermocline towards the equator and eastward along the equator have not been observed. Thus, much of the present day research is focused in the western equatorial Pacific.

Some researchers, sceptical of the self-contained oscillator theory, believe that ENSO is triggered by changes in the atmosphere over the Indian Ocean which spill over into the western Pacific via the Walker circulation. Others feel that large amounts of rain in the western Pacific have a large effect on changing conditions as the additional fresh water increases the vertical stability of the water column and leads to an increase in the sea surface temperature.

9.2.2 Teleconnections, patterns and oscillations

The substantial variations in the atmosphere and ocean that are part of the El Niño-southern oscillation appear simultaneously with abnormal meteorological and/or oceanographic conditions in regions far removed from the equatorial Pacific. These remote associations are termed teleconnections but the processes linking the remote anomalies to the equatorial Pacific are still under study. The terms pattern and oscillation have also been commonly used to specify large long-term changes in the atmosphere which may or may not be part of the recently discovered ENSO teleconnections.

In the ocean, the best known remote connections with El Niño are changes in the conditions off the west coasts of both North and South America. In the Gulf of California, for example, interannual variations in the sea level are correlated with the occurrences of El Niño and the California current is observed to slow or reverse direction during strong El Niños (Mysak 1986; Enfield 1989). These changes are transmitted along the coast by the baroclinic Kelvin waves travelling north along the west coast from the equator where they are generated when the thermocline along the equator is either increasing or decreasing its downward tilt towards the west.

The unusually large El Niño in 1982–1983 led to unusually warm conditions along the western coast of North America as far north as Alaska. Mysak (1986), however, presents two reasons why baroclinic Kelvin waves could not propagate as far north as Alaska. The thermocline, on which they travel, is too deep near the equator and farther north the thermocline rises to the surface which stops such waves. He argues that the warm conditions in the northeast Pacific were caused by the intensification of the Aleutian low. This led to stronger than normal winds from the southwest along the western coast of North America bringing an unusually long period of downwelling, high coastal sea levels, and a strong northward transport of warm water along the coast from California to Alaska. Mysak (1986) also makes the point that although intensification of the Aleutian low often follows an ENSO event, the historical records show that weak ENSO events have been followed by strong anomalies in the Aleutian low, and that ENSO events have occurred without being followed by intensification of the Aleutian low. Hence, the correlation between the two is strong but not by any means complete.

In the atmosphere there are strong correlations between the warm period of ENSO and weather changes along the equator. Droughts in Australia, Brazil and Africa occur at this time along with unusually high amounts of rain along the equator east of 160° E. Higher rainfall is also common in Ecuador and northern Peru. Descriptions of some of the disasters inflicted by the altered weather caused by the 1982–1983 El Niño are given by Glantz (1984) and Canby (1984). The abnormal weather is thought to be primarily due to the changes in the position and intensity of the atmospheric convection which moves eastward during the warm phase of ENSO from Indonesia towards the central Pacific. This affects the adjacent Walker cells resulting in changes in the equatorial weather remote from the central Pacific.

In the temperate zones, teleconnections with ENSO also exist but the correlations are usually not as strong as they are with the abnormal events near the equator. The pattern of anomalous atmospheric pressure related to the warm phase of El Niño (suggested by Horel and Wallace 1981) is given in Fig. 9.03. These authors first constructed indices for

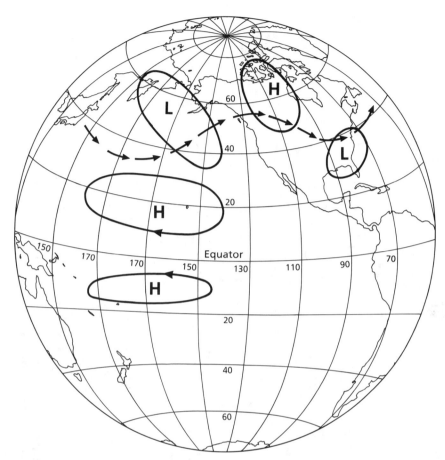

Fig. 9.03 Pattern of anomalous pressure distribution in winter during the warm phase of El Niño. Arrows indicate mean path of westerlies. After Horel and Wallace (1981).

various features of the southern oscillation over the 28 years between 1951 and 1978 and then calculated the correlations between these indices and various atmospheric parameters throughout the northern hemisphere during winter. During the warm phase of El Niño, in the western Pacific on either side of the equatorial region of high rainfall are found regions of higher than normal pressure. Towards higher latitudes a sequence of alternating low and high pressure anomalies form along a curving eastward path. These anomalies which are strongest in the winter cause the path of the mid-latitude westerlies to be distorted as indicated by the line of arrows in Fig. 9.03. The distortion brings westerlies further south in mid-ocean causing a decrease in the rainfall at Hawaii but at the west coast of North America the westerlies are further north than usual and bring warmer air to Alaska and the Canadian prairies.

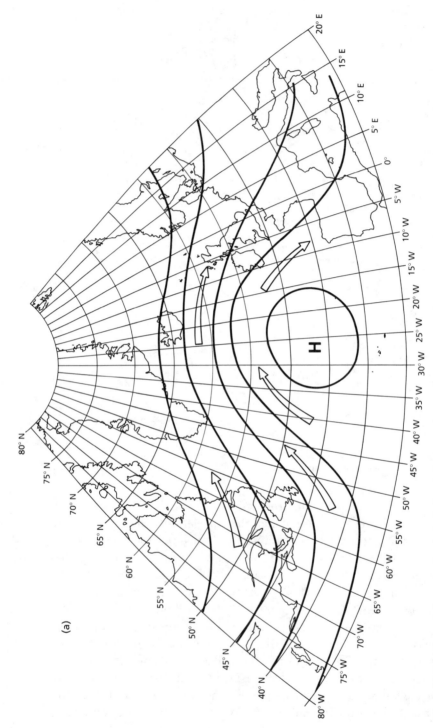

Fig. 9.04 Positions of the westerly winds during the two extremes of the North Atlantic oscillation. Adapted from Dole (1989). (a) Pressure difference between Iceland and the Azores is less than average; westerlies further north.

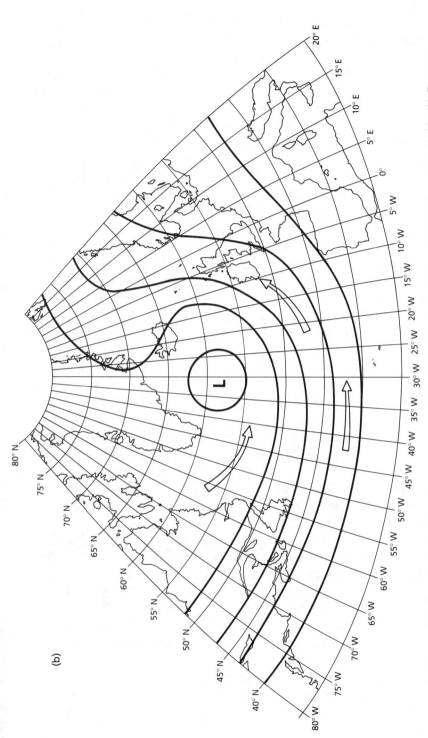

Fig. 9.04 Positions of the westerly winds during the two extremes of the North Atlantic oscillation. Adapted from Dole (1989). (b) Pressure difference between Iceland and the Azores is greater than average; westerlies further south.

On the eastern side of North America the air flow is more northerly than usual and brings more of the cold arctic air to the eastern USA.

The pattern in Fig. 9.03 is an average over many realizations and does not accurately describe the teleconnection pattern for each unique El Niño episode. For example, the 1976–1977 El Niño was associated with severe drought in the western USA and cold in the southeast but during the 1982–1983 episode the west coast was hit by an unusual number of storms and the east was unusually mild (Namias and Cayan 1984).

The alternating series of anomalous highs and lows shown in Fig. 9.03 constitutes one of the better known long term weather patterns. It is called the Pacific/North American pattern and has been used successfully for many years as an aid in long range forecasting. Another pattern, well known since the whaling days in Baffin Bay, is the North Atlantic oscillation described by van Loon and Rogers (1978).

The best-known effect of the North Atlantic oscillation is that warm winters in Europe coincide with cold winters in Labrador and West Greenland and vice versa. How this happens is illustrated (Fig. 9.04) by the positions of the westerlies at the two extremes of the oscillation. In Fig. 9.04(a) the westerlies are pushed far north of their average position bringing higher than normal pressures to the northern central North Atlantic. The flow of air over the western side of the ocean is more southerly than normal and thus warmer while the flow on the eastern side is more northerly and colder than normal. In the opposite extreme (Fig. 9.04b) the westerlies are pushed farther south than their average position bringing west to southwest winds over Europe and cold northerly air out of the arctic to the Labrador Sea and West Greenland.

The state of the North Atlantic oscillation is often indicated by the atmospheric pressure difference at sea level between the Azores and Iceland. High values of this difference occur when the track of the westerlies is particularly far south and the sea level atmospheric pressure at Iceland is lower than normal corresponding to Fig. 9.04(b). Low values indicate the track of the westerlies is farther north in the mid-Atlantic and cold northerly winds are common over Europe, while the Labrador Sea is warmer than normal. A time series of this difference for the years between 1895 and 1983, shown in Fig. 9.05, indicates that both the anomalous high and low pattern can exist for many years at a time such as from the mid-1950s to the early 1970s when the pressure difference was well below average bringing cold air to Europe. The early 1970s and early 1980s on the other hand had a positive index with warm European winters but some of the severest on record off Labrador and Greenland.

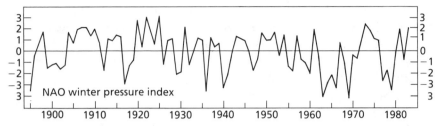

Fig. 9.05 The North Atlantic oscillation winter index based on the average pressure difference between the Azores and Iceland during the winters of 1895–1983. Adapted from Rogers (1984).

9.3 BIOLOGICAL VARIABILITY IN THE PACIFIC OCEAN

The biology of the Peruvian upwelling system and its interannual variability were treated in Chapter 5. We saw that at the height of the 1982–1983 El Niño event the upwelling waters off Peru were at 29 °C instead of the usual 16–18 °C, had very low nutrient concentrations and supported a primary productivity that was only about 5% of normal. It is well established that El Niño years are times of greatly reduced catches of anchoveta, although the exact cause is uncertain. A proportion of the stock migrates south and finds cooler waters less affected by the abnormal surface warming. Others remain in the affected area and die from thermal shock, a lack of suitable food, or perhaps a combination of both. Since the economy of Peru is tied quite strongly to the anchoveta industry, the social consequences of ENSO events are serious, and there is a great deal of interest in trying to predict their occurrences. This is perhaps the clearest example in the world of the influence of physical oceanography on fish productivity. It underlines the need for a better understanding of physical–biological interactions in the sea. Meantime, evidence is accumulating from other areas that year-to-year variability in ocean circulation contributes to the variability in biological events.

For example, McGowan (1985) reported the anomalies present in the southern California bight during the 1983 El Niño. Beginning in the autumn of 1982, sea level began to rise, there were anomalously high sea surface temperatures and pelagic red crabs (*Pleuroncodes planipus*), tuna, marlin and other warm-water fish normally found far to the south began appearing nearshore. By March 1983 the coastal waters had been invaded by anomalously warm water of low salinity, of a type normally found about 600 km offshore. At this time, the biological response was quite modest, but by August the nutricline was very deep and the distribution of phytoplankton and zooplankton very different from the long-term mean. Chlorophyll concentrations in the upper 50 m were very much reduced. The normal summer bloom of macro-zooplankton close to shore was much reduced, and the offshore maximum was entirely missing. This pattern remained approximately constant over four months of autumn and early winter.

Further north, in Oregon, Miller et al. (1985) reported that during the 1983 El Niño a thick blanket of warmer-than-usual water lay over the Oregon shelf and slope, out to more than 100 nautical miles from the coast. Although there were upwelling-favourable winds during 1983 they did not 'dig through the lens of warm water', except once during August. The isotherms lay almost flat across the shelf, with surface temperatures up to 7 °C above normal. The overall biomass of zooplankton was about 30% of that found in non-El Niño years. Normally, there is a seasonal change in species composition of zooplankton, with southern species appearing in the autumn and persisting through the winter, but with northern species replacing them from May to October. In 1983 the northern species did not appear in the coastal water, though some appeared offshore late in the summer. There were many occurrences of the euphausid *Nyctiphanes simplex* which normally appears no further north than central California. The offshore stock of northern anchovy, *Engraulis mordax*, spawned inshore and spawned early.

Juvenile salmon were less abundant in the nearshore waters of Oregon and Washington in 1983 than in previous years, and the catch of adult coho salmon were the lowest since records began in 1952 (Pearcy et al. 1985). Reproductive success and survival of several fish-eating birds (Brandt's cormorants, pelagic cormorants and common murres) were sharply reduced along the Oregon coast in 1983 (Graybill and Hodder 1985). All of these observations are explicable in terms of the intensification of the Aleutian low and its oceanographic consequences, as discussed in section 9.2.2.

9.3.1 Events at the larger scale: evidence from zooplankton

It was mentioned in the previous section that deepening and intensification of the Aleutian atmospheric low pressure system, which often follows an ENSO event, is associated with a rise in temperature and in sea level along the west coast of North America and a reduced flow of the California current. Some years before this mechanism was understood Chelton et al. (1982) found that the prime determinant of interannual changes in the biomass of zooplankton in the California current was the volume of flow of the current itself, plankton volumes being high when the flow was high. They also documented the correlation between elevated sea levels at the coast and reduced flow of the current. It therefore appears that the complex of factors associated with the deepening and intensification of the Aleutian low pressure system, loosely referred to as 'El Niño north', leads to a reduction in secondary production in the California current. In considering the probable mechanism for reducing zooplankton volume, Chelton *et al.* (1982) suggested either reduced advection from northern waters or reduced upwelling of nutrients associated with geostrophic tilting of isopycnals. They finally concluded that both elements were important.

Fulton and LeBrasseur (1985) reported on zooplankton studies in the northwest Pacific. Sub-Arctic water at the periphery of the Alaskan gyre, at the peak period after the spring phytoplankton bloom, has a mean biomass of 147 mg m^{-3} of net zooplankton, and contains good numbers of the large copepods *Neocalanus plumchrus* and *Neocalanus cristatus*. The California current system, part of the subtropical Pacific gyre to the south, has a mean spring biomass of only 72 mg m^{-3}, and has many species of small copepods belonging to the genera *Calanus, Mesocalanus, Paracalanus, Clausocalanus, Acartia*, etc. There is a transition zone between the two, that has an even lower mean biomass, 35 mg m^{-3}. The authors found that between Cape Mendocino and the Queen Charlotte Islands (Fig. 9.06) there was an area in which the sub-arctic boundary, marking the transition between high and low zooplankton biomass fluctuated from year to year by as much as $15°$ latitude at the coast, being strongly deflected to the north in warm-water years (often associated with ENSO), and to the south in more normal cold-water years. Fulton and LeBrasseur (1985) pointed out that the large zooplankton of the sub-Arctic water mass provide good food for juvenile pink salmon, while the small species characteristic of the California current do not, so that growth and perhaps survivorship of the salmon would be affected when warm water invaded their feeding area. Fulton and LeBrasseur (1985) also provided evidence that the California current is not entirely independent of the Alaska sub-Arctic gyre, for in years when the sub-Arctic boundary is far south, the California current contains some of the large species of zooplankton characteristic of the sub-Arctic waters.

To summarize to this point, it seems that there is a North Pacific interannual variability which frequently appears to be related to ENSO. The Aleutian low pressure system intensifies, and there are unusually strong winds from the southwest along the west coast of North America. This piles up warm water inshore, elevates sea levels, and forces the boundary between the sub-Arctic water of the Alaskan gyre and the warmer waters of the California current section of the subtropical gyre to move northwards, sometimes as much as 1500 km.

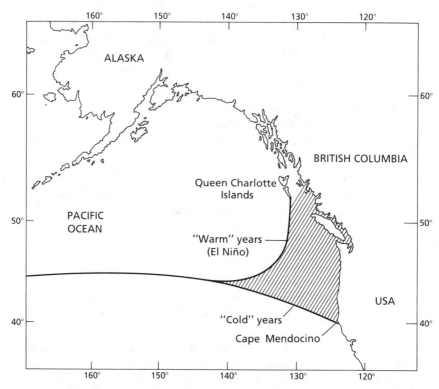

Fig. 9.06 Diagram illustrating the interannual variation in the position of the boundary between the sub-Arctic gyre and the California current portion of the subtropical gyre, in the North Pacific. Adapted from Fulton and LeBrasseur (1985).

9.3.2 Effects on fish stocks

Sinclair *et al.* (1985) noticed an apparent correlation between coastal warming events off California and good survival of Pacific mackerel (*Scomber japonicus*). They constructed a survival index from the estimated year-class size at age 1 year, divided by the estimated spawning biomass. They then ran correlation analyses using coastal sea-level data and a suite of biological data from the California current. There was a highly significant positive correlation between the survival of the mackerel and elevated sea level (Fig. 9.07) and several significant negative correlations with primary productivity and zooplankton biomass in the California current. Since others had shown that coastal warming and elevated sea level were accompanied by a reduction in volume of the California current and a reduction in zooplankton volume, the various patterns of correlation obtained by Sinclair *et al.* (1985) are not surprising, except for the strange fact that Pacific mackerel appear to survive better under conditions of reduced food supply. The authors did not

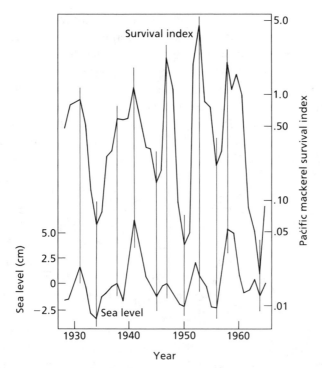

Fig. 9.07 Survival index for Pacific mackerel (upper line) and sea level anomalies along the California coast. High sea levels indicate El Niño events. After Sinclair *et al.* (1985).

have evidence for or against the various possible explanations, but they appeared to favour the idea that since the years of high sea level are also years of reduced offshore Ekman transport, the eggs, larvae and juveniles may survive better in the coastal waters because they are not advected out into the open ocean in such large numbers. In 'normal' years there may be a heavy loss from the coastal population by offshore advection. There is no direct evidence that young fish so advected fail to survive, but it is postulated that they are lost to the parent stock that breeds in coastal waters.

Similar results were obtained for Pacific hake (*Merluccius productus*) off the coast of California. In years of strong upwelling the larvae occurred at a greater mean distance from shore, and the year-class strength was low (Bailey 1981). Herring off southeast Alaska show strong year-classes in El Niño years (Pearcey 1983), suggesting that high sea levels and onshore convergence may be conducive to high retention of the herring larvae in inshore nursery grounds, and to good survival.

It is possible that the interannual variations in the northward flow of coastal waters, associated with variation in the development of the

Aleutian low, are responsible for part of the year-to-year variation in recruitment of Pacific cod. Tyler and Westerheim (1986) noted that eggs and bathypelagic larvae of cod are present in Hecate Strait, BC (inshore of the Queen Charlotte Islands) during January to March each year. They postulated that unusually strong northward transport carried the larvae out of the optimum feeding area and reduced recruitment of that year-class. A model which incorporated northward transport, water temperature and size of spawning stock was a reasonable fit to recruitment data for the years 1958–1980, accounting for 67% of the variance in the time series, with transport accounting for 44%.

For salmon which ascend the Fraser River to spawn, there are two possible migration routes from the open sea. One takes them south of Vancouver Island, through Juan de Fuca Strait, where they are equally available to US and Canadian fishermen. The other takes them north of Vancouver Island and into the Fraser River by way of Canadian waters. Prediction of the percentage following each route is thus of considerable economic interest to the two countries. Xie and Hsieh (1989) found that they could predict the percentage diverted to the northern route from a quadratic equation having terms for mean March temperature at Kains Island, Fraser River run-off during the same month, and an autocorrelation term representing the index of diversion from 2 years earlier. The percentage taking the northern route was higher in years having high surface temperatures in March and high river run-off. Predictions from this equation had a correlation coefficient $r^2 = 0.80$ with the actual data. When we remember that El Niño-type conditions involve higher than normal surface temperatures and heavy rainfall along the coast; that high river run-off increases stratification and spring warming; and that fresh water discharge in the northern hemisphere normally turns to the right and drives a coastal current, it is clear that there is a possible relationship with the Aleutian low and El Niño.

A second influence on the migration route of salmon has been proposed by Hamilton and Mysak (1986). In some years, but not all, there is a large anticyclonic eddy about 300 km in diameter located off Sitka, an island off the southeast Alaska coast (Fig. 9.08). The Sitka eddy is thought to be driven by interactions of bottom topography with northerly flowing water of a certain range of current speed. Near-surface currents in the eddy can exceed 50 cm s^{-1}, which is approximately the swimming speed of a mature sockeye. Hamilton and Mysak (1986) proposed that when the eddy is well developed salmon returning to the Nass and Skeena Rivers in northern British Columbia avoid the eddy and are diverted southwards. A southerly diversion carries the fish more into Canadian coastal waters and less into US (Alaskan) waters. Catch

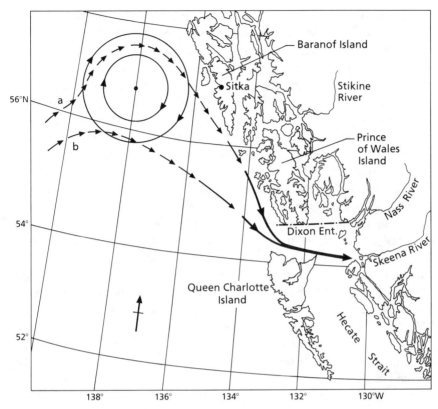

Fig. 9.08 Diagram showing position of the Sitka eddy and its postulated effect on the migration of salmon returning to the Nass and Skeena Rivers. (a) Weakly developed eddy, salmon follow the coast of Alaska; and (b) strongly developed eddy, salmon pass more directly to the coast of British Columbia. After Hamilton and Mysak (1986).

statistics tend to confirm this. Since the occurrence of the eddy is related to the strength of the northward flowing coastal current, and since this changes with the strength of the Aleutian low, this interannual variability seems once again to be a climatic effect.

So far we have spoken mainly of the elevated temperature and sea level of the eastern margin of the Pacific, from Peru to Alaska, that accompanies an ENSO event. There is also a converse effect, namely a negative sea surface temperature (SST) anomaly in the western Pacific but the peak of the anomaly appears to occur 6–9 months later than in the east. In June 1983 when the SST anomaly off Peru was up to + 7 °C, in the western Pacific it was only – 1 °C, but between January and May 1984, SST deviations off Japan were between – 2 °C and – 7 °C. The associated biological effects have been briefly reported by Yamanaka (1985). Bluefin tuna (*Thunnus thynnus*) spawn southwest of

Japan and in normal years large quantities are caught off the coast of Japan in purse seines. During their feeding migrations the fish are believed to follow the Kuroshio current and the North Pacific current out towards the mid-Pacific, and considerable quantities are caught by long-lining. As the time for breeding approaches, the fish migrate south into the north equatorial current and return with it to the western Pacific.

During El Niño years the purse seine catches close to Japan are much decreased. The centre of distribution of the tuna, as reflected in longline catches, appears to move eastward and purse seine catches off the California coast tend to increase. It therefore seems that the movement eastward of the SST anomaly is accompanied by an eastward shift in the bluefin tuna population.

During the 1983–1984 El Niño a number of species which normally stay well north of Japan made appearances along the Japanese coast. These included euphausids, glow squid, pollack, cod, walrus and fur seals. In the sea of Japan, heavy mortality of sea bass, sea bream, and parrot fish were believed to have resulted from the low water temperatures. One could say that in the North Pacific basin as a whole, an ENSO episode is associated with a counter-clockwise extension of the range of many organisms, so that they are found further north than usual on the west coast of North America, and further south than usual on the coast of Japan.

9.3.3 Effects on coral reefs

It appears that the high water temperatures associated with the 1982–1983 El Niño caused heavy mortality among corals off the coast of Costa Rica, Panama, Colombia and the Galapagos Islands, and that the destruction was amplified by associated biological interactions. According to Glynn (1985) sea surface temperatures were at a mean of 30–31 °C for 5–6 months and the mixed layer was 100 m deeper in some areas than in previous years. After the event, there were patches of 0.1–1.0 hectare of dead reef surface. Predominant among the reef-building corals in this area are the branching *Pocillopora* species, which are resistant to the attacks of the sea star *Acanthaster planci*. Their defences consist of effective nematocysts and of symbiotic crustaceans (*Trapezius* and *Alpheus*) which attack and repel sea stars.

On a number of reefs, coral species that are not resistant to *Acanthaster* have been able to persist because they were growing centrally on a reef that was protected by an outer ring of *Pocillopora*. When the El Niño warming killed the *Pocillopora* it also allowed the sea stars to move in and take the vulnerable species. Cores taken through some large colonies of *Pavona* which is vulnerable to sea stars, showed them to be nearly 200 years old, with no evidence of sea-star attack during that period. Glynn (1985) therefore concluded that reef growth had proceeded continuously in Panama during this period, in spite of the presence of sea stars. Thus, a disturbance of the magnitude of the 1983 El Niño had not been experienced in the area for nearly 200 years.

9.4 SUMMARY FOR THE PACIFIC OCEAN

The cycle of physical events comprising an El Niño–southern oscillation is summarized in section 9.2.1. The most dramatic biological event is failure of the production system supporting anchoveta off Peru. The southern oscillation causes the thermocline to move much deeper, with the result that water upwelled in the coastal zone is warm, nutrient-poor water. Primary production is greatly reduced, and with it most kinds of secondary production. In extreme conditions, such as those experienced in the 1983 El Niño, tropical coral reefs may suffer severe damage from the elevated water temperature.

While the changes just described are first seen close to the equator, off the Ecuador–Peru coast, they gradually spread to higher latitudes. A rise in sea level and anomalously high sea-surface temperatures have been recorded along the North American coast from California to Alaska. They are loosely referred to as 'El Niño north', but it seems that their occurrence is not always correlated with an El Niño. Intensification of the atmospheric low pressure system over the Alaskan subarctic gyre (the Aleutian low), causes an intensification of onshore winds from the southwest, a rise in sea level and an increase in strength of the northward-flowing coastal current. This usually, but not always, happens in El Niño years. As in Peru, coastal upwelling in the California current system during El Niño tends to bring to the surface warmer, nutrient-poor water and biological productivity decreases. Paradoxically, some species of fish, such as mackerel, hake and herring seem to have better survival under these conditions, producing bigger year-classes. It is suggested that in non-El Niño years they lose a large proportion of the newly hatched stock through offshore Ekman transport. During 'El Niño north' these species benefit from the reduced offshore transport. Similar arguments have been produced for the improved recruitment of crab stocks during El Niño years.

Changes in the strength of the northward-flowing coastal current are correlated with changes in the survival of cod and in the migration routes of salmon. Strong currents appear to carry a high proportion of the cod away from their nursery grounds, leading to poor year-classes.

Factors associated with 'El Niño north', such as warmer surface waters and higher rainfall along the coast (hence more run-off from the rivers) are thought to cause salmon to change their migration routes, and hence change their vulnerability to fishing fleets along different parts of the coast.

We thus arrive at a concept of oscillations in the atmosphere–ocean system which benefit some species of fish in El Niño years and other species in non-El Niño years. This is important to fisheries managers. For a long time overfishing was most frequently blamed for declines in fish stocks. Later, biological interactions such as changing food availability or predation by one species on the young stages of another were implicated. It is only recently that it has become apparent that major physical changes on the scale of whole ocean basins occur routinely and have drastic effects on the year-class success of a wide variety of fish species.

A desirable economic aim of a fishing industry is to have a relatively constant annual yield from a stock, so that capital equipment and staff can be used to the best advantage year after year. Study of the large-scale physical processes and their coupling to biological events reveals that major year-to-year fluctuations in fish stocks are a natural phenomenon. Studies of the abundance of fish scales preserved in anaerobic basins in the California current system confirm this (Soutar and Isaacs 1974).

There is no complete agreement on the physical mechanism responsible for the alternation of El Niño and non-El Niño conditions, but one of the more plausible theories suggests that after the end of an El Niño episode a train of Rossby waves moves slowly westward across the Pacific and when the waves reach the western boundary they cause changes which lead to the onset of a fresh ENSO episode. This is clearly not the whole story, for it does not take into account the atmospheric terms of this ocean–atmosphere phenomenon. Nevertheless the main outlines of the southern oscillation mechanism in the Pacific Ocean are sufficiently well understood to enable us to say with confidence that biological productivity, all the way from phytoplankton to fish, birds and mammals will continue to undergo strong oscillations.

9.5 BIOLOGICAL VARIABILITY IN THE NORTH ATLANTIC

9.5.1 Evidence from long-term plankton records

In the years between the two world wars Alister Hardy developed a 'continuous plankton recorder' (CPR). It consisted of a towed body with a tunnel through it, along which sea water flowed. Across the tunnel was stretched a fine gauze mesh that unrolled continuously from a spool below and wound onto a propeller-driven spool above, that was immersed in formalin. In this way, plankton was sampled and preserved continuously so long as the body was in forward motion. After World War II a mechanism was put in place to have these recorders towed by merchant ships and ocean weather ships on regular routes (Glover 1967). Samples were collected at a standard depth of 10 m and an attempt was made to obtain records at monthly intervals from twelve different geographical areas (Colebrook 1986). For zooplankton, the record extends from 1948 to the present day, and recent analyses are based on 24 species in 247 series. For phytoplankton, a change in the method of enumeration was made in 1958, so analyses start from that time. Nevertheless, records are available for 24 species in 263 series.

This study showed that there are strong differences between the open ocean, the shelf and the North Sea in the seasonal patterns of phytoplankton biomass. Stratification and bloom development begin very early in the North Sea, where the water column is shallower and the salinity is lower, a little later on the shelf and latest of all in the open ocean (see Chapter 3). However, the open ocean is warmer in winter than is the North Sea, and once stratification is established, the phytoplankton grow faster. Thus the region of maximum phytoplankton abundance showed a regular seasonal shift from coastal waters, out onto the shelf and then into the open sea.

The seasonal patterns of zooplankton distribution were very different. In general, the highest densities appeared on the shelf at all times of year, the least in the offshore waters. The densities of summer populations in each area appeared to be mainly a function of the densities of overwintering populations in those same areas, and not a function of the food supply or the dynamics of the water column. As Colebrook (1986) expressed it, phytoplankton achieve something approaching steady state in their response to seasonal changes in temperature and stratification but zooplankton do not. Instead, they appear to be limited by their population growth rates. As has been suggested elsewhere, zooplankton on the shelf tend to be unable to fully utilize the phytoplankton produced during a bloom. Instead, much of the phytoplankton biomass sinks to the bottom and nourishes the benthos. Davies and Payne (1984) documented this for the North Sea.

Against this background information we may now look at long-term trends in the records. For the zooplankton, almost all species in all types of habitat, from shallow coastal to deep open ocean, showed a downward trend in numbers from 1948 to about 1980 (though an upward trend became apparent after 1980). Principal components analysis was used to obtain an expression of the common pattern for all areas using data standardized for zero mean and unit variance. The data for *Pseudocalanus elongatus* are shown in Fig. 9.09. The same technique was used to show the trends in total zooplankton, in each of the 12 areas sampled. In every case the 36-year trend was towards lower numbers, although there was an upward turn in the last few years which has since proved to be important. When the time series for the various taxa were compared on a seasonal basis it was found that the similarity was greatest in the winter data. From this Colebrook (1985) concluded that the factor responsible for the decline in zooplankton numbers probably operates primarily in winter.

In spite of the quite different seasonal and geographic patterns displayed by phytoplankton and zooplankton in the CPR data, it turned out that total abundance of phytoplankton and zooplankton followed the same 20-year trend of decreasing abundance from about 1960 to 1980 (Fig. 9.10). It would be very surprising if the decrease in zooplankton were not a consequence of the decrease in availability of phytoplankton as food.

Referring to Fig. 9.05 we see that the winter index of the North Atlantic oscillation was negative from 1955 to 1970. This means that the track of the westerlies is expected to be further north than usual in the mid-Atlantic and cold northerly winds should be more common

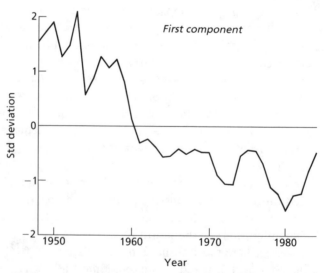

Fig. 9.09 Abundance data for *Pseudocalanus elongatus* from 12 different areas, combined. The data were normalized to zero mean and unit variance. The plot is of the first principal component, after principal components analysis. From Colebrook (1986).

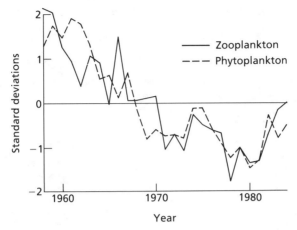

Fig. 9.10 Time series of the abundance of zooplankton and phytoplankton for 1958–1984, derived from principal components analyses of the data for all species and all areas. From Colebrook (1986).

over the eastern Atlantic. Dickson *et al.* (1988a) investigated the situation by calculating a slightly different index, the mean atmospheric pressure difference between 20° W and 10° E over the latitude interval 35–65° N, reporting monthly averages for the years 1873–1981. They found a regular seasonal pattern, with the pressure difference increasing during the first half of the year, as the Azores–Bermuda high became stronger. They then showed that during the decade of the 1970s the strengthening of that high started earlier and was more intense than it has been during the 1950s. This resulted in much stronger northerly winds over the coastal waters of Western Europe.

This analysis was confirmed by the records of light vessels in the North Sea, many of which showed, between the 1950s and the 1970s, increases of 20–55% in the incidence of gale force winds during the springtime. The regions of maximum increase were close to the east coast of England, and were the same areas in which the plankton biomasses showed the most strong decline.

When the evidence from the winter index of the North Atlantic oscillation and the analysis of Dickson *et al.* (1988a) are considered together, it seems that there was a period from about 1950 to 1980 when the tendency for cold northerly winds to blow over the eastern Atlantic in winter and spring was progressively increasing. This leads to the conclusion that the most probable cause of decline in the phytoplankton stocks was strong wind mixing of the surface layers of the ocean in spring which, as we saw in Chapter 2, leads to a delay in the formation of the spring phytoplankton bloom because the phytoplankton are mixed to below the critical depth. Figure 9.11 shows contours of

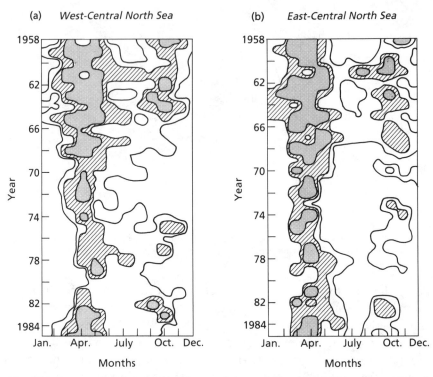

(a)   West-Central North Sea   (b)   East-Central North Sea

Fig. 9.11 Contour plots by month and year of the total biomass of 12 'spring species' of phytoplankton from continuous plankton recorder data. The west-central North Sea area was more subject to gale force northerly winds in winter and spring, than the east-central North Sea area. Dark shading indicates maximum biomass. From Dickson *et al.* (1988a).

phytoplankton density by month and by year for an area where the spring gales increased greatly, and for an area where the increases in northerly winds were less marked. The former shows both a decline in spring biomass and a 1-month delay in onset of the bloom during the mid-1970s, while the latter shows a decline in biomass but not a delay. Both areas showed a partial recovery during the 1980s. All this amounts to powerful evidence in support of the view that climatic factors caused the decline in phytoplankton biomass between 1958 and 1980, by delaying or diminishing the extent of spring stratification and hence the ability of phytoplankton stocks to increase their biomass by growth and multiplication.

The close correlation between the trend of zooplankton biomass and that of the phytoplankton makes it difficult to dismiss the view that zooplankton biomass is controlled by phytoplankton production. A delay in the spring bloom and a reduction in the phytoplankton biomass present throughout the growing season of the zooplankton would be likely to reduce the rate of growth and reproduction of the zooplankton.

Lamb (1969) developed an analysis of climatic trends in Britain using an index which might be regarded as roughly the complement of the index used by Dickson *et al.* (1988a). He called it the 'frequency of westerly weather'. Westerly weather in Britain is characterized by surface winds that are predominantly southwest to west and by rapidly changing weather produced by a series of depressions that travel across the country bringing frontal rain. In general the temperature is higher than when northerly winds predominate. Lamb (1969) found that the number of days per year on which this type of weather occurred increased from the year 1890 to 1925, decreased for 20 years, peaked again in 1950 and decreased again for more than 30 years. Thus westerly weather was decreasing during the period that the incidence of northerly winds was increasing. In general the changes represent a period of climatic warming in the northeast Atlantic in the first 50 years of this century, and a cooling trend in the period 1950–1980. Perhaps not coincidentally, the first 50 years of this century showed an increase in mean global temperature, and the next 30 years a decrease (Fig. 10.01). We shall see that this is reflected in a number of changes in the fish stocks.

9.5.2 Evidence from fish stocks

Fish stocks are noted for their variability, but some of it is attributable to overfishing and some to man's destruction of habitat (for example, salmon spawning beds). However, the view is gaining ground that a very large proportion of the variability of the stocks is related to natural variability of the marine environment. Many examples of correlations between changes in fish stocks and physical changes in the North Pacific were given in section 9.3.2, and historical evidence that such changes have occurred in the past, independently of man's fishing efforts, were obtained by counting fish scales from the sediments of anoxic basins off the California coast (Soutar and Isaacs 1974). For the North Atlantic, the leader in the search for evidence of physical factors influencing fish stocks has been Dr David Cushing (Cushing and Dickson 1976; Cushing 1979; Cushing 1982).

North Sea gadoid stocks

One of the more striking and consistent changes in the fish stocks of the North Sea has been the sharp increase in stocks of cod, haddock, whiting, coalfish and Norway pout (collectively known as the gadoids), which took place during the decade of the 1960s (Fig. 9.12). Total stocks increased by a factor of almost five, as a result of a series of very successful year-classes. The proportionate increase was about the same for each species. After discussing various possibilities, Cushing (1982) concluded that the most probable explanation was that the climatic changes that took place during the period of enhanced northerly winds, 1950–1980, and which led to the delay in events in the plankton (see section 9.5.1) also led to an improvement in conditions for gadoid larvae.

To understand this, we must make a brief excursion into Cushing's (1975) match/mismatch hypothesis. In essence it states that a year-class

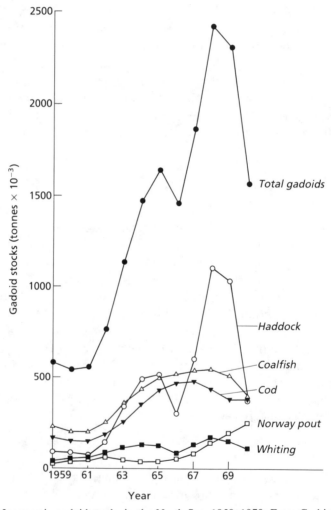

Fig. 9.12 Increase in gadoid stocks in the North Sea, 1959–1970. From Cushing (1982).

of fish is strong if the timing of the zooplankton maximum matches the timing of the appearance of the larvae requiring that zooplankton as food. In the case of a mismatch, the year-class is weak. When Cushing plotted the recruitment of cod against the time interval between cod spawning and *Calanus* peak production, he obtained a highly significant correlation. When the zooplankton peak was delayed by 3 months after peak cod spawning the year-classes were on average stronger. It appears that late blooming plankton is favourable to cod larvae, creating a better match.

From this, it seems that the meteorological events of the 1960s, with a prevalence of strong northerly winds in the North Sea which had the effect of delaying the onset of the spring bloom, were unfavourable for the biomass of zooplankton but were favourable to the survival of cod larvae. As Cushing (1982) put it: 'In many ways the events in the North Sea reflect the climatic changes as clearly as the more dramatic ones, such as El Niño or the rise and collapse of the West Greenland cod fishery.'

Portuguese sardine stocks

An interesting link with the story about increasing strength and frequency of northerly winds is found in the Portuguese sardine fishery. It was mentioned briefly in section 5.7.2 that northerly winds produce upwelling off the coast of Spain, near Cape Finisterre. Similar upwelling also occurs further south, off the coast of Portugal, and Dickson *et al.* (1988a) found that an April–September upwelling index at Porto showed an increasing trend from 1950 to 1980. They attributed it to the same increase in northerly winds that is believed to have caused the decrease in plankton in the northwest Atlantic. One might have thought that increased upwelling meant increased plankton production and increased sardine production, but the story is not that simple. The joint sardine catch of Spain and Portugal (for which statistics are available) shows a *negative* correlation with the upwelling index. Moreover, the negative correlation is strongest with the April upwelling index over the previous three years. Since April is one of the spawning seasons of the sardine, it appears that weak upwelling may permit the formation of a planktonic population suitable in quality and distribution as food for the newly hatched sardines. In other areas there are examples of *positive* correlations between upwelling index and sardine catches, and this may depend on the range of strength of upwelling that occurs. Whatever the explanation, it is clear that changing weather patterns have exerted a marked effect on the sardine stocks of Portugal.

Deep-water stocks and the great salinity anomaly

The trend towards cold northerly winds in winter over the coasts of western Europe turns out to have been part of a larger pattern of weather anomaly in the North Atlantic during the 1950s and 1960s (Dickson *et al.* 1975). When the average atmospheric pressure over Greenland during the period 1956–1965 was compared with the average for 1900–1939, it was found that there was a positive pressure anomaly over Greenland during the later period. It was particularly pronounced during winter months, when it amounted to an average difference of 7 mbar. The pressure ridge first formed in the early 1950s and persisted and intensified through to the late 1960s. It was accompanied by abnormally strong and cold northerly winds, especially in winter. Sea ice in the Greenland Sea expanded its range each year and reached a maximum in 1968.

One result of these changes was that abnormal amounts of polar water were brought south to join the East Greenland and East Icelandic currents so that they became cooler and fresher (Dickson *et al.* 1988b; Cushing 1988). The East Icelandic current, normally ice-free, began to transport and preserve drift ice and the East Greenland current increased greatly in volume. Furthermore, the surface salinities north of Iceland decreased below a critical value so that further cooling made them lighter and they did not mix with underlying layers. Below the surface ice there formed a layer of cold, fresher water about 200–300 m deep which was isolated from underlying layers. Although the effect was first seen in 1962, this water mass reached its minimum temperature in 1967 and minimum salinity in 1968. It retained its identifiable characteristics for nearly two decades, while making an enormous journey round the sub-Arctic gyre (Fig. 9.13). Christened by Cushing 'the great slug', it travelled round the southern tip of Greenland (1969–1970), south along the Labrador coast to the Grand Banks (1971–1972), turned eastward along the sub-Arctic front to pass the weather ship Charlie in mid-Atlantic in about 1974, and on to the northwest of the British Isles in 1975 and 1976. Minor portions of the slug are thought to have penetrated many parts of European coastal waters, but the signal rapidly became confused with local salinity variations. By 1977 the main anomaly reached the ocean weather ship Metro off the Norwegian coast and in 1978 and 1979 it was off the North Cape and to the west of Spitzbergen. Finally, although the evidence is incomplete, it appears that part of the same slug of water of anomalously low temperature and salinity entered the southward-flowing East Greenland current in 1981–1982. At about the same time these changes also appear (Lazier 1988)

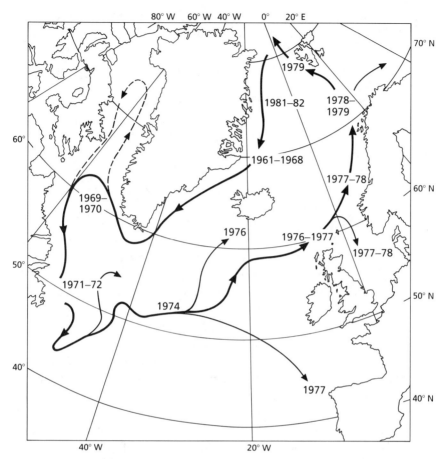

Fig. 9.13 Route followed by the great salinity anomaly, from its formation north of Iceland in 1961–1962 to its return to the east coast of Greenland in 1981. Modified from Dickson *et al.* (1988b).

to have found their way into the dense water which flows over Denmark Strait to form the bottom water in the Labrador Sea.

Naturally, the slug did not retain absolute integrity, its signal was reduced by mixing with adjacent waters as it travelled. Nevertheless, from an original water mass just over 2 °C cooler and 0.55‰ less saline when it left the north of Iceland in 1968, it was still 0.7 °C cooler and 0.11‰ less saline off the North Cape a decade later. It is possible to calculate that there was a 'salt deficit' of approximately 72×10^9 tonnes as it passed south along the Labrador coast, and a deficit of 47×10^9 tonnes as it passed through the Faroe–Shetland channel to the north of Britain (Dickson *et al.* 1988b). Since the slug originally approached Iceland from the Arctic Ocean, it is assumed that high latitude regions of the North Greenland Sea became correspondingly saltier. The total distance travelled between 1968 and 1979 is of the order of 10,500 km, which indicates a mean speed of about $3\,cm\ s^{-1}$, which is consistent with other estimates of the mean circulation speed of the subpolar gyre. What are the biological consequences of this major perturbation of the North Atlantic?

There were marked changes in the plankton community north of Iceland during the formation of the great salinity anomaly. Astthorsson *et al.* (1983) showed that on various sections off the north coast of Iceland the primary production measured at 10 m depth in spring averaged $2.6\,mg\ C\ m^{-3}\ h^{-1}$ during 1958–1964, but between 1965 and 1971 it was only $0.7\,mg\ C\ m^{-3}\ h^{-1}$. During the same periods the volume of zooplankton obtained from a standard net haul filtering 21 m^3 changed from 10 to 30 ml per haul to less than 5 ml. They speculated that the presence of the cold, low-salinity water caused a year-round stratification which prevented the renewal of nutrients in the surface layers prior to the spring bloom. This in turn led to greatly reduced primary production and consequently a greatly reduced build-up of zooplankton populations.

On the Grand Banks during the passage of the 'great slug' in 1973, the continuous plankton recorder network indicated that the biomass of phytoplankton was reduced to less than one-third of the average from the previous 13 years, and that similar reductions occurred in the biomasses of copepods and euphausids.

Cushing (1988) examined the spawning success of 15 stocks of fish whose breeding grounds were thought to lay in the path of the salinity anomaly. He did this by noting the distribution of year-classes in the catches. In eleven of those stocks, the year-classes spawned during the low salinity years were significantly below normal abundance. In Table 9.1 we see that Icelandic summer herring had anomalously low abundance from 1965 to 1971. A Wilcoxon rank test of 36 year-classes showed that those associated with the salinity anomaly were significantly different from those not so associated. Similarly, Icelandic spring herring had poor year-classes over an even longer period. The effect was visible in cod stocks of east and west Greenland, and the southern Grand Banks, but not in the stocks off Labrador and the northern Grand Banks. Similarly, many stocks in the northeast Atlantic and northeast Arctic showed the effect, but some did not, and the reason is not clear.

Low salinity of itself is not thought to cause mortality of young stages, but lowering of temperatures would extend the period of larval development and increase the risk of mortality by predation, etc. Cushing (1988) was inclined to look to reduced food supply as

Table 9.1 Wilcoxon rank tests of the significance of the difference in level of recruitment to year-classes produced during the passage of the great salinity anomaly, compared with normal years. From Cushing (1988).

Fish stock	Years of anomaly	Year-classes (No.)	p
Icelandic summer herring	1965–71	36	0.01
Icelandic spring herring	1962–71	45	0.01
E Greenland cod	1965–71	13	0.05
W Greenland cod	1969–72	15	0.01
Labrador, N Grand Banks cod	1971–73	20	n.s.
S Grand Banks cod	1971–73	22	0.05
W Scotland saithe	1974–78	16	0.01
North Sea saithe	1975–77	18	0.05
Faroe saithe	1975–77	18	0.01
Faroe plateau cod	1975–77	18	n.s.
Faroe plateau haddock	1975–77	18	n.s.
NE Arctic saithe	1978–81	19	n.s.
NE Arctic cod	1978–81	21	0.01
NE Arctic cod, first year-class	1978–81	15	n.s.
NE Arctic haddock	1978–81	22	0.01
NE Arctic haddock, first year-class	1978–81	16	n.s.
Blue whiting	1978–81	6	0.01

the chief cause of reduced breeding success, and in seeking a connection between the physical characteristics of the slug and the reduced plankton production, he emphasized the possibility that the cooler water would lead to a delay in onset of summer stratification and hence a delay in the onset of the spring bloom. He suggested that cool surface waters would be more subject to intermittent downward mixing by strong surface winds during the spring. In other words there would be an extended period of transient thermoclines, and this would have the effect of delaying the spring bloom and reducing the biomass of both phytoplankton and zooplankton during the growing season.

Assembling the relevant data to relate physical with biological events on such a large scale is a difficult task, and it is not surprising that the story has many loose ends. Nevertheless, the case for the formation of the great salinity anomaly, its journey of 10,500 km around the subarctic gyre, and its depressing effect on the recruitment of fish stocks in its path is a persuasive one.

The rise and fall of the west Greenland cod fishery

Most of the biological evidence reviewed so far in this chapter deals with the period of the 1960s and 1970s when the North Atlantic winter pressure index (Fig. 9.05) was predominantly negative, so that cold northerly winds were prevalent in winter in the northeast Atlantic and plankton productivity was depressed. If we now look at earlier periods in the record, we find that the index was mostly positive in the 1920s and 1930s, and again in the 1940s and 1950s. This implies (Fig. 9.04b) that westerly weather rather than northerly winds predominated over the northeast Atlantic and the climate was warmer. There are several patterns in the cod fisheries that appear to reflect these changes. Templeman (1972) developed indices of recruitment to the main cod stocks of the northwest and northeast Atlantic and found that for each region separately, and for the North Atlantic as a whole, recruitment tended to increase during the period 1940–1950 and to decrease during 1950–1970, thus apparently following the climatic trend.

The history of the cod fishery in Greenland indicates that the stocks spread northward along the West Greeland coast between 1917 and 1936 (Cushing 1982). From Julianehaab, near the southern tip of Greenland, the fishable stocks spread north about 14° latitude, until by 1936 they were being caught far above the Arctic circle. The changing pattern of the North Atlantic weather from the 1950s onward seemed not to affect them at first. There were record catches in the early 1960s, but by the end of the decade catches were at a low level, and have not so far recovered.

The story is given added interest by the fact that fish were tagged in West Greenland from 1924 onwards, and in the decade 1930–1940 significant numbers of tagged fish were found on the spawning grounds south of Iceland. When the West Greenland cod stock was at its most northerly distribution and the biomass of fish was approaching its maximum, large numbers of adults were returning to Iceland to breed. Although the tagging program continued in West Greenland in the 1950s and 1960s, very few tagged fish were recovered in Iceland after 1945. In 1963 developing cod eggs were traced in the Irminger current which flows from the southwest of Iceland, across the Davis Strait to Greenland. This led to the hypothesis that under favourable circumstances the life history of cod in this area was as follows (Fig. 9.14).

Of the cod eggs produced on the major spawning ground south of Iceland, a proportion were carried in the Irminger current to join the East Greenland current, which carried them to the southern tip of Greenland. Here the developing larvae and juveniles were carried north in the West Greenland current. In warm years these larvae and juveniles successfully grew to maturity as far north as Disko Bay and beyond, well into the Arctic circle. In cool years successful development was confined to the southern regions. At sexual maturity, the adults made a

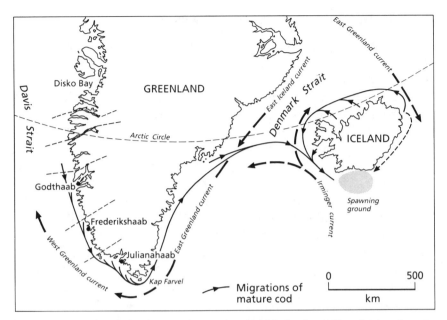

Fig. 9.14 Chart showing major ocean currents around Iceland and Greenland and possible migration routes of cod. From Cushing (1979).

return journey in excess of 1500 km to the spawning grounds south of Iceland. This pattern appeared to reach its peak of effectiveness in the 1940s, but when the weather patterns changed during the 1950s, fewer and fewer fish seemed to return from Greenland to the Icelandic spawning ground. Yet as late as 1963, eggs were being carried to East Greenland by the Irminger current.

At this point we may recall that the 1960s and 1970s were the time of strong northerly winds to the east of Greenland. Cushing (1982) suggested that these winds may have disrupted the drift of eggs, larvae and juvenile cod from Iceland to Greenland. This may have led to the final collapse of the West Greenland fishery in the late 1960s.

As in previous examples, one is struck by the large number of assumptions required to construct a coherent story. Nevertheless, the meteorological data and the fish catch data are well established facts. They show clearly the nature of interannual variability in both physical and biological systems. It remains to establish the connections between the two. Until we have more detailed information on, for example, the connection between the incidence of strong northerly winds and levels of plankton productivity, we can only invoke such understanding as we have of the mechanism of the spring bloom and its relationship to the formation of a shallow mixed layer, as discussed in Chapter 3.

9.5.3 Evidence from distributions of organisms

Let us recall atmospheric events in the first 25 years of this century. The North Atlantic oscillation pressure index was predominantly positive, so that the track of the westerlies across the Atlantic was well to the south and relatively warm air was brought to the shores of western Europe. It is probable that there was a corresponding warming of coastal waters, or movement of warm water masses to more northerly locations. Cushing (1982) has documented the northward movement of animals in the eastern Atlantic during this period. There were three main groups of events. First, an unusually large number of surface-living and pelagic subtropical animals were carried to western France and the British Isles. These included the Portuguese man-of-war (*Physalia physalis*), two species of goose barnacle (*Lepas*), loggerhead turtles (*Caretta*), and the octopus. Secondly, Atlantic, as opposed to Arctic, species of fish and bottom-living animals reached further north than they had previously been recorded. Sword-fish, pollack and ray appeared in Icelandic waters, cod and haddock appeared in the Barents Sea, and the hermit crab *Eupagurus*, together with three species of benthic molluscs appeared on the Murman coast for the first time. Thirdly, in waters off Nova Scotia there were, during this period, an unusually large number of new records of subtropical species (borne, presumably, by the Gulf Stream) but also of Arctic fish species. We have seen that when the North Atlantic oscillation pressure index is positive, winds tend to blow from a northerly direction over Labrador, and the weather is colder. The presence of Arctic species off Nova Scotia may reflect a cooling of coastal waters of eastern Canada, with perhaps a strengthening of the Labrador current.

All these events reached their peak in the period 1925–1935, which Cushing has dubbed 'the dramatic decade'.

9.5.4 Events in the English Channel: the Russell cycle

The end of the dramatic decade marked the beginning of a major change in the biota of the English Channel. Most of the observations on this change were made by staff of the Marine Laboratory at Plymouth, and the most important evidence was collected and summarized by Sir Frederick Russell (Russell 1973), so the event is known as the Russell cycle.

The story begins with a decline in recruitment to the herring stock in the western part of the English Channel, beginning in 1925. The last recorded year-class entered the fishery in 1931, and the Plymouth herring stock collapsed in 1936. Throughout this period, macro-plankton was being regularly monitored at International Station E1, off Plymouth, and samples were taken at the surface and near the sea bed for phosphate analysis. It had previously been established that two species of the arrow worm *Sagitta* tended to occur in different water masses and could be taken as indicators of those water masses. *S. setosa* characterized water that had been found to be resident in the English Channel while *S. elegans* characterized water occurring to the west of the channel, south of Ireland. In the autumn of 1931 the quantity of macroplankton declined by a factor of four, and *S. elegans* was replaced by *S. setosa*. During the next decade the number of non-clupeid fish larvae (flat-fish, cod family, etc.) declined drastically (Fig. 9.15), but the number of pilchard eggs increased. During the same period, the amount of phosphorus present in the water column during winter also declined. It seemed as if there had been some shift in a key environmental parameter that had caused a total change in the structure of the ecosystem. The old system had indicator organisms which it shared with a water mass to the west. It had abundant larvae of flat-fish and gadoids, and in the winter, when phytoplankton activity was minimal, it had a good supply of dissolved nutrients (as indexed by phosphorus). Suddenly, a persistent shift occurred, with different indicator organisms, less nutrients, less macro-plankton and far fewer fish larvae. The change was from one dominated by herring, macro-plankton and bottom-living fishes to one dominated by pilchard and small plankton.

For the next 25 years only a few comparable observations were made, but these indicated that numbers of non-clupeid fish larvae remained low. However, between 1965 and 1979 the situation reversed

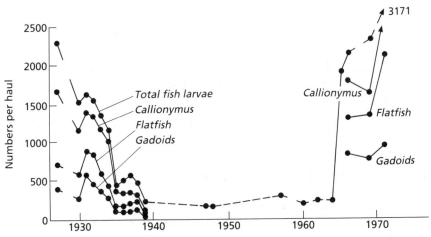

Fig. 9.15 Changes in the abundance of fish larvae in the English Channel 1926–1972. These changes are part of the Russell cycle. After Russell (1973).

itself. In 1965 the total number of fish larvae increased by an order of magnitude. Over the next 5 years the winter level of phosphorus rose to pre-1930 levels. In 1970 *S. elegans* came to predominate over *S. setosa* and by 1978 its abundance was back to earlier levels. During the 1970s the number of copepods returned to normal levels and the number of pilchard eggs decreased.

There were interesting changes along the shore lines during the same period. An unusual seaweed, *Laminaria ochroleuca* appeared in the Plymouth area in the late 1940s, spread widely in the south and west of England by 1960, but declined sharply between 1965 and 1975. The barnacle *Chthalamus stellatus* is better adapted to higher temperatures than *Balanus balanoides*. The warm-water species spread eastward in the English Channel during the 1930s and 1940s, but in the 1960s retreated quite sharply.

All of these events point to a tendency for the English Channel to have been invaded by more southerly species during the pre-1940 period of positive values for the North Atlantic oscillation index, and for these to have retreated during the period when the index became negative. The special feature of the Russell cycle is that it appears to have involved a major flip from one fairly stable configuration of organisms in the ecosystem to another quite different mix of species. In ecosystem theory it has long been recognized that multiple stable states may exist for a particular community of organisms, and that relatively small perturbations may cause a flip from one stable state to the other. The point of separation between the two states is referred to as a bifurcation. Discussion of these ideas is to be found in May and Oster (1976), Allen (1985) and Mann (1988).

A full understanding of the mechanisms of change involved in the Russell cycle still elude us. However, the English Channel is a well-studied area close to the zone of transition between the subtropical gyre and the subarctic gyre in the North Atlantic. When changes in the North Atlantic oscillation pressure index caused changes from colder, more northerly weather to warmer more westerly weather off the coast of western Europe, the biological community of the English Channel was induced to flip from one stable state to another.

9.6 SUMMARY FOR THE NORTH ATLANTIC

The North Atlantic oscillation winter pressure index (Fig. 9.05) enables one to divide the years of the 20th century into three main groups: (i) before 1935 the index was predominantly positive, so that the prevailing winds approaching western Europe were from the south and west, and were warmer than average; (ii) between 1935 and 1950 the index oscillated about the long-term mean; and (iii) from 1950 to 1965 the index was falling, and remained negative until 1971. This was a period of increasing cold northerly winds over the waters of the northeast Atlantic.

The biological events associated with the three periods are illustrated in Figs 9.16 and 9.17. Beginning about 1917 cod spread north along the coast of western Greenland, probably as a result of receiving a good supply of larvae from Iceland in the Irminger current. Soon afterwards, subtropical species began to appear off the coast of western Europe. The peak of these incidents was in the period 1925–1935, the 'dramatic decade'. In the mouth of the English Channel, herring were replaced by pilchard, and the numbers of young stages of gadoid fish and flat-fish dropped sharply. The whole balance of the plankton in the English channel shifted in the early 1930s, and remained changed for more than 30 years. The decade 1940–1950 saw a marked improvement in recruitment of cod in many parts of the North Atlantic, and between 1960 and 1970 the cod landings from West Greenland reached a high level, then collapsed.

It is interesting to note the lag between the onset of the cooling trend, in about 1950, and the downturn in biological events. Cod recruitment in the North Atlantic did not begin to decline until well into the 1950s, the pilchard-dominated community in the English Channel

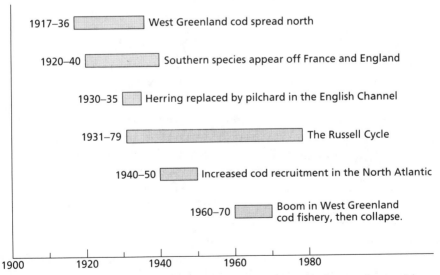

Fig. 9.16 Diagram summarizing biological events associated with the warming trend in the eastern North Atlantic, when the North Atlantic oscillation winter pressure index tended to be positive and the westerlies took a southern route.

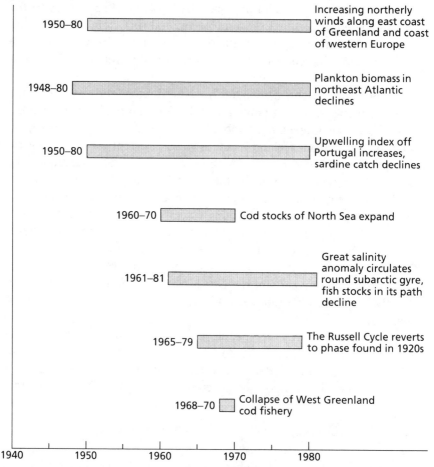

Fig. 9.17 Diagram summarizing biological events associated with the cooling trend in the eastern North Atlantic, when the NAO winter pressure index tended to be negative and northerly winds predominated.

did not begin to decline until about 1965, and the collapse of the West Greenland cod fishery took place between 1968 and 1970.

Once the cooling trend was well established, with weakening of westerlies and a marked increase in north wind along the east coast of Greenland and across to the coast of Western Europe, a well-documented decline in the biomass of phytoplankton and zooplankton was seen in the North Sea and the northeast Atlantic. Interestingly, the delay in the onset of the spring bloom which accompanied this decline seemed to be advantageous to young cod in the North Sea, for their numbers increased between 1960 and 1970.

The northerly winds blowing down the east coast of Greenland created a major slug of cold, low salinity water which travelled all the way round the subarctic gyre, taking about 15 years. Most fish stocks in its path showed poor recruitment while it was present, possibly because the cold water delayed the surface warming required to establish the spring bloom of plankton.

In about 1980 the phytoplankton and zooplankton stocks of the northeast Atlantic began an upward trend in biomass that seems to have been maintained.

There is a long history of biological observations off the coasts of western Europe, and there is little doubt that many of the events summarized here are linked to large-scale changes in weather patterns. However, these patterns are not as clearly delineated as the southern oscillation in the Pacific Ocean, so the links between physical and biological phenomena in the Atlantic are rather more tenuous, and much remains to be done to establish the mechanisms linking the two.

10

The Oceans and Global Climate Change: Physical and Biological Aspects

10.1 INTRODUCTION

The oceans have a deep circulation, the thermohaline circulation, which we have so far mentioned very little. Water is heated in equatorial regions then moves polewards in major currents, giving off heat to the atmosphere. In sub-Arctic regions cooling and ice formation cause water to become more dense and it sinks to form the 'deep water'. This sinking is the beginning of a long journey close to the ocean floor. Some of the deep water travels south in the Atlantic basin, moves across to the Pacific basin and there moves slowly northward in a journey that may take a thousand years.

At the regions of deep water formation large quantities of carbon dioxide dissolved in the water sink to great depth and are removed from contact with the atmosphere. Conversely, at regions of upwelling, especially the large upwellings at the tropical divergence, heating of the cold upwelled water causes it to give off billions of tonnes of carbon dioxide. These are the major physical mechanisms by which the oceans exchange

384

carbon dioxide with the atmosphere, and there is no particular reason to think that they are not approximately in balance. However, there are in addition important biological processes that remove carbon dioxide from the atmosphere and transfer it to the deep ocean. Over 99% of the carbon dioxide added to the earth's atmosphere throughout its history has been taken up by phytoplankton and sedimented to the sea floor to form the calcareous rocks and the fossil fuels. This biological mechanism is known as the biological pump.

For at least 150 years the carbon dioxide concentration of the atmosphere has been rising as a result of man's activities in cutting down the forests and burning fossil fuel. It is now expected that there will be a global rise in atmospheric temperature — the 'greenhouse' effect. Before the magnitude of this important change can be predicted, it will be necessary to understand the extent to which the oceans may be a reservoir for excess carbon dioxide. In this chapter we seek first of all to describe the mechanism of global warming and the present day global carbon cycle. We shall then explore the relative importance of oceanic sources and sinks compared with industrial activities and terrestrial biota. Finally, we shall attempt to peer into the future, and consider what changes might be expected to occur in physical and biological mechanisms for circulating carbon in the ocean, if the expected rise in global atmospheric temperature occurs.

10.2 PHYSICAL ASPECTS

10.2.1 The greenhouse effect

(a) The process

The average surface temperature of the earth and the lower atmosphere is about 15 °C. But if there were no water vapour, carbon dioxide or methane in the atmosphere the surface temperature would be below freezing by $\simeq 18$ °C and all the rivers, lakes and oceans would be frozen solid. The reason for the higher, more habitable, temperature is the fact that these greenhouse gases delay heat from leaving the earth by trapping it in the lower atmosphere.

As pointed out in Chapter 3 all the heat received on the earth comes originally from the sun's surface ($\simeq 6000$ °C) via electromagnetic radiation with wavelengths between 0.2 and 4.0 μm, often called the shortwave radiation. Of this incoming radiation $\simeq 31\%$ is reflected back into space, $\simeq 23\%$ is absorbed by the ozone, water vapour, clouds and dust in the atmosphere, and 46% is absorbed by the land and water at the earth's surface (Mitchell 1989). All these absorbers in turn radiate heat in the form of electromagnetic radiation but the wavelength of this radiation is much longer than the incoming radiation in accordance with Planck's radiation law which dictates that cooler bodies radiate at longer wavelengths. The radiation from the surface of the earth and the atmosphere is thus at wavelengths between 5 and 100 μm, the long-wave radiation.

The atmosphere is quite transparent to the short-wave radiation from the sun as indicated above by the fact that only 23% of the incoming radiation is absorbed by the atmosphere. The long-wave radiation is another story. Roughly 90% of the long-wave radiation leaving the earth's surface is absorbed in the atmosphere by the greenhouse gases. This heat eventually reaches the upper layers of the lower atmosphere through convection and is lost to outer space, but the absorption by the greenhouse gases delays the loss and keeps the lower atmosphere warmer than if the atmosphere were transparent to the long-wave radiation. The amount of heat trapped and the resulting temperature of the atmosphere clearly varies directly with the concentrations of these gases. If they are very concentrated, as on the planet Venus, the temperature is very high (+ 400 °C) and if they are at low concentrations, as on Mars, the temperature is very low (– 50 °C).

On the earth there are seven or so gases which contribute to warming the atmosphere, ranging from the naturally occurring carbon dioxide, water vapour, methane, ozone and nitrous oxide to the human-produced chlorofluorocarbons (Freon). The warming effect of each of these gases is different because their concentrations are different and because they absorb radiation with different efficiencies at different wavelengths. At present, about 65% of the warming effect is due to water vapour and 32% to carbon dioxide. Between them they absorb almost all the long-wave radiation with wavelengths longer than 15 μm and shorter than 8 μm. The other gases each contribute $\simeq$1% or less to the total effect but an increase of one molecule of chlorofluorocarbon is about 10,000 times more effective at trapping heat than one molecule of carbon dioxide (Mitchell 1989). This difference arises because the chlorofluorocarbons strongly absorb energy in the 'window' between 8 and 12 μm where water vapour and carbon dioxide do not absorb. A small increase in the chlorofluorocarbons thus makes a big difference in the total absorption because the band is still full of unabsorbed radiation whereas a similar increase in carbon dioxide can only produce a small increase in absorption in a band where most of the radiation is already absorbed.

For the future of life on the earth, the important point is that the concentrations of these gases in the atmosphere are slowly increasing. Carbon dioxide, for example, has been monitored since 1958 and these data along with determinations of the concentrations in the 'old' air trapped in the polar ice sheets clearly indicate that the concentration in the atmosphere has increased by about 25% since 1850 (Schneider 1989). Most of this increase is due to the burning of wood, coal, oil and gas. Deforestation in the northern hemisphere also plays an important role as the reduced biomass takes up less carbon dioxide from the atmosphere and the decomposing litter gives off carbon dioxide. It is many years before new growth begins to take up carbon dioxide faster than the old forest. The trace gases: methane, nitrous oxide, etc., have also been increasing. Although their contribution to the total budget is small, at the present time they together produce about the same *increase* in the greenhouse effect as does the increased level of carbon dioxide (Hansen *et al.* 1988). These well documented increases raise the question which is today of great concern throughout the world, that is, what effects are these increases having on the world's temperature, precipitation, ice cover, biological processes, etc,?

(b) Trends in air temperature

Finding the answers, which is of course not easy, involves two main parts; first the monitoring of the important variables to determine if and how they are changing with time and second the construction of a sophisticated model of the climate to determine which processes are causing the observed changes. The main variable to monitor is the surface air temperature which has been measured accurately throughout the past 150 years over land and sea. But as can be imagined the temperature records are very uneven in quality and coverage and creating a useful data set for the necessary length of time is prone to some unusual problems. For instance, the air in a city can be $\simeq 0.2\,^\circ$C higher because of the activity in the city and the shape of the buildings. This 'heat island' effect must be removed to get a true picture of the long-term temperature trends (Hansen and Lebedeff 1988). At sea other problems arise. Folland *et al.* (1984) analysed the available sea surface temperature data but only used the night-time values because the sun's warming of the ships during the day produces falsely-high values. They also point out that the temperatures obtained from ships during the second world war appear to be high because the thermometers had to be taken inside the ship to be read, as no lights were allowed on the open deck. These and other spurious signals have to be carefully weeded out of the record as they can create systematic errors that are as big as the signal that is being sought.

A clean data set of the world's temperature is now available and updated analyses appear regularly. Two of these, Jones *et al.* (1988) and

Hansen and Lebedeff (1988), are reviewed by Schneider (1989) (Fig. 10.01). Each analysis uses a different technique but both use roughly the same raw data except that Jones *et al.* (1988) include temperatures over the sea. The two graphs agree in that they both indicate that the temperature of the earth's surface as a whole has increased by $\simeq 0.5$ °C over

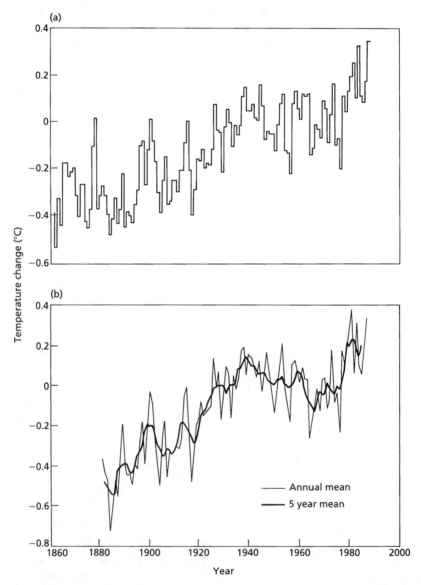

Fig. 10.01 Two analyses of the surface temperature of the earth over the past 100 years relative to the average temperature between 1951 and 1981, adapted from Schneider (1989). (a) from Jones *et al.* (1988); includes temperatures from continents, islands and the sea surface; and (b) from Hansen and Lebedeff (1988); approximately the same data set as in (a) except no sea surface temperatures are included.

the past 100 years with the decade of the 1980s being the warmest on record. For a long time, the existence of a temperature increase has been regarded as unproven, but the evidence becomes more convincing with each new analysis. Even now, the cause of the rise is uncertain. Some invoke a change in solar radiation (the solar constant) but none doubt that increasing atmospheric carbon dioxide eventually leads to a rise in temperature at the earth's surface, so the most probable cause of the present rise in temperature is the observed increase in atmospheric carbon dioxide.

As indicated above, it has been possible to determine from direct measurement the increases in the concentrations of the greenhouse gases. If their concentrations are known it is a relatively straightforward calculation to determine the consequent atmospheric heating rates (Mitchell 1989). The difficult thing to determine is the temperature change due to the changed heating rate. This is because the atmosphere and/or ocean may react to the increased heating with processes which produce positive or negative feedback. For example, the content of water vapour in the atmosphere is usually assumed to remain in equilibrium with the oceans and varies with the air temperature. If the atmosphere warms up because of an increase in the greenhouse gases the amount of water vapour in the air will automatically increase because of the higher temperature. This leads to a further increase in the absorption of long-wave radiation and causes a further increase in temperature; a positive feedback. Another positive feedback is related to the snow and ice cover over land. If the atmosphere warms up, the area of the earth's surface covered with snow and ice decreases while the area of soil and water increases. These darker-coloured surfaces absorb more of the sun's heat than the white snow and ice, causing a further increase in temperature. On the other hand, an increase in temperature and water vapour results in more clouds. This has two opposing effects. The clouds reflect back into space a proportion of the incoming solar radiation, resulting in lower temperatures. The clouds also cut down the amount of back-radiation escaping from the earth to outer space. This effect is particularly prominent at night, and has a warming effect. It is the total effect of all such processes that results in the observed temperature variations. One way of integrating all the processes is by mathematical models.

(c) Climate models and predictions

The climate models are computer programs in which the atmosphere is represented by a three-dimensional grid (see Mitchell 1989, for review). One cell in the grid typically represents 500–1000 km in the east–west and north–south directions and 10–20 km in the vertical. At the grid points forming corners of each cell variables such as temperature, humidity, pressure and wind are specified by the average conditions thought to exist at some time in the past. Changes in these variables over a short time interval (hours) are then calculated according to the equations governing thermodynamics, momentum balances and mass conservation. The calculated changes are then used to update the values at the grid points. The process continues until the program has modelled the required time interval under study (decades).

The aim is to create a model sophisticated enough to accurately predict the consequences of the increases in the greenhouse gases. It is improved by comparing the results to the observations until it can predict the observed changes accurately, after which it can be used to estimate future trends with some confidence. This of course is no different than the usual scientific method of comparing observations to a model, but the climate system of the earth is too complicated to be modelled with a few linear equations and it is only with the largest and fastest computers that even rudimentary models are possible. Even then, when variables have been changed to make a model fit observations, it is often impossible to know whether the right variables have been adjusted.

All the recent models (Hansen *et al.* 1988; Mitchell 1989) predict that the increases in the greenhouse gases will cause the temperature of the earth to rise. The amount varies with the model but lies in the range of 1.5–5.5 °C for a doubling of the concentration of carbon dioxide. Such a doubling is expected, if the present growth rates continue, in about the year 2030. The effects of such a temperature rise and the accuracy of the models are now both widely discussed topics. There are many, for example, who believe that the model predictions are far too high and that the models contain too many uncertainties to produce accurate values. Some (Roberts 1989) believe that the expected decrease in the energy output from the sun over the next century will counteract the warming due to the increased greenhouse effect. Others (Kerr 1989) believe that the models are missing too many of the important negative feedback processes which would serve to counter the warming.

The effects of a world-wide increase in temperature will depend of course on the size of the increase and will be beneficial for some areas and processes and disastrous for others. If the water in the oceans increases in temperature it will expand and cause a rise in sea level which will be added to the rise in sea level caused by the partial melting of the polar ice caps. In the extreme case the oceans will rise by a metre or so and inundate heavily populated low-lying areas which also tend to be rich agricultural areas. Millions of people will be affected. The higher temperatures are also expected to change patterns of agriculture as some areas will become too hot or too dry to grow the present crops while other areas which are now too cold become warm enough to produce crops.

One of the biggest problems with the present climate models is the fact that the oceans are not completely included. This is partly because the circulation of the oceans is less well known than the circulation of the atmosphere and partly because the computers are not yet big enough to include both the atmosphere and the oceans in a coupled model. A typical method of including the oceans in the models (Hansen *et al.* 1988) is to assume the oceans do not change appreciably during the time period being modelled. Thus the oceanic heat transport, the seasonal changes in temperature, the mixed layer depth, etc., are assumed to remain the same for each year. The global climate models therefore are not yet capable of predicting or incorporating large-scale sea-surface temperature anomalies as were found to be so important in the discussion of the El Niño in Chapter 9.

In the next sections our purpose is to first indicate why the oceans are important in the global carbon dioxide budget and then describe the deep ocean currents which are responsible for moving the carbon dioxide from the surface layers down into the deeper layers of the various oceans.

10.2.2 The carbon cycle

The main source of carbon throughout earth's history has been in the form of carbon dioxide ejected into the atmosphere by volcanoes. Revelle (1982, 1983) estimates that over time volcanoes have added $\simeq 50 \times 10^6$ giga tonnes (1 giga tonne = 1 Gt = $10^9 \times 10^3$ kg = 10^{12} kg) of carbon dioxide to the atmosphere and add a further 0.04 Gt y^{-1} while the continued burning of fossil fuel adds about 5.3 Gt y^{-1}, Fig. 10.02. Some of the carbon dioxide in the atmosphere becomes incorporated into the structure of terrestrial plants and some enters the ocean by dissolving in the surface layers. At present the atmosphere holds about 720 Gt and has a cycle of $\simeq 10$ Gt y^{-1} reflecting the summertime transfer of CO_2 from the atmosphere to the land plants by photosynthesis, and the wintertime transfer back to the atmosphere through respiration and decomposition. The terrestrial plants and the soil contain about 2000 Gt, the mixed layer of the ocean (0–80 m) 2300 Gt and the deep ocean 35,000 Gt. The rocks of the world that have been formed from ocean sediments, together with present sediments contain about

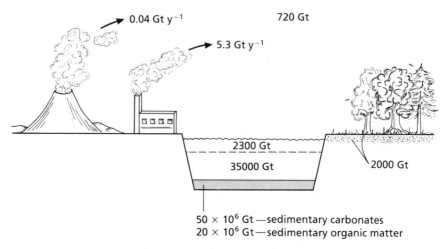

$0.04\ Gt\ y^{-1}$ 720 Gt

$5.3\ Gt\ y^{-1}$

2300 Gt

35000 Gt

2000 Gt

$50 \times 10^6\ Gt$ —sedimentary carbonates
$20 \times 10^6\ Gt$ —sedimentary organic matter

Fig. 10.02 Quantities of carbon in the various reservoirs of the world and the annual flux due to volcanoes and burning of fossil fuels. Units are giga tonnes.

2000 times the total amount in all these reservoirs or 70,000,000 Gt. Roughly 50,000,000 Gt of these sediments is in the form of carbonates and 20,000,000 Gt is organic matter of which 5000 Gt is in the form of recoverable fossil fuels. The sediments therefore contain the vast majority (99.95%) of all the carbon ever issued into the atmosphere by volcanoes. If there were no oceans to form sediments the concentration of carbon dioxide in the atmosphere would be far higher than it is and the average temperature of the earth would be closer to that found on Venus which does have a high concentration of CO_2 and where it is too hot for water to remain liquid.

From direct measurements it is observed that the concentration of CO_2 in the atmosphere is now about 350 ppm and increasing by about 0.35% or 1.2 ppm y^{-1} (Takahashi 1989). In absolute terms this is about 2.1 Gt y^{-1} which is less than half the estimated 5.3 Gt that is being added each year to the atmosphere through human activities. The fate of the missing CO_2 is not yet fully understood. Revelle (1983) argues that the increasing concentration of CO_2 leads to increased photosynthesis and therefore increases carbon storage in plants and Takahashi (1989) suggests that the extra CO_2 is taken up partly by an increase in the size of the northern forests and partly by an increase in biological activity in the oceans.

Since the oceans and their sedimentary rocks contain almost all the CO_2 ever issued into the atmosphere and they possibly absorb a significant fraction of the additional CO_2 being put into the atmosphere each year it is obvious that the processes controlling the disposition of the gas are vital to understanding the climate system of the world.

Carbon dioxide enters the ocean by dissolving in the surface waters of the ocean at a rate determined by the difference in the partial pressure of the gas in the air and the water. Once in the ocean only a small amount of the gas remains in the dissolved form as most of it reacts with the water to form carbonic acid (H_2CO_3), bicarbonate ions (HCO_3^-) and carbonate ions (CO_3^{-2}) (Takahashi 1989). In the surface layer the carbon is incorporated into organic compounds, skeletons and shells, especially during the spring bloom, but when the organisms die the remains sink. Some decompose, releasing the carbon back into the water column, but some are buried in sediments. This flux of carbon out of the euphotic zone into the deeper layers is called the 'biological carbon pump' and is explored in more detail in section 10.3. The result of the downward flux is an increase of the concentration of carbon dioxide from $\simeq 2.0$ nmol kg^{-1} (n = nano = 10^{-9}) in the surface layer to about 2.2–2.4 nmol kg^{-1} at 1000 m (Fig. 10.03). From 1000 m to the bottom the concentration increases or decreases only slightly. As is evident in Fig. 10.03 the concentration of CO_2 below 1000 m is about 10%

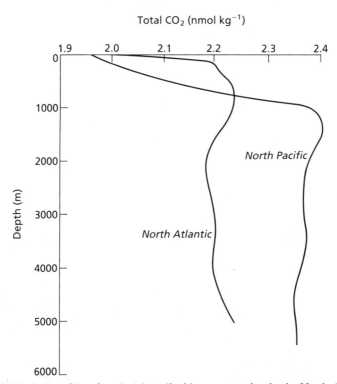

Fig. 10.03 Vertical profiles of total carbon dioxide concentration in the North Atlantic at GEOSECS station 37 (12° N 51° W) and in the North Pacific at GEOSECS station 214 (32° N 177° W), adapted from the GEOSECS atlases Vols 2 an 4 (1981), respectively.

higher in the North Pacific Ocean than in the North Atlantic. This difference is due to the differences in the rates of the deep circulation between oceans. Deep water passes more rapidly through the Atlantic than the Pacific (Takahashi 1989). This gives the deep water in the Pacific Ocean more time to accumulate the CO_2 being pumped down from the upper layer. As indicated in the next section the deep water originates in the North Atlantic and works its way into the Pacific which suggests that the difference in the CO_2 concentrations between the two oceans of $\simeq 0.2$ nmol kg^{-1} indicates an increase above the base concentration of 2.2 nmol kg^{-1} contained in the North Atlantic deep water when it begins its journey. The formation and flow of these deep waters is part of the thermohaline circulation as distinct from the wind-driven circulation.

10.2.3 The deep thermohaline circulation

The fact that the deep and bottom waters of the oceans are colder than the near surface waters is now common knowledge but when it was first discovered in 1791 the implications were far from understood (Warren 1981). Scientists at that time considered the ocean to be a relatively stagnant body of water in which the temperature at any location was in equilibrium with the atmosphere. In such an ocean heat absorbed from the sun would slowly diffuse downward into the deep waters which would become much warmer than they are observed to be. The cold water, it was finally realized, must originate in the cold regions of the world and sink to the bottom under the warm water. This process was first suggested by Count von Rumford in 1797 as a consequence of laboratory experiments on thermally-driven convection in fluids.

Rumford and subsequent authors envisaged large convectively-driven circulations in which the surface waters in the polar regions cooled to a high enough density to sink to the bottom. This new deep water would then flow away from the high latitudes and fill the deep basins of the oceans. As more cold water was added, the deep layer would become thicker, thus pushing cold water up towards the surface throughout the warmer regions of the oceans. The rate of this deep water upwelling has been calculated to be about 10^{-5} cm s^{-1} which is too small to be measured, but high enough to have important effects on the structure of the ocean.

When the upward-moving cold water reaches a sufficiently shallow depth it is warmed by the downward diffusion of heat from the surface layer. Over the long term, a balance is achieved in which the downward diffusion of heat equals the rate that heat is being moved back up by the upward motion in the water. This advection–diffusion balance in the vertical flow of heat leads to the ocean being divided into a warm upper layer and a cold lower layer separated by the permanent thermocline (as distinct from the seasonal thermocline). The rising deep water slowly becomes part of the warm upper layer and eventually flows back towards the poles where it is again cooled to renew the deep water and complete the convection cell. This convection circulation, because it depends on changes in density related to changes in temperature and/or changes in salinity, is called the thermohaline circulation. It exists alongside and in addition to the wind-driven circulation which interestingly is more or less confined to the upper layer of the ocean (Chapter 8) created by the thermohaline circulation (Gill 1982).

The reasons for the existence of the thermohaline circulation are not, however, as straightforward as originally perceived. The temperature decrease in the water between the equator and the polar regions creates horizontal density and pressure gradients. In an idealized world without irregularly shaped continents and ocean basins the rotation of the earth would simply lead to geostrophically balanced eastward currents perpendicular to the poleward pressure gradient. In such a system, any decrease in the temperature of polar water or increase in the equatorial water would increase the pressure gradient and increase the speed of the eastward flow. It would not lead to polar water sinking under the equatorial water. The formation and sinking of dense waters occurs in particular locations where the conditions are quite different from those in the ideal world.

Because much of the deep water is formed near the surface in high latitudes, the formation of ice is often an important feature of the process. When ice is formed, there is a partitioning between the water and the salt. The ice is made from the water; the salt and other impurities are excluded from the ice and trapped in small channels or cavities within the ice. With time the salt slowly drains out of these channels into the water below and thereby increases the density of the underlying water.

One of the regions where dense water is formed is over the continental shelf around Antarctica especially in the Weddell Sea. This sea is situated from 15 to 60° W and from 65 to 75° S, bounded on the west by the Antarctic Peninsula and on the east by Queen Maud Land. In the winter, new ice is formed on the western side of the sea. As we have seen, salt is excluded from the ice and added to the water below. Prevailing winds blow the ice offshore, leaving behind the more saline water. Thus the water flowing into the sea from the east is both cooled and made more saline by the formation of ice. The resulting water has a salinity of about 34.65 and a temperature of $-0.7\ °C$ and flows off the shelf down to the bottom of the ocean. This along with similar processes on the other shelves around Antarctica produces $2–5 \times 10^6\ m^3\ s^{-1}$ (Warren 1981) of dense water which subsequently spreads northward at the bottom of the Atlantic, Pacific and Indian Oceans as the Antarctic bottom water.

Another important source of dense water is found in the Greenland and Norwegian Seas where relatively saline water from the North Atlantic current is cooled to reach a density high enough to form bottom water. Clarke *et al.* (1990) describe the processes in detail and point out that one of the important features is the fact that the water circulates in a large well-defined cyclonic gyre in which the warmer saltier water from the Norwegian current and the colder fresher waters from the East Greenland current are mixed together to create the water which when cooled will form the denser deep water. Within this cyclonic flow, which is partly forced by the bathymetry, the water is confined to remain in a region of intense cooling for a considerable length of time. Consequently the water continues to lose heat and gets progressively denser, with the result that the upper mixed layer becomes deeper and deeper until it reaches the bottom. The role of ice, if it is important in this area, is different than in the Weddell Sea. The extraction of fresh water to increase the salinity of the underlying water is not important because there is no net gain of salt in the formation area. This is because the amount of salt added to the water from the freezing process is balanced by the amount of fresh water added by melting ice. If ice is important Clarke *et al.* (1990) suggest it will be due to the leads between the ice floes. This is where the heat loss from the water is greatest and where the convection will begin. The spacing between the leads may therefore control the spacing of the convection cells.

The dense water formed in the Greenland and Norwegian Seas is important to the world's oceans because after it flows into the North Atlantic over the sills between Iceland and Scotland it forms one of the main components of the North Atlantic deep water (Harvey and Theodorou 1986). Similar processes in the sea to the north of Denmark Strait, between Greenland and Iceland, produce a slightly different form of dense water which flows over the sill in Denmark Strait to form the bottom water in the Labrador Sea (Swift *et al.* 1980).

There are other areas in the ocean where deep penetrative convection occurs in winter to form homogeneous dense water but the density is not generally as high as the water formed in the polar seas. In the Labrador Sea for instance there exists a cyclonic circulation, as in the Norwegian Sea, bringing together the relatively warm saline water from the Irminger current and the cold and fresher water from the Labrador current. The slow circular flow maintains the water in the region while it attains the properties that are just right to lead to deep convection in the depth of a cold winter. The depth of this convection has been observed to 2000 m so it does not form bottom water but remains at an intermediate depth. The depth of the convection in this region is also very dependent on the atmospheric conditions, and these can be very variable from one year to the next (Lazier 1980). When the winter winds are from the northwest they are exceptionally cold and the convection is the deepest, but in years when the wind patterns cause warmer air to persistently flow over the area the convection is quite shallow. Intermediate water masses are also formed in the northwest Pacific Ocean but no deep or bottom water is formed there (Warren 1983).

The densest water having a major influence on any of the major oceans is formed in the Mediterranean Sea and flows into the Atlantic through the Straits of Gibraltar (Armi and Farmer 1988). This water differs from the other dense waters because it is formed primarily by evaporation and the consequent increase in salinity, rather than by the decrease in temperature found in the dense polar waters. One of the interesting things about the Mediterranean water is that even though it is the densest water flowing into the Atlantic, having a σ_t of $\simeq 29.0$ as compared to 27.8 for the Antarctic bottom water, it does not descend to the bottom of the ocean. Killworth (1977) pointed out that this is because the relatively warm Mediterranean water at 12–13 °C is less compressible than the cold (-0.7 °C) Antarctic bottom water. The more highly compressible cold water becomes even more dense than the surrounding water as it descends towards the bottom because of the increase in pressure. The less compressible 12 °C water does not increase in density as rapidly as the cold water as it descends to higher pressures, so it becomes entrained into the surrounding water more readily than the Antarctic bottom water.

The movement of the various intermediate, deep and bottom waters is detected by tracking anomalous properties, and this is known as the core layer method of water mass analysis. The dense water which flows off the shelves around Antarctica, for example, is followed by its characteristic low temperature. The North Atlantic deep water is traced by its high salinity and high dissolved oxygen content relative to the layers above and below. A suggestion of some of these water masses and their supposed flows is illustrated in Fig. 10.04. In the north the section begins in the Norwegian Sea and the flow of Norwegian Sea water into the North Atlantic over the Iceland–Faeroe rise. This is one of the main

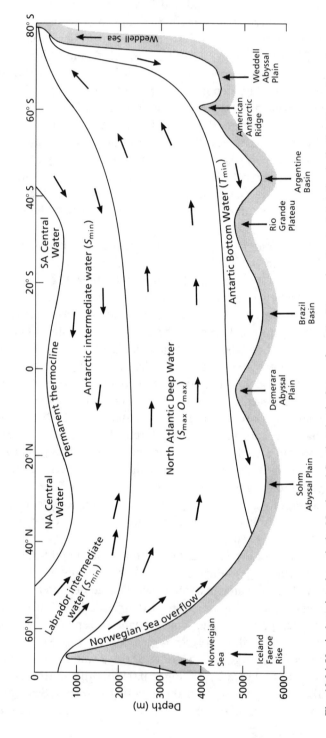

Fig. 10.04 Vertical section through the North and South Atlantic Oceans from the Norwegian Sea to Antarctica indicating the major water masses, their principal characteristics and their directions of flow. S_{min} = low salinity, S_{max} = high salinity, O_{max} = high oxygen content, T_{min} = low temperature. Adapted from Dietrich et al. (1980).

components of the North Atlantic deep water which flows into the western basin of the ocean through the Charlie Gibbs fracture zone in the mid-Atlantic ridge at 53° N. South of this point the diagram illustrates distributions west of the ridge.

The central waters of the subtropical gyres are outlined by the base of the permanent thermocline. Its depth is shallower at the equator than in the centres of the gyres and significantly deeper in the North Atlantic than in the South Atlantic. Beneath the permanent thermocline lie the intermediate waters which are formed in the high latitudes of both oceans. The Antarctic intermediate water renewed around Antarctica occupies a much larger volume than does the Labrador Sea water formed between Canada and Greenland in the Labrador Sea. The Mediterranean water (not shown) exhibits a strong salinity maximum at intermediate depths especially on the eastern side of the mid-Atlantic ridge. Below the intermediate waters are the deep and bottom waters. The principal deep water is the North Atlantic deep water characterized by relatively high salinity and oxygen values. It is observed throughout the length of the Atlantic Ocean between $\simeq 2000$ m and $\simeq 4500$ m. The water has also been shown to provide most of the deep water in the Indian and Pacific Oceans. The largest volume of bottom water originates on the shelves around Antarctica and flows north in all the major ocean basins. Its northern limit in the Atlantic is about 40° N where it gives way to the Denmark Strait overflow water (not shown) which lies at the bottom in the Labrador Basin.

The flow of water depicted in Fig. 10.04 assumes that the north–south convection cells driven by the thermohaline circulation are broad ocean-wide features. A north–south section at any other longitude should show the same pattern of flow. There are however good dynamical reasons to doubt that such a simple pattern is realistic. As pointed out by Stommel (1958) and discussed by Warren (1981) the deep slow flows associated with the thermohaline circulation must, with very little error, be in geostrophic balance. Also, because the deep waters are being supplied continuously, there must be a slow upward movement in the upper layers of the ocean. This upward velocity is like the Ekman pumping we discussed in Chapter 8 and the balance between the vertical motion and the north south–flow is the same as described by Eq. 8.10 in association with the discussion of the Ekman pumping and the Sverdrup flow. That equation is:

$$V_S = \frac{f}{\beta H} W_E, \qquad\qquad (10.01)$$

where β is the rate of change with latitude in the Coriolis parameter f, and V_S and W_E are the north–south and vertical velocities respectively. In Chapter 8 the vertical velocity W_E was the Ekman pumping and was downward or negative leading to a negative or equatorward Sverdrup velocity V_S. In the present case the vertical velocity is upward or positive leading to a positive or poleward interior flow. This strange result leads to the conclusion that the deep waters which are formed in the polar regions are flowing towards their sources in the interior of the oceans. Because of this restriction Stommel (1958) postulated that the deep waters must flow from the north to the south or from south to north in the case of the Antarctic bottom water in narrow western boundary currents much like those associated with the wind-driven circulation in the upper layer. These boundary currents have indeed been found and Warren (1981) gives a good description of them.

The dense waters thus flow away from their sources in narrow western boundary currents and move laterally into the interior of the oceans through eddy diffusion. An illustration of the flow of one of the most important deep waters, that is the flow of the North Atlantic deep water (NADW) is illustrated in Fig. 10.05 (Warren 1981; Gordon 1986). The flow begins at the sill between Iceland and Scotland which separates the North Atlantic from the Norwegian Sea. The overflow water mixes with the resident deep water in the eastern basin of the North Atlantic then moves into the western basin of the North Atlantic through the gap in the mid-Atlantic ridge at $53°$ N known as the Charlie Gibbs fracture zone. The flow continues in an anti-clockwise fashion around the Irminger and Labrador Seas and then moves south towards the equator and the South Atlantic Ocean along the western boundary of the ocean. In the southern ocean the NADW is incorporated into the eastward flowing Antarctic circumpolar current from which it makes its way into the Indian and Pacific Oceans. The water enters the remainder of each ocean basin from the western boundary currents by eddy diffusion. The regions where convection renews the deep and intermediate waters of the Norwegian and Labrador Seas are indicated in Fig. 10.05 by the circled crosses. Water must slowly upwell throughout all the ocean basins to provide for the continuous supply of deep and bottom water.

The return flow to the North Atlantic starts in the western Pacific Ocean north of Australia (Gordon 1986) and follows the path indicated by open arrows in Fig. 10.05. The flow out of the Pacific into the Indian Ocean transports about $8.5 \times 10^6 \, m^3 \, s^{-3}$ and continues across the southern Indian Ocean into the Aguhlas current reaching the South Atlantic by flowing round the southern tip of Africa. At this point the transport

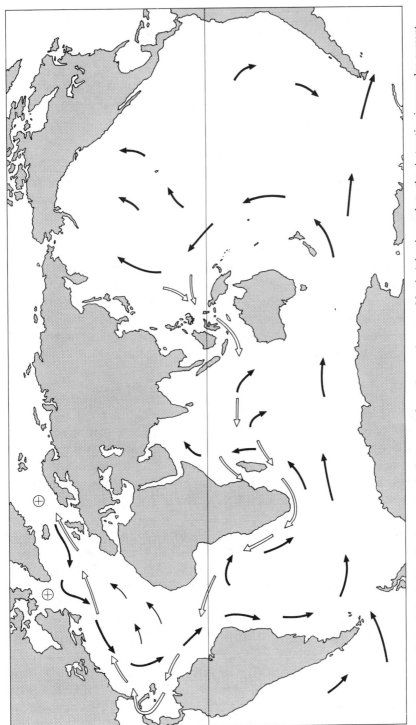

Fig. 10.05 The flow of North Atlantic deep water (solid arrows) from the northeastern North Atlantic south to the Antarctic circumpolar current thence to the Indian and Pacific Oceans. The return flow (open arrows), after Gordon (1986), is in the thermocline and mixed layers. Slow upwelling of deep water is assumed to occur where the permanent thermocline exists.

is $\simeq 13.5 \times 10^6 \, \text{m}^3 \, \text{s}^{-3}$. The flow then becomes part of the equatorial current system in the Atlantic and eventually joins the Gulf Stream system which delivers $\simeq 18.5 \times 10^6 \, \text{m}^3 \, \text{s}^{-1}$ of upper layer water back to the Norwegian Sea ready to be turned into deep water. This apparently continuous circulation connecting the deep and surface flows throughout the world's oceans is sometimes loosely referred to and depicted as a conveyor belt (Steele 1989).

10.3 THE BIOLOGICAL PUMP

Given the existence of a very large, slowly moving deep ocean circulation that may take a thousand years to traverse the major oceans, we need to look at the mechanisms that will take carbon from the atmosphere and transfer it either to the sea floor or to the deep parts of the ocean, for in either event it will be unable to return to the atmosphere for hundreds of years. We saw earlier that, at the limited areas that are sites of deep water formation, carbon dioxide dissolved in the water is carried to depth. We will now examine mechanisms operating over the entire surface of the oceans by which carbon is fixed in photosynthesis and introduced into the deep ocean by the sinking of dead organisms. It is called the biological pump.

10.3.1 The magnitude of carbon fixation in organic matter

The biological fixation of carbon is taking place continuously in the world ocean, but the pattern is extremely complex as may be seen in the satellite images (Plate 1). At any one place it changes with the alternation of light and dark. In most places this is a 24 h cycle but within the Arctic and Antarctic circles there is a period of continuous darkness in winter and light in summer. It also changes seasonally, especially in temperate waters where there is an alternation of shallow mixed layer in summer and deep mixed layer in winter. Waters of the continental shelves tend to be more productive than the open ocean, and sites of upwelling more productive than sites where upwelling is absent. These phenomena have been reviewed in previous chapters.

Until now, all estimates of global rates of carbon fixation by phytoplankton have been based on compilations of the world's literature on field measurements made at particular points at particular times. They range from 20 to 55 Gt y^{-1} of carbon. The literature has been reviewed by Walsh and Dieterle (1988) who themselves suggest a figure of about 30 Gt y^{-1} of carbon for total photosynthesis by phytoplankton less respiration — the net fixation (Table 10.1). Of this, they estimate that more than half is fixed in the open ocean, which has the lowest rate

Table 10.1 Aquatic photosynthesis and sedimentation of organic carbon. From Walsh and Dieterle (1988).

Region	Area (km^2)	Net primary production ($\times 10^{12}$ kg C y^{-1})	Sediment organic carbon sink ($\times 10^{12}$ kg C y^{-1})
Open ocean	3.1×10^8	18.6	0.19
Continental shelf	2.7×10^7	5.40	0
Continental slope	3.2×10^7	2.24	0.50
Fresh water marshes	1.6×10^6	1.51	0.15
Estuaries/deltas	1.4×10^6	0.92	0.20
Salt marshes	3.5×10^5	0.49	0.05
Rivers/lakes	2.0×10^6	0.40	0.13
Coral reefs	1.1×10^5	0.30	0.01
Seaweed beds	2.0×10^4	0.03	0
Totals	3.75×10^8	29.89 (C input)	1.23 (C output)

of fixation per unit area, but has an enormous area compared with the other habitats listed. As we discussed in Chapter 8, there is still a great deal of uncertainty about the magnitude of primary production in the major ocean gyres. Within the last decade estimates have been put forward that vary by an order of magnitude (see Chapter 8; Platt *et al.* 1989 and references therein). With our newly acquired ability to monitor the whole ocean by satellite sensors, it is hoped that the confidence limits of estimates will be greatly decreased.

Satellite colour images now make it possible to make estimates of the distribution of chlorophyll in the surface waters of the world ocean, but this gives, at best, an indication of the distribution of phytoplankton biomass. To convert from this to rates of carbon fixation is no easy task and it has hardly begun. Platt and Sathyendranath (1988) have put forward an algorithm that combines a spectral and angular model of submarine light with a model of the spectral response of algal photosynthesis. They point out that one needs to know from field measurements the geographical and temporal variation in the physiological rate parameters of the algae and the biological structure of the water column, and they suggest that such data be collected, after which it should be possible to use satellite data to routinely estimate rates of carbon fixation anywhere that has the appropriate satellite coverage.

We also discussed in Chapter 8 that there are two categories of primary production: regenerated production, in which the algae use as a nitrogen source the ammonia excreted by the consumers in the system, and new production, in which the algae use nitrate upwelled from below the nutricline. The ratio of new production to total production is called the f ratio. In an idealized situation in which the pelagic zone of the ocean is in a biochemical steady state, regenerated production can be considered a closed loop with no net gain or loss of carbon or nitrogen. In this situation the flux of nitrogen leaving the pelagic zone in sedimenting organic particles is balanced by the upward flux of nitrate–N into the zone. Similarly, the downward flux of carbon in organic matter is balanced by uptake from the atmosphere. (There is also an upward flux of carbon by eddy diffusion, but the vertical gradient in carbon is so slight that this component can be ignored.) This brings us to the realization that when considering the magnitude of the biological pumping of carbon from the mixed layer down to the interior of the ocean, it is not the total production that we need to know, but the new production. Hence, even if a routine method is developed for using satellites to determine total net production of phytoplankton over wide areas, it will also be necessary to know the f ratio for each area under study. Fortunately, there is less disagreement among different observers about the levels of new production than about the levels of total production.

One of the most difficult problems facing those trying to determine the true magnitude of new production is that much of it is believed to occur as a result of intermittent bursts of upwelling of nitrate, resulting from eddy activity, internal waves, etc. (Goldman 1988). Any method of observation that is not continuous is likely to miss a proportion of these intermittent events. In a permanently stratified tropical ocean with a deep nutricline, the new production may take place at considerable depth and escape detection by satellites. In places where this occurs, the lower part of the euphotic zone will have a relatively high proportion of new production (high f ratio) while the upper part may function mainly on regenerated nitrogen (low f ratio). Hence there is the added difficulty of a non-uniform f ratio at a given geographic location.

In spite of all these complications, the role of the phytoplankton in pumping carbon down into the interior of the ocean is of such importance in the context of rising atmospheric CO_2 that a great deal of effort will be made to evaluate the magnitude of the new production.

10.3.2 Sinking of organic matter into the interior of the ocean

Primary production may be by organisms of a wide range of sizes, from large diatoms down to picoplankton, and their consumers may be small ciliates, copepods in the middle size ranges, larger euphausids, or large gelatinous zooplankton that nevertheless filter very fine particles. These consumers produce faecal pellets with a great range of sizes and sinking rates. Calculation of the downward transport of carbon by evaluating all the terms in such a complicated food web may well be a nearly impossible task, and it is not surprising that there have been a number of attempts to obtain the value of the sinking flux by direct measurement, using particle traps. For example, Martin *et al.* (1987) reported on the material intercepted in traps placed at depths ranging from 100 to 2000 m in the northeast Pacific off California. They estimated that at 100 m depth a station close to the coast had a downward flux of 7.1 mol C m^{-2} y^{-1} while at the same depth a station far out in the subtropical gyre had a downward flux of only 1.2 mol C m^{-2} y^{-1}. These figures yielded f ratios estimated at 0.14 for the open ocean and 0.17 for the coastal zone. In an area of active upwelling it was 0.20. Applying these percentages to some of their own primary production data, they tentatively extrapolated to the world ocean, concluding that global total production was of the order of 50 Gt of carbon, of which 7.4 Gt was new production.

Martin *et al.* (1987) were also able to reconstruct the vertical profiles of sinking organic matter, down to 2000 m. As expected, the deeper a trap was placed, the less material it collected, and this gave a measure of the rate at which organic matter was being decomposed during the journey downwards. Flux data from six open ocean stations were combined and fitted rather well the expression

$$F = 1.53\,(z/100)^{-0.858}, \qquad (r^2 = 0.81) \qquad\qquad (10.02)$$

where z is depth in metres.

As Fig. 10.06 shows, the flux at 500 m is only about 0.4 mol C m^{-2} y^{-1} and at 2000 m it is of the order of 0.1 mol C m^{-2} y^{-1}. The difference is accounted for by the oxidation of the organic carbon to CO_2 during its downward journey. It works out that 50% of the C removed from the surface is regenerated in less than 300 m, 75% by 500 m and 90% by 1500 m. The authors conclude that this rapid regeneration in near-surface waters means that the majority of the CO_2 removed from the upper 100 m will be available for exchange with the atmosphere within 10 years. That may be true for the area studied (between northern California and Hawaii) but in areas where the permanent thermocline is normally less than 100 m from the surface, much of the sinking carbon would be removed to the interior of the ocean and cut off from the atmosphere for at least several decades.

The question then arises, is this situation typical for oligotrophic gyres in the world ocean? Jenkins (1982) and Jenkins and Goldman (1985) studied oxygen utilization rates at a range of depths in the north Atlantic subtropical gyre. Martin *et al.* (1987) converted their carbon oxidation rates to oxygen utilization rates and compared the two sets of results. While the rates at 100 m in the Atlantic were somewhat lower

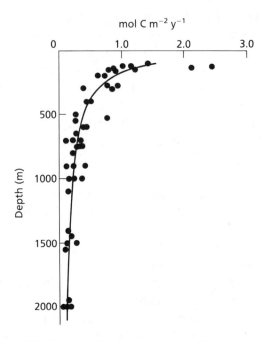

Fig. 10.06 Downward flux of carbon in the northeast Pacific as determined by sediment traps. Composite of six open ocean stations. From Martin *et al.* (1987).

than in the Pacific, the rates measured from 200 to 500 m were all higher in the Atlantic, indicating that less carbon was oxidized above 200 m and more was oxidized at 200–500 m than in the Pacific.

Interpretation of differing patterns of sinking fluxes is difficult, complex and far from being resolved. One quickly gets into a mass of biological detail. Consider, for example, the timing of the spring bloom of phytoplankton in relation to the ability of zooplankton to consume it. If zooplankton are present in good numbers in surface waters and are able to keep up with the consumption of phytoplankton as fast as it is produced, more phytoplankton carbon will be oxidized in the near-surface waters. We saw in section 8.6 that the sub-Arctic gyres of the North Atlantic and North Pacific differ in this respect. On the other hand, common forms of zooplankton such as copepods produce large quantities of faeces which sink rapidly and would have the effect of accelerating downward movement of carbon.

Bacteria have the opposite effect. If a dead organism begins to break into smaller particles and release dissolved organic matter, both the particulate and the dissolved organic matter provide food for bacteria, which then multiply rapidly. But bacteria are so small that they have a negligible sinking rate, and large quantities of carbon fixed in surface waters may remain in suspension and be converted back to CO_2 by bacterial respiration. The whole question of the operation of the biological pump is a matter of active debate. See, for example Karl *et al.* (1988) and Cho and Azam (1988).

10.3.3 Sedimentation on the sea floor

The last stage in this biological pumping is sedimentation on the sea floor. As far as the major ocean gyres are concerned, the rate of carbon fixation by phytoplankton is relatively low and a large proportion of that which sinks is regenerated to CO_2 in the water column, so that deep ocean sediments receive only a few g m^{-2} of organic carbon annually. As Table 10.1 shows, the whole of the deep ocean floor accounts for only about 11% of the world's carbon storage.

The situation on the continental shelves is quite different. They have much higher rates of primary production than the open ocean, and budgets show that perhaps half of this production is not grazed by the zooplankton, but sinks to the bottom (Walsh *et al.* 1981; Walsh 1983). One might therefore expect large accumulations of sedimentary carbon on the shelves. Instead, one finds large accumulations on the continental slopes, and the authors postulate that there is cross-shelf transport and export to the continental slopes which world wide amounts to 2.7 Gt C y^{-1}. Perhaps two-thirds of this is oxidized to CO_2 and returned to shelf and slope waters, but from studies of the organic carbon content of the sediments, their porosity, and ^{210}Pb measurements of sediment accumulation rates, it is concluded that 0.9 Gt of carbon is stored annually on the shelves and slopes, along with another 0.2 Gt in the estuaries and deltas.

10.3.4 The significance of biological fixation of carbon in calcium carbonate skeletons

Reef-building activities by corals and formation of tests by such planktonic forms as foraminifera are clearly mechanisms for the removal of carbon from the mixed layer of the ocean. World-wide geological deposits of calcium carbonate attest to the long-term and pervasive influence of this activity by marine organisms. The question arises, should this downward flux of inorganic carbon be counted as a term of the biological pump, when considering ocean–atmosphere interactions? The short answer is no. Formation of calcium carbonate removes CO_2 from the mixed layer (and indirectly from the atmosphere) but it also removes calcium. The removal of calcium affects the total alkalinity by the liberation of hydrogen ions:

$$Ca^{2+} + HCO_3^- \Rightarrow CaCO_3 + H^+ .$$

This shifts the equilibrium between bicarbonate and carbon dioxide in the water, leading to the liberation of carbon dioxide to the water and eventually back to the atmosphere:

$$H^+ + HCO_3^- \Rightarrow CO_2 + H_2O .$$

The addition of CO_2 by this process more than compensates for its removal in the formation of carbonate, so there is no net downward pumping from the atmosphere to the interior of the ocean. One may then ask, 'But surely the sinking of calcium carbonate skeletons represents a net downward flux of carbon, a biological pumping?' The answer is, 'This flux is compensated for by a change in the alkalinity of the surface waters, but not by a flux of carbon from the atmosphere. In the steady state situation, the change in alkalinity is compensated by advection of some neutralizing material from other sources, such as the land, and not by a flux of carbon dioxide from the atmosphere.'

10.3.5 Summary for the biological pump

Among those attempting to model carbon dioxide fluxes between the atmosphere and the ocean, there is agreement that biological processes lead to a net flux of carbon into the ocean. While part of the primary production in the mixed layer is consumed and respired in that layer so that the carbon dioxide released by respiration is returned to the atmosphere, and therefore leads to no net loss to the atmosphere, another part (equivalent to the 'new production') sinks out of the mixed layer carrying with it fixed carbon that is withdrawn from contact with the atmosphere for some considerable period. At present, the details of the biological pump are very uncertain. We do not know the total amount of carbon fixed world-wide on an annual basis, we do not know what proportion of that total is new production, and we do not know the details of the sinking process, or of the regeneration of carbon that takes place in the water column. Finally, we have only the roughest of estimates for the amount of carbon that is buried in the sediments.

Nevertheless it is frequently useful to construct models that represent a synthesis of knowledge at a particular time about the total flux of carbon dioxide between the atmosphere and the ocean. Takahashi (1989) has reviewed a number of models and arrived at the following synthesis. Carbon dioxide transfer across the sea surface is dependent on the difference between the partial pressure of carbon dioxide in surface waters and in the air, multiplied by a gas transfer coefficient. The partial-pressure difference is strongly influenced by biological carbon fixation, since vigorous photosynthesis by algae reduces the carbon dioxide concentration in surface waters. On the other hand if cold water is upwelled and warmed at the surface, the change in temperature will increase the partial pressure, because warm water saturates at a lower gas concentration than cold water.

The gas transfer coefficient is dependent on such factors as wind strength, degree of surface turbulence and the shape of the waves. Attempts to quantify these variables in wind tunnel experiments have led to widely differing results, but some cross-checking against the values used to simulate conditions in the ocean can be carried out by checking the rate at which radon gas and radioactive carbon have moved between the ocean and the atmosphere.

Using the best estimates for the values of partial-pressure differences and gas transfer coefficients in different parts of the world ocean, calculations have been made of net carbon flux between the ocean and the atmosphere which Takahashi (1989) claims are accurate within ± 0.3 Gt y^{-1}. The greatest flux from the ocean to the atmosphere is believed to result from warming of upwelled water at the equatorial divergences. Although strong biological pumping occurs there, its magnitude is believed to be substantially less than the outgassing resulting from warming of upwelled water. The major sinks for carbon dioxide are found in polar and subpolar regions. Cooling of water as it flows from the equator towards the poles causes it to take up CO_2 from the atmosphere, and deep water formation, mentioned earlier, leads to the downward transport of large amounts of the absorbed CO_2. In addition, the biological pumping that occurs in temperate waters transports large quantities of carbon to the interior of the oceans. The overall estimate according to Takahashi (1989) is a net flux of 1.6 ± 0.3 Gt y^{-1} of carbon from the atmosphere to the oceans. Many would say that the uncertainties in these estimates are much larger than claimed by Takahashi. The figure may be compared with the estimate of 5.3 Gt y^{-1} liberated from the burning of fossil fuels and the increase of 3.0 Gt y^{-1} remaining in the atmosphere. Takahashi proposes that the oceans absorb 1.6 Gt of the difference leaving 0.7 Gt to be absorbed by terrestrial processes. There is as yet no general agreement about this.

10.4 EVIDENCE FROM PALAEOCLIMATE STUDIES

The magnitude of the recent rise in atmospheric CO_2 from about 280 ppm in the 17th century to over 350 ppm at present should be viewed in the context of long-term changes. From studies of the carbon dioxide content of bubbles formed in the ice caps thousands of years ago, we now know that during the last ice age CO_2 was as low as 180 ppm. Some of the most complete evidence comes from a Vostok ice core (Lorius *et al.* 1988). Vostok station is located on the central part of east Antarctica, where the present mean annual temperature is $-55.5\,^{\circ}C$. The ice core was drilled to a depth of 2200 m and the deepest part is 160,000 years old (Fig. 10.07). Ambient temperature through the life of the core (expressed as difference from present surface temperatures) has been estimated from the isotope composition (both deuterium and ^{18}O, which give good agreement). There was a rise of about 10 $^{\circ}C$ about 15,000 years ago, marking the end of the last ice age. The previous interglacial period was from about 140,000 to 116,000 years BP and its beginning was also signalled by a relatively rapid rise in mean temperature of about 10 $^{\circ}C$. Temperatures during the glacial period fluctuated considerably, and this is often attributed to variations in the earth's orbit (Hays *et al.* 1976). Turning attention to the record of CO_2 in the bubbles trapped in the ice, there is a remarkable correspondence with the temperature, each rise in temperature being accompanied by a corresponding rise in atmospheric CO_2. At the end of each of the last two

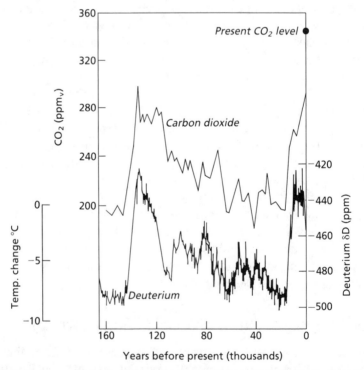

Years before present (thousands)

Fig. 10.07 Data on carbon dioxide content and deuterium δD levels (indicative of atmospheric temperature) of gas bubbles in the Vostok ice core at various dates back to 160,000 years BP. From Barnola *et al.* (1987).

glacial periods, when the temperature rose about 10 °C, the carbon dioxide content of the atmosphere rose by about 40%. Interpretation of this result is controversial, an example of the classic 'chicken and egg' type of situation. Was the temperature change brought about by changing CO_2 levels or did the changing CO_2 levels result from the climate change?

This is not the place to debate theories of the causes of ice ages, but it seems to be agreed that a glacial period is brought on by a triggering factor which does not by itself cause the large drop in temperature. The triggering factor leads to a series of climatic feedbacks (such as increasing albedo as a result of the increasing ice cover) which lead to the observed drop in mean temperature, formation of greatly enlarged ice caps, drop in sea level, and so on.

Many regard changes in the earth's orbit ('orbital forcing') as the prime initiating factor. The direction that the earth's axis points, in relation to the stars, cycles every 23,000 years. The obliquity of the earth's axis in relation to its tilted position changes on a cycle of 41,000 years, and its orbital eccentricity has a period of 100,000 years. Evidence that past ice ages have corresponded with these periodicities comes from ocean sediment cores going back 800,000 years (Hays *et al.* 1976; Kerr 1983). Three parameters were measured: (i) oxygen isotopic composition of planktonic foraminifera (a measure of temperature); (ii) a statistical analysis of radiolarian assemblages, previously shown to be correlated with temperature; and (iii) percentages of another radiolarian, not used in the previous data, thought to be an indicator of stratification in the water column. They found by spectral analysis that climatic variation in these records peaked every 23,000, 42,000 and 100,000 years, strongly supporting the theory that orbital forcing is a fundamental cause of quaternary ice ages.

Glacial periods can be satisfactorily accounted for by various models involving orbital forcing, climatic feedbacks and atmospheric carbon dioxide (reviewed in Lorius *et al.* 1988), but falling CO_2 levels are responsible for more than half of the effect. This would not be expected from the direct radiative effect of the CO_2 changes, which therefore require amplification. Nevertheless, the CO_2 changes are essential to the models. This brings us to the question: What caused them?

10.4.1 Theories concerning rapid changes in atmospheric CO_2

We have seen that the ending of the last two ice ages was marked by a roughly 40% rise in atmospheric CO_2. If we could understand the causes and consequences of these changes, we should be in a better position to attempt to forecast the effect of further increases in atmospheric CO_2 as a result of man's activities. Obviously, two classes of factors are involved, physical and biological. The ocean is divided into two concentric spheres: the surface mixed layer whose CO_2 partial pressure is a little above or below atmospheric value, and the deep ocean that is supersaturated with CO_2 and contains a huge reservoir, 50–60 times that of the atmosphere. Any physical mechanism that increases the transport from the deep ocean to the surface mixed layer will cause a rise in atmospheric CO_2. Alternatively, since the biological pump is responsible for the downward transport of carbon dioxide into the deep ocean, any factor that decreases biological activity, while upward transport from the deep ocean continues at its normal rate, will lead to increasing atmospheric CO_2. Most explanations put forward to date have both physical and biological elements.

Mix (1989) proposed a mainly biological model, and reviewed a number of preceding models. Several early workers in the field had suggested that during glacial periods the supply of nutrients to the phytoplankton was increased, especially at areas of upwelling, thus stimulating biological productivity and drawing down the atmospheric CO_2. At the end of the glaciation the flux of nutrients reverted to something approaching present-day levels and the atmospheric CO_2 rose sharply. Fortunately, it is possible to test this hypothesis. The concentrations of dissolved cadmium and phosphate in the sea are closely correlated, and from the cadmium to calcium ratio in the shells of benthic foraminifera it is possible to calculate the ocean's phosphate content in times past. There is no evidence for high nutrient concentrations in upwelling areas in glacial periods, or of a sharp change at the end of those periods.

Several modellers have concentrated on processes in the Antarctic. As we discussed earlier, deep water enriched in nutrients rises to the surface, but extensive biological pumping also occurs and the net result is that the southern ocean is a major sink for carbon. It has been suggested that iron is a limiting factor in the biological use of nutrients, and that a higher input of iron-rich dust during the last ice age could have stimulated biological productivity, reducing atmospheric CO_2. Another suite of models has invoked increased nutrient upwelling at the tropical divergence, with increased biological production and a decrease in atmospheric CO_2 (even though outgassing of CO_2 at the equator would presumably continue).

Attempts have been made to calculate changes in ocean productivity between glacial and interglacial times by comparing the organic content and the carbon isotope ratios of the appropriate parts of ocean bottom cores. A recent example is Sarnthein *et al.* (1988). They concluded that there was a massive decrease in new production at low and mid-latitude upwelling areas at the termination of the last ice age with a consequent increase in atmospheric CO_2. They attributed it to orbital forcing, leading to an increase in high latitude insolation, reduction in ice cover and a reduction in the strength of meridional trades.

Mix (1989), while not disagreeing with the conclusions of Sarnthein *et al.*(1988) cited numerous difficulties with interpretation of organic matter data from cores. He proposed instead to carry out a major reconstruction of ocean productivity in times past, using the community structure of foraminifera assemblages as his index. Beginning with a global ocean productivity map, he developed for the North and South Atlantic Ocean a relationship between foraminifera assemblages on the contemporary sea floor ('core tops') and the level of productivity of the waters above. This was done using standard transfer function techniques (Imbrie and Kipp 1971). The result was a regression of measured productivity against estimated productivity with an error envelope of $\pm$ $12\,g$ C $m^{-2}y^{-1}$. Then, turning to the foraminiferan assemblages present at the glacial maximum 18,000 y BP, he constructed surface productivity maps for that time. They were markedly different from contemporary productivity maps. The largest increases were seen at the equator and in subtropical regions, the least in eastern boundary currents. Overall, Mix (1989) estimated an increase of 18% in total primary productivity of the North Atlantic. Using the arguments of Eppley and Peterson (1979) that show that higher rates of production tend to have a higher proportion of new production, Mix (1989) estimated that new production would have been 38% higher overall, and 87% higher in equatorial regions during the coldest period of the last ice age.

It seems that during the last ice age atmospheric CO_2 was reduced by stronger biological pumping, particularly at the equator. There is evidence for this both in the foraminiferan communities from that period, and from the amounts and types of organic carbon in the sediments. The earlier suggestion that this was caused by an increase in dissolved nutrients in the deep ocean is rejected. Instead, modifications of the present-day atmosphere–ocean interactions are invoked. During the ice ages, it is suggested, there was a stronger thermal gradient between the poles and the equator, and this resulted in stronger trade winds and stronger upwelling of nutrient-rich water, particularly at the equator. There may well have been enhanced upwelling in Antarctic waters, but the evidence from palaeobiology is less convincing.

The current scenario for the life of an ice age now appears to have the following components:

1   changes in the earth's orbit led to reduced solar radiation at high latitudes, lower temperatures and spread of the polar ice caps;

2   increase in the thermal gradient between the poles and the equator led to strengthening of trade winds and increased upwelling of nutrient-rich water at the equator;

3   increased upwelling caused enhanced biological pumping, and a decrease in atmospheric CO_2, which in turn led to further cooling; and

4   after an appropriate period the orbital position of the earth changed enough to begin a reversal of the process.

10.4.2 Postulated changes in the rate of deep water formation

We saw in section 10.2.3 that the world's major source of deep water formation is in the North Atlantic. Broecker *et al.* (1985) pointed out that as polar water sinks, water from the south moves north to take its place. While doing so, it cools and about 5×10^{21} cal of heat are released into the atmosphere each year. This modifies the climate of the North Atlantic basin and adjacent land areas. The water that sinks in the Norwegian and Greenland Seas is relatively deficient in nitrogen and phosphorus. Foraminifera in surface sediments taken from the floor of the North Atlantic have a low cadmium : calcium ratio, indicative of this low nutrient status.

Broecker *et al.* (1985) reviewed the evidence that in glacial times the formation of deep water in the North Atlantic was severely reduced. It has been found that the Cd : Ca ratio in benthic foraminifera was higher in glacial times than today, indicating a more nutrient-rich water. It has also been shown that their stable carbon isotope signature in glacial times was indicative of deep water much more like that found in the Pacific during the same period. All this suggests that the organisms living deep in the North Atlantic were in water that had a different origin than today, and this opens the possibility that in glacial times the thermohaline circulation of the world was radically different than it is today. Even the possibility of reversal, with deep water formation in the Pacific and upwelling in the Atlantic, should not be ruled out. Just how this would relate to the scenario reviewed above in 10.4.1 remains unclear.

10.5 PHYTOPLANKTON AND DIMETHYLSULPHIDE

In addition to their role in fixation of carbon dioxide, phytoplankton are thought by some workers to influence the earth's climate through the production of dimethylsulphide, DMS (Charlson *et al.* 1987). Most species of phytoplankton excrete DMS, and some of it escapes to the atmosphere where it reacts to form a sulphate and a methane sulphanate aerosol, MSA. The latter substance stimulates the formation of clouds over the oceans, thus increasing albedo and reducing temperature at the earth's surface.

Dimethylsulphide is the only volatile sulphur compound excreted by phytoplankton and its biological function is unclear. The concentration of DMS in surface waters is only weakly correlated with the biomass or productivity of the phytoplankton. Once released into the water, DMS may be broken down photochemically, removed by bacteria or released into the atmosphere. Rates of emission from the sea to the atmosphere are in the range 2.2–5.7 mmol m^{-2} y^{-1}, with the highest values being found in warm, highly productive coastal regions. Coccolithophorids, which are most abundant in tropical, oligotrophic waters have a high rate of DMS production per unit biomass, while diatoms have a low rate of production.

Cloud condensation nuclei, CCN, are essential to the formation of clouds of liquid water droplets. It is thought that at the present time the air over continents has an abundance of particles that could act as CCN, but that air over the oceans remote from land has a relative shortage of suitable particles. Water-soluble sulphates are believed to be the natural CCN in the marine air. Sea-salt particles are possible candidates, but concentrations at cloud height are much too low for them to be important. Attention therefore focussed on 'non-sea-salt-sulphate', and DMS was identified as the most likely precursor of such sulphate particles.

Charlson *et al.* (1987) then developed at considerable length the technical arguments which led them to believe that an increase in DMS in the atmosphere (with water content held constant) would lead to a larger number of smaller droplets, and that this would increase the earth's albedo. This in turn would have a cooling effect. Since DMS is released into the atmosphere at the highest rate in tropical waters, the authors argued that it serves a useful function in moderating the extremes of temperature over the tropical oceans. They see this as part of the global feedback control to which Lovelock (1979) drew attention in his book *Gaia*.

The importance of DMS as a climatic influence was challenged by Schwartz (1988). He argued that if DMS emissions have a strong influence on global albedo, an even stronger effect should be detectable from man-made sulphur dioxide emissions, for these are thought to release twice as much sulphur into the atmosphere as the phytoplankton. Moreover, the release is mostly in the northern hemisphere, and has increased to its present level during the last 100 years. Schwartz searched the records for any evidence of higher albedo in areas most under the influence of industrial sulphur emissions, and found none. He also found no evidence of reduced temperatures in the northern hemisphere compared with the southern. His conclusion was that anthropogenic sulphates do not have an important effect on the cloud component of albedo. By extension, he suggested that neither do the products of DMS emission.

An important body of data pertinent to this subject was published by Prospero and Savoie (1989). They reported on analyses of methane sulphonate aerosol, MSA, and non-sea-salt sulphate, NSSS, on eight Pacific islands. They took Fanning and American Samoa as sites totally remote from industrial activity and found that the ratio MSA/NSSS was 0.065. They then used this ratio to calculate what was the proportion of non-sea-salt sulphates derived from DMS at other Pacific sites. They concluded that about 80% of all cloud condensation nuclei over the North Pacific were attributable to DMS from phytoplankton. They claimed that most of the industrial sulphur emissions emphasized by Schwartz are deposited on land, locally or regionally, so that, for example, only 20–25% of those produced in North America are exported from the continent. In the light of these conclusions, the original hypothesis of Charleson *et al.* (1987) may still be valid. The possibility that, in the context of global warming, DMS production by phytoplankton could be a negative feedback, moderating some of the effects of increasing atmospheric CO_2, cannot be ruled out.

10.6   SOME SPECULATIONS ABOUT THE FUTURE

In almost everything discussed in this chapter there are large areas of uncertainty. Yet we are confronted with urgent global problems, and to meet them we need all the insight we can muster about the interactions of the atmosphere, the terrestrial biota and the oceans.

The following is a summary of current understanding of the global warming problem, taken from Hare (1988). About half of the predicted warming is due to increased carbon dioxide levels in the atmosphere, the other half to increasing chlorofluorocarbons, nitrous oxide, methane and ozone (the latter in the lower layers only). In combination they are expected to lead to the equivalent of a doubling of carbon dioxide by the year 2030. According to existing ocean–atmosphere models this should lead to an increase of between 1.5 and 4.5 °C in mean global surface temperatures, with the high latitudes warming the most, especially in autumn and winter. Sea level may rise between 20 and 140 cm, chiefly because of an expansion of the ocean's water column and only secondarily from melting of glacial ice.

The most important aspect to note is that the predicted rate of change exceeds anything that has happened since the end of the last glacial period. The changes will be too fast for many plants and animals, or for economic systems, to adapt. It will amount to a destabilization of the global ecosystem. There is heated controversy about whether the process is already under way and measurable. Climate is notoriously variable in time and the problem is to discern a signal amidst the environmental noise. At the time of writing (1989) many leading authorities are convinced that the warming process has begun.

What, then, are the expected responses of the oceans to such a change? A preliminary study commissioned by the Department of Fisheries and Oceans of the Canadian Government (Wright *et al.* 1986) came to the following conclusions (some of which are general while others are particular to the North Atlantic).

1 Sea surface temperatures generally will rise, with the largest changes expected near 60° N. This will lead to greater amounts of evaporation and hence a more vigorous hydrological cycle.

2 The expected balance will be that mid-latitudes will have the greatest increase in evaporation, with a corresponding increase in precipitation in polar and tropical regions. As a result, meridional (i.e. north–south) gradients in surface salinity will increase, creating fresher northern conditions and saltier subtropical conditions.

3 The thickness, areal extent and duration of ice cover will decrease.

4 Enhanced surface warming at high latitudes will mean a reduced meridional temperature gradient and probably a reduction in wind stress of the order of 10% over the entire North Atlantic. This in turn will lead to reduced flow in the major wind-driven gyres so that the Gulf Stream and the Kuroshio current, for example, will be less strong.

5 Increased precipitation in the higher latitudes should lead to more river run-off, greater stability in coastal water columns and a strengthening of buoyancy-driven currents. For the western North Atlantic this should mean that while the Gulf Stream will relax, the Labrador current will be enhanced.

6 Increased buoyancy in surface waters, combined with reduced wind stress should lead to a reduction in rate of intermediate water formation in the Labrador Sea. In the Norwegian and Greenland Seas the inflow of fresh water is expected to be less, and the amount of wind mixing may increase as a result of the retreat of the ice. For this area it is

difficult to predict whether, in the short term, deep water formation will increase or decrease.

7　In the long term, global warming should lead to increased evaporation in the North and South Atlantic basins, thus increasing average salinity. According to the theories of Broecker *et al.* (1985) this will probably increase the deep water formation and thermohaline circulation. However, at the same time, less cooling is expected at the surface, and this will tend to reduce deep water formation. The balance between these two tendencies cannot yet be predicted.

What are the likely biological consequences of these changes? This is a difficult, almost impossible question to answer. If, as suggested, there is a general reduction in wind stress there will probably be less upwelling of new nitrogen at the equator and other major upwelling areas, leading to less biological production. Reduced biological pumping would be a positive feedback towards increasing atmospheric CO_2. But until we are in a position to make predictive models of ocean circulation, we really do not know whether the upwelling of nutrient-rich water will increase or decrease overall in the world ocean. Planktonic organisms, with their relatively short life histories, may adapt much more rapidly to changing climate than would longer-lived organisms like trees and mammals. After a few years of changed environmental conditions, the planktonic community may have changed its species composition and adapted well to the new regime.

A second report commissioned by the Canadian Government (Frank *et al.* 1988) called for a preliminary assessment of the changes to be expected in the fisheries of Atlantic Canada. The authors reiterated many times that their report was no more than informed speculation. Nevertheless, their ideas have interesting links with some of the material dealt with in earlier chapters in this book. For example,. it has been suggested that global warming will lead to increased precipitation in high latitudes, which means increased freshwater run-off and greater buoyancy in coastal waters. This can be expected to lead to more rapid warming and stabilization of coastal waters in spring. Among the species that might be expected to benefit from this change are lobsters and scallops. Lobster landings have shown interesting positive correlations with river run-off in the first year of the lobster's life (section 4.7.2), and scallop year-class success has been positively correlated with the water temperature in the year in which they were recruited.

Species such as cod and haddock, on the other hand, might be adversely affected. It was found in the Gulf of Maine area that the ctenophore *Pleurobrachia* (a jellyfish-like animal), when present in its peak abundance (>1 m^{-3}), can graze down the zooplankton on which larval cod and haddock rely. Normally, the larval fish occur in spring and the ctenophores in mid-summer so competition for food is minimal. However, in 1983 the sea surface was unusually warm between January and April and the time of peak ctenophore abundance advanced from summer to spring. As a result, the ctenophores and larval haddock coincided, and the 1983 haddock year-class is the lowest on record (Frank 1986). Hence, there is a possibility that freshening and warming of the coastal waters off Nova Scotia and in the Gulf of Maine might lead to reduced success of cod and haddock recruitment.

It would naturally be expected that if there was a general warming of the ocean, the geographic range of fish species would shift northward. There is a certain amount of observational data from previous warming events to indicate which fish might be most affected. For example, during the warming trend of the 1940s there was a northward shift in the abundance and distribution of mackerel, lobster and menhaden, and green crab established a resident population north of any previously recorded location. During the cooling trend of the 1960s American plaice and butter-fish retracted their range southward, and capelin and spiny dog-fish extended their migrations southward. Other species, such as haddock, yellowtail flounder, winter flounder and winter skate appeared not to change their range in response to temperature changes, and it is suggested that their distribution is determined more by the presence of an appropriate bottom type. From all of this it is concluded that in the Gulf of Maine, for example, many species might be displaced northward by global warming, and their place taken by an assemblage of fishes more characteristic of the middle Atlantic bight. This would include menhaden, butter-fish, red hake, silver hake and herring.

From Chapter 9 we may recall that the warming event of the 1940s was associated with a strong northward extension of the range of cod in west Greenland, and we may speculate that a similar event might be precipitated by global warming in the 1990s. Similarly, the changes in the whole biological community in the English Channel, which took place in association with the warming of the 1940s, might be expected to recur in the years to come.

In section 8.4.4 it was shown that when a warm-core ring interacts with the continental shelf, there may be adverse effects on year-classes of fish because the larvae are swept off the shelf and dispersed into the open ocean. One prediction that could be made is that this will occur less frequently because reduction in strength of the Gulf Stream as a result of global warming will lead to the formation of fewer warm-core rings.

In the discussion of tidal currents (section 7.3.2) it was shown that the distribution of herring spawning grounds in the North Atlantic often coincides with tidally-mixed areas, and that the young stages often seem to be retained in those areas for a considerable period. It was further suggested that the upper limit of the size of a spawning stock might be determined by the limits of the tidally-mixed area. Under the conditions of global warming, with a freshening of coastal waters, a small decrease in the extent of tidally-mixed areas might be expected and this might have some consequence for the herring stocks.

Turning from species of commercial importance to more broad ecological principles, the freshening of coastal waters might be expected to lead to earlier onset of stratification in areas that are not tidally mixed, hence a lengthening of the period of stratification. This will shift the growth advantage from larger phytoplankton such as diatoms to smaller species such as dinoflagellates. In general, this should lead to a lengthening of the food chain between the primary producer and the larger consumer species, with a consequent loss of secondary productivity.

One might have hoped that it would be possible to offer an opinion as to whether ocean productivity would be higher or lower as a consequence of global warming. Certainly many features, such as the reduced wind stress and greater water column stability in north temperate coastal waters, suggest less upwelling of nutrient-rich water and hence reduced productivity. On the other hand, waters that are now ice-covered for a substantial part of the year will become ice-free and have increased productivity. We can say for certain that things will be different, and that those who exploit the ocean resources in a particular place will need to pass through a painful period of adjustment, but it is not possible to say what the overall effect will be on global ocean productivity.

PART D
DISCUSSION AND CONCLUSIONS

11

Questions for the Future

11.1 INTRODUCTION

We have reviewed the dynamics of marine ecosystems from three perspectives. In Chapter 2 we concentrated on the turbulent flow of the water surrounding the organisms in order to understand better how algae obtain inorganic nutrients and microscopic animals their algal food. Chapters 3–8 reviewed the biological and physical processes associated with formation of phytoplankton concentrations in spring blooms, at fronts, in coastal upwelling areas, at shelf edges, in rings and in estuaries. The final perspective was of long-term variations in phytoplankton productivity due to climate changes.

In all of this we reviewed information on consumer dynamics where this was available, but by far the greater part of the text concerns phytoplankton. The reason for this bias is that chlorophyll can be measured continuously by means of a fluorometer, and this serves as a useful index of phytoplankton biomass. Instruments for the continuous measurement of zooplankton biomass are still in the development stage. Future advances of marine ecology require major improvements in automated methods for collection of data on zooplankton.

As marine ecology emerges as an integrated discipline concerned with the physics, chemistry and biology of the oceans, there is a search for generalities around which to organize the multiplicity of observations. We see this search revolving around three questions.

1 Is there a common mechanism to account for the occurrence of high biological productivity in a variety of physical environments?

2 To what extent are events in marine ecosystems determined by the physical processes? To paraphrase Hartline (1980), do physical factors feed fish?

3 How can we develop concepts and models that span the enormous range of scales in marine ecology, from the microscopic to the global and from seconds to geological ages?

11.2 A BASIC MECHANISM OF PHYTOPLANKTON PRODUCTIVITY

It was pointed out by Legendre (1981) and is amply confirmed by our review, that there is one sequence of events that occurs in a variety of physical settings and on time scales ranging from a few hours to a year, that normally leads to an increase in phytoplankton production. The essence of it is strong vertical mixing followed by stratification of the water column. As first described by Gran (1931) and presented as a quantitative model by Sverdrup (1953), it was offered as the explanation of the spring bloom in temperate waters. The vertical mixing brings nutrients from depth to surface waters, and the formation of stratification confines the phytoplankton to a well-lit zone in which daily photosynthesis exceeds daily respiration. The driving force for vertical mixing is wind stress at the surface and the chief agent of stratification is solar heating.

In estuaries and parts of the continental shelf vertical mixing may be driven by tidal currents interacting with the bottom, and stratification may be mainly a function of freshwater run-off, but the effect on the phytoplankton is the same. The major difference is the time scale. The strength of tidal currents varies with a diurnal rhythm and with the fortnightly cycle of spring and neap tides. When tidal currents are strong, vertical mixing is also strong, and when tidal currents are weak stratification is more marked. We have reviewed examples of high phytoplankton productivity corresponding with each period of stratification.

At tidally-mixed fronts, the area of tidal mixing increases with the spring tides and decreases on the neap tides, causing the front to move back and forth horizontally. At some geographical locations there is an alternation of tidal mixing and stratification which leads to enhanced phytoplankton production and biomass.

Similarly, in areas noted for their coastal upwelling, the prevailing wind brings nutrient-rich water to the surface and a relaxation of that wind permits surface warming and stratification. The high productivity is associated with the relaxation of the winds, as the phytoplankton utilize the upwelled nutrients. Even in areas of more or less continuous winds, a horizontal succession of events can be detected. Close to the coast, nutrients are abundant in the freshly upwelled water, but as it streams away from the area of upwelling stratification sets in and phytoplankton biomass increases. Still further from the coast is a zone in which the zooplankton become more abundant as they feed on the phytoplankton. A similar zonation is found in relation to equatorial upwelling.

Here, then, is a mechanism of biological productivity that is found in situations as diverse as a temperate estuary and a mid-ocean band of upwelling running parallel with the equator. The alternation of vertical mixing and stratification is surely one of the most important sequences in marine ecology.

11.3 THE RELATIONSHIP BETWEEN PHYSICAL AND BIOLOGICAL PROCESSES

In the majority of situations we have reviewed, biological processes appear to be strongly influenced by the physics, while the physical processes are largely independent of the biology. This perception led Hartline (1980) to write an article entitled: 'Coastal upwelling: physical factors feed fish'. In estuaries, on continental shelves and in the open ocean, the physical processes appear to set the stage (albeit a constantly changing stage) on which the biological play is enacted. The organisms are adapted to the physical conditions that occur in a given place, but, in the short term, the physical conditions are very little affected by the biology.

Lewis *et al.* (1983) have provided one of the few documented examples of short-term feedback from biological to physical processes. They showed that at an oligotrophic ocean station the chlorophyll maximum was sufficiently strong that the heat absorbed from the downwelling radiation caused local heating of the water and this caused increased vertical mixing and deepening of the mixed layer. The chlorophyll maximum also decreased the rate of heating of the water immediately below it, thus increasing the stability of that part of the thermocline. The phenomenon is probably widespread in the ocean.

On the other hand, there are numerous instances in which, in the short term, the physical processes are the prime determinants of the dynamics of marine ecosystems. However, when we turn to a time scale of thousands of years, the feedback from biology to physics is far from trivial. The present composition of the earth's atmosphere and the present average temperature at the surface of the earth have been profoundly influenced by living organisms. Volcanoes have been constantly adding to the carbon dioxide content of the atmosphere, but 99.5% of that carbon dioxide has been transported to the ocean sediments by marine organisms. They have fixed carbon dioxide in their calcareous skeletons and they have incorporated carbon dioxide in their organic matter. As they sank to the bottom of the ocean, they gave rise to the carbonaceous rocks and to the deposits of fossil fuel. Without these processes it is likely that the carbon dioxide content of the atmosphere would be very high and the temperature of the surface of the earth would be about 400 °C. Life as we know it would be impossible.

Biological processes are also thought to have played a major part in starting and ending the ice ages. It has been found that the rise in mean temperature (about 10 °C) associated with the end of the last glaciation was accompanied by a rise in atmospheric carbon dioxide. There is not yet agreement about the mechanisms involved. Changes in the earth's orbit are believed to have had a triggering effect, but it seems that there was also a sharp decrease in the fixation of atmospheric carbon dioxide by the phytoplankton. If this turns out to be correct, biological processes are responsible for profoundly influencing the composition of the atmosphere, the temperature of the earth's surface and as a consequence, the temperature and circulation of the whole global ocean.

11.3.1 The concept of auxiliary energy

In our review we have seen that physical processes in the ocean are driven by solar energy which is transmitted to the water mass as heat, wind energy or freshwater run-off, and results in currents, turbulence and patterns of stratification. Solar energy is also captured by phytoplankton, passed from one organism to the other through the feeding process and finally dissipated in the heat of respiration. Physical oceanography deals with the first of these pathways and biological oceanography with the second. In marine ecology we become concerned with both, and we then notice that the two pathways are inter-related. Water movement breaks down the boundary layers round organisms, transports nutrients and waste products, and influences the rate of encounter between planktonic predators and their prey. Organisms use currents to assist their migrations and sessile organisms occupy sites with a range of current speeds appropriate to their way of life. These physical energies assist the transfer of energy in the biological food web without themselves taking part in the process. From a biological point of view they have been labelled auxiliary energy.

The concept was first elucidated by Odum (1967a, b) who drew attention to the way in which the ebb and flow of the tide assists the productivity of a salt marsh, and the way in which agricultural productivity (as measured by energy flow) is enhanced by the use of auxiliary energy to prepare the ground, remove weeds, spread fertilizers, and so on. In the history of agriculture the auxiliary energy has changed from being human labour (the farmer and his family), through the use of domestic animals to the present-day use of fossil fuels. Odum drew attention to the fact that the auxiliary energy for a salt marsh is provided free by nature, while the auxiliary energy of a farm is costly.

Margalef (1974, 1978b) pointed out that the physical energy that upwells nutrient-rich water in the ocean is an energy subsidy to the phytoplankton. He showed that a model relating photosynthesis to energy of upwelling is, in many situations, a better predictor of phytoplankton production than the more complex models involving light, nutrients and temperature. He also developed detailed explanations of the morphology and physiology of various phytoplankton species in terms of their adaptations to high or low levels of auxiliary energy. For example, diatoms tend to be characteristic of turbulent water in which nutrients are being upwelled vigorously, while flagellates are better suited to low-turbulence, low-nutrient waters.

The topic was further explored by Legendre (1981), Legendre and Demers (1984, 1985) and Legendre *et al.* (1986). He and his colleagues pointed out that high biological productivity often occurs at places where there is a sudden change of auxiliary energy, such as fronts, thermoclines, sediment–water interfaces, the underside of ice, etc. They also pointed out that for there to be an effective physical–biological interaction there must be a matching of scales in time and space. For example, since the doubling time for phytoplankton cells is in the range of hours to days, favourable physical conditions must persist for at least a few days for there to be a marked increase in phytoplankton populations. A good example of the importance of matching time scales was reviewed in Chapter 6. High phytoplankton productivity on tidally-mixed fronts is stimulated by changes in the strength of tidal mixing and has a periodicity of 14 days, whereas high productivity at the shelf break is stimulated by internal waves that form on the ebbing tides with a semi-diurnal rhythm. Since the generation time of meso-zooplankton is of the order of weeks, they cannot reproduce fast enough to use the pulses of primary production occurring at fortnightly intervals on the tidally-mixed fronts, but they are able to build their populations by using the more continuous supply of phytoplankton generated in association with the internal lee waves at the shelf break.

One of the problems inherent in trying to build the concept of auxiliary energy into numerical models is that it is difficult to determine what proportion of the energy of physical processes impinges on biological processes. Margalef (1978a) made some rough estimates and made

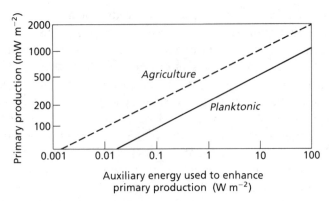

Fig. 11.01 Auxilliary energy (W m^{-2}) used to enhance primary production in agricultural systems and in the phytoplankton. From Margalef (1978a).

comparative plots of the regressions of primary production in the plankton and in agriculture on auxiliary energy (Fig. 11.01). The diagram shows that a given amount of primary production in the plankton requires about ten times more auxiliary energy than the same production on land. This is not surprising when we recall that in the ocean nutrient regeneration often occurs at great depth, far removed from the euphotic zone, whereas on land nutrients are regenerated in the soil, close to the roots of the plants.

It will be interesting to see whether the concept of auxiliary energy leads to better models of primary production. Now that it is possible to obtain vertical profiles of turbulence (see section 3.3.3) data on primary production as a function of turbulent energy may begin to accumulate.

11.4 THE QUESTION OF SCALE

The cartoon in Fig. 11.02 helps to illustrate some problems of scale that confront us in the future. The figure represents the 1000 km domain of a fictitious fish which is shown moving from place to place on its annual migration. From its spawning grounds in the early spring the fish swims to the upper layer of the open ocean to feed on the zooplankton which are feeding on the early phytoplankton blooms. Later they move to a tidal front over the shelf before swimming on to feed on a small spawning fish. They next move on to the outer region of the continental shelf where internal tidal waves have enhanced the productivity and finally they end up back at their own spawning grounds.

The fish swim thousands of kilometres over the year to feed at a series of areas of enhanced productivity which are each about 50 km across. From the size of territory associated with each species it is clear

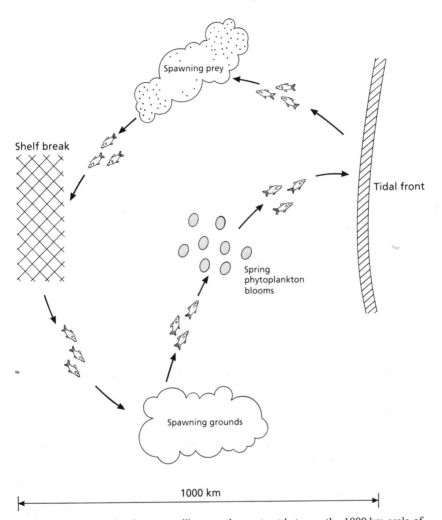

Fig. 11.02 A schematic diagram to illustrate the contrast between the 1000 km scale of the fish's domain and the 50 km scale of the various features, such as fronts and plankton blooms, upon which the fish depend.

that complete studies of fish will need to cover a much larger area than studies of phytoplankton. The time scales associated with the two organisms are also much different. Individual phytoplankton live for a few days and the population of a patch can change significantly over the same length of time. Fish live for 5–10 years and measurable changes in their population occur over a year or so. Thus besides covering a large area the study of fish will also have to take a long time in order to understand the causes of any variations in time. The length and time scales associated with the populations are sometimes plotted jointly on a chart such as the one shown in Fig. 11.03 (Steele, 1981).

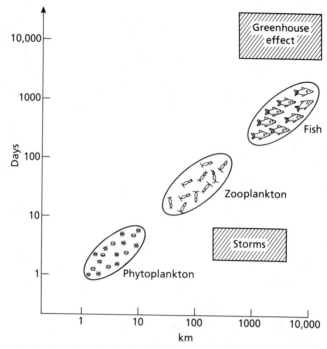

Fig. 11.03 The horizontal scale of phytoplankton, zooplankton and fish populations plotted against the time required for a significant change in the population. The time and length scales of storms and climatic effects such as a change in the greenhouse effect are also plotted for comparison.

The fishes domain, in the upper right, is thousands of kilometres across and the population takes years to change significantly. Phytoplankton at the other extreme change their numbers quickly and are concentrated over small scales. The length and time scales of the zooplankton lie in between those of the phytoplankton and the fish. They also lie in between the other two as prey to the fish and predators to the phytoplankton. The health of the fish population is obviously dependent on the health and availability of the zooplankton which in turn is dependent on the phytoplankton but these links between the groups can be rather 'elastic'. That is, the zooplankton population covers a sufficiently large area and is sufficiently mobile to survive short term fluctuations in the phytoplankton population at any one location. The fish in turn are mobile enough to not be strongly affected by a decline of zooplankton in one area. The larger organisms therefore, because of their mobility, are able to smooth out the effects of fluctuations in the population of the prey populations. The population of the smaller organism, however, will tend to fluctuate inversely with the population of the predator. Thus a large fish population may reduce the zooplankton population and lead to a higher phytoplankton population.

Changes in the environment will also cause large variations in the populations of the smaller organisms. Violent storms can disrupt and disperse spawning fish and increase the turbulence in the upper layer to disperse plankton patches. Longer term changes in temperature such as those expected from the greenhouse effect will change the phytoplankton species of a given region and possibly the locations of phytoplankton concentrations will change. A plankton population is said to absorb the variations due to storms and adapt to the multi-year variations (Steele 1988).

Not shown in Figs 11.02 and 11.03 are important relationships between the fish and its predators, and between the fish and its prey. It is also necessary to understand fluctuations in the other fish populations inhabiting the same space as the one illustrated because they may strongly interact. The diagram in Fig. 11.03 may then sprout other axes to depict these other relationships such as the different trophic levels and the different levels of hierarchy of each particular species. Four axes is too many to show on one diagram so for simplicity the time/length scale plane is sometimes collapsed to a line with the assumption that time scales increase directly in proportion to the length scales. Such a figure is given in Fig. 11.04 but it must be remembered that the simplification comes at the cost of ignoring some important effects such as short- and long-term meteorological forcing.

Diagrams like those in Figs 11.03 and 11.04 help to organize marine ecology and aid in separating the important and tractable relationships that have to be understood. One initiative aimed at understanding the interactions between the passively floating phytoplankton, the actively swimming zooplankton and the turbulence in the water has recently been discussed at a workshop entitled 'Small-scale bio-physical coupling' (Bucklin *et al.* 1989). The participants in this workshop were experts in acoustics, numerical modelling, physical oceanography and biological oceanography. They agreed that progress in the field is most likely to be made through study of small-scale physical and biological processes with the object of understanding how the small organisms, both passive and active, interact with each other and with the small-scale physical processes.

To achieve this understanding they propose to obtain observations over a series of length and time scales. At the one metre scale optical and acoustic instruments will obtain high resolution time series observations of currents and the movements of particles and organisms from

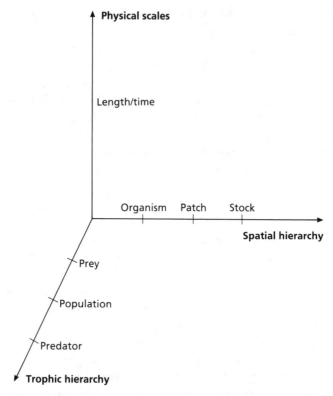

Fig. 11.04 The three main ecological variables plotted on a three-dimensional grid as suggested by Steele (1988).

$5\,\mu$m to 5 mm in size. Results from these fine-scale observations will lead to the development of accurate parameterizations of the fine-scale physical and biological processes which can then be used in the models connecting the fine-scale observations to the large-scale ones.

The observations over the larger scales will include the use of submersibles to observe zooplankton, moored instruments to record temperature, velocity, light intensity, temperature and velocity microstructure profilers, drifters, and an acoustic imaging array. In short, a high resolution multidiscipline experiment using the most advanced instrumentation.

Hats off to the dedicated, hard working individuals who will write the next chapters in this vital and fascinating field!

Appendix

Table A1 Variables, constants and non-dimensional numbers

Symbol	Quantity or variable	Typical value and/or units
A	Area of cross-section	m^2
α	Thermal expansion coefficient (water)	$10^{-4}\ °C^{-1}$
α, α_λ	Absorption coefficient	$\simeq 0.1\ m^{-1}$ for red light $\simeq 0.001\ m^{-1}$ for blue light
β	Variation of Coriolis parameter with latitude	$2 \times 10^{-11}\ m^{-1}\ s^{-1}$ ($\phi = 45°$)
C	Concentration of a substance	$kg\ m^{-3}$
C_D	Drag coefficient	varies with shape and flow
c	Specific heat of water	$4.2\ kJ\ kg^{-1}\ °C^{-1}$
D	Molecular diffusivity	$1.5 \times 10^{-7}\ m^2\ s^{-1}$ (heat) $1.5 \times 10^{-9}\ m^2\ s^{-1}$ (salt)
D_c	Compensation depth	m
D_E	Ekman depth	$\simeq 80\ m$
D_t	Tidal energy dissipation	$W\ m^{-2}$
d	Typical diameter	m
δ	Boundary layer thickness	m
ε	Turbulent energy dissipation	$W\ kg^{-1}$
F	Flux	$kg\ m^{-2}s^{-1}$
F_D	Drag force	N
f	Coriolis parameter ($2\Omega \sin\phi$)	$10^{-4}\ s^{-1}$ ($\phi = 45°$)
ϕ	Latitude	degrees
g	Gravitational acceleration	$9.98\ m\ s^{-2}$
g'	Reduced gravity $\left(\dfrac{\Delta\rho}{\rho}g\right)$	$0.01\ m\ s^{-2}$
h	Height of water column	m
I	Intensity of light	$W\ m^{-2}$
J	Energy, in Joules	$N\ m$
K	Half saturation constant	$kg\ m^{-3}$
K_h	Horizontal eddy diffusivity	$500\ m^2\ s^{-1}$
K_v	Vertical eddy diffusivity	$10^{-4}\ m^2\ s^{-1}$
l	Typical length	m
L_b	Buoyancy length scale $(\varepsilon/N^3)^{1/2}$	$1–60\ m$
L_d	Diffusive length scale $2\pi\,(\nu D^2/\varepsilon)^{1/4}$ (Batchelor scale)	$0.2–1.0\ mm$
L_v	Viscous length scale $2\pi(\nu^3/\varepsilon)^{1/4}$ (Kolmogoroff scale)	$6–35\ mm$
M	molar concentration	$mol\ l^{-1} = g\text{-at}\ l^{-1}$
μM	micro-molar concentration	$\mu g\text{-at}\ l^{-1}$

Table A1 Variables, constants and non-dimensional numbers (*Continued*)

Symbol	Quantity or variable	Typical value and/or units
M_E	Ekman transport	kg m s^{-1}
N	Force, in newtons	kg m s^{-2}
N	Brunt–Väisälä frequency $(g/\rho\, d\rho/dz)^{1/2}$	$10^{-3}\text{–}10^{-2}\,\text{s}^{-1}$
P_a	Pressure, in Pascals	N m^{-2}
R_e	External Rossby radius $(gh)^{1/2}/f$	1000 km
R_i	Internal Rossby radius $(g'h)^{1/2}/f$	30 km
Re	Reynolds number ud/v	
ρ	Density	$1000\ \text{kg m}^{-3}$ (fresh water)
		$1026\ \text{kg m}^{-3}$ (salt water)
σ_t	Sigma-t (density anomaly)	$(\rho\text{–}1000)\ \text{kg m}^{-3}$
τ	Stress	N m^{-2}
u,v,w	Components of velocity along x, y, z	m s^{-1}
$\bar{u},\bar{v},\bar{w}$	Average velocity	m s^{-1}
u',v',w'	Fluctuating component of velocity	m s^{-1}
u_*	Friction velocity $\left(\dfrac{\tau}{\rho}\right)^{1/2}$	m s^{-1}
v	Kinematic viscosity	$10^{-6}\ \text{m}^2\,\text{s}^{-1}$ (water)
W	Power, in Watts	J s^{-1}
Ω	Angular velocity	rad s^{-1}
x,y,z	Position coordinates	m
ζ	Relative vorticity	s^{-1}

Table A.2 Basic units

Quantity	Name	Symbol
mass	gram	g
length	metre	m
time	second	s
amount of substance	mole	mol
temperature	degree Celsius	deg C, °C

Table A.3 Prefixes

Prefix	Factor	Symbol
pico	10^{-12}	p
nano	10^{-9}	n
micro	10^{-6}	μ
milli	10^{-3}	m
centi	10^{-2}	c
deci	10^{-1}	d
deca	10	da
hecto	10^2	h
kilo	10^3	k
mega	10^6	M
giga	10^9	G

References

Alldredge, A.L. and Hamner, W.M. (1980) Recurring aggregations of zooplankton by a tidal current. *Estuar. Coast. Mar. Sci.* **10**: 31–37.

Alldredge, A.L. and Cohen, Y. (1987) Can microscale chemical patches persist in the sea? Microelectrode study of marine snow, fecal pellets. *Science* **235**: 689–691.

Allen, P.M. (1985) Ecology, thermodynamics and self-organization: towards a new understanding of complexity, pp. 3–26. In: R.E. Ulanowicz and T. Platt. (Eds) *Ecosystem Theory for Biological Oceanography. Can. Bull. Fish. Aquat. Sci.* 213.

Anderson, S.M. and Charters, A.C. (1982) A fluid dynamic study of water flow through *Gelidium nudifrons. Limnol. Oceanogr.* **27**: 399–412.

Andrews, W.R.H. and Hutchings, L. (1980) Upwelling in the southern Benguela current. *Prog. Oceanogr.* **9**: 1–81.

Ansa-Emmin, M. (1982) Fisheries in the CINECA region. *Rapp. P.-v. Reun. Cons. Int. Explor. Mer.* **180**: 405–422.

Apel, J.R. (1987) *Principles of Ocean Physics.* Academic Press, New York.

Apel, J.R., Byrne, H.M., Proni, J.R. and Charnell, R.L. (1975) Observations of oceanic internal and surface waves from the Earth Resources Technology Satellite. *J. Geophys. Res.* **80**: 865–881.

Armi, L. and Farmer, D.M. (1988) The flow of Mediterranean water through the Strait of Gibraltar. *Prog. Oceanogr.* **21**: 1–105.

Armstrong, D.A., Mitchell-Innes, B.A., Verhaye-Dua, F., Waldron, H. and Hutchings, L. (1987) Physical and biological features across the upwelling front in the southern Benguela. In: A.I.L Payne, J.A. Gulland and K.H. Brink. (Eds) The Benguela and Comparable Ecosystems. *South African J. Mar. Sci.* **5**:171–190. Sea Fisheries Research Institute, Cape Town.

Astthorsson, O.S., Hallgrimsson, I. and Jonsson, G.S. (1983) Variations in zooplankton densities in Icelandic waters in spring during the years 1961–1982. *Rit. Fiskideildar (J. Mar. Res. Inst. Reykjavik)* **7**: 73–113.

Atkins, W.R.G. (1928) Seasonal variation in the phosphate and silicate content of sea water during 1926 and 1927 in relation to the phytoplankton crop. *J. Mar. Biol. Assoc.* **15**: 191–205.

Atkinson, L.P., O'Malley, P.G., Yoder, J.A. and Paffenhöfer, G.A. (1984) The effect of summertime shelf break upwelling on nutrient flux in southeastern United States continental shelf waters. *J. Mar. Res.* **42**: 969–993.

Atkinson, L.P. and Targett, T.E. (1983) Upwelling along the 60-m isobath from Cape Canaveral to Cape Hatteras and its relationship to fish distribution. *Deep-Sea Res.* **30**: 221–226.

Backus, R.H. and Bourne, D.W. (Eds) (1987) *Georges Bank.* M.I.T. Press, Cambridge, Mass. 593 pp.

Backus, R.H. and Craddock, J.E. (1982) Mesopelagic fishes in Gulf Stream cold core rings. *J. Mar. Res.* **40** (Suppl.): 1–20.

Bailey, K. M. (1981) Larval transport and recruitment of Pacific hake *Merluccius productus. Mar. Ecol. Prog. Ser.* **6**: 1–9.

435

Bakun, A. (1973) Coastal upwelling indices, west coast of North America 1946–71. NOAA Tech. Rep. NMFS SSRF-671. US Dept. Commerce, Seattle.

Bakun, A. and Parrish, R.H. (1982) Turbulence, transport and pelagic fish in the California and Peru current systems. *CalCOFI Rep.* **23**: 99–112.

Banse, K. (1987) Clouds, deep chlorophyll maxima and the nutrient supply to the mixed layer of stratified water bodies. *J. Plankt. Res.* **9**: 1031–1036.

Barber, R.T., Chavez, F.P. and Kogelschatz, J.E. (1985) Biological Effects of El Niño. pp. 399–438. In: M. Vegas (Ed.) *Seminario Regional Ciencas Tecnologia y Agression Ambiental: El Fenomeno 'El Niño'.* Contec Press, Lima, Peru.

Barber, R.T. and Smith, R.L. (1981) *Coastal Upwelling Ecosystems*, pp. 31–68. In: A.R. Longhurst (Ed.) *Analysis of Marine Ecosystems.* Academic Press, New York, 741pp.

Barnola, J.M., Raynaud, D., Korotkevich, Y.S. and Lorius, C. (1987) Vostok ice core provides 160,000-year record of atmospheric CO_2. *Nature* **329**: 408–412.

Barton, E.D., Huyer, A. and Smith, R.L. (1977) Temporal variation observed in the hydrographic regime near Cabo Corbeiro in the northwest African upwelling region, February to April 1974. *Deep-Sea Res.* **24**: 7–24.

Berg, H.C. and Purcell, E.M. (1977) Physics of chemoreception. *Biophys. J.* **20**: 193–215.

Bernal, P. and McGowan, J.A. (1981) Advection and upwelling in the California current. *Coast. Estuar. Sci.* **1**: 381–399.

Bienfang, P.K., Syper, J. and Laws, E. (1983) Sinking rate and pigment responses to light-limitation of a marine diatom: implications to dynamics of chlorophyll maximum layers. *Oceanol. Acta* **6**: 55–62.

Bigelow, H.B. (1927) Physical oceanography of the Gulf of Maine. *Bull. U.S. Bureau Fisheries* **40**: 511–1027.

Biggs, D.C. (1982) Zooplankton excretion and NH_4^+ cycling in near-surface waters of the Southern Ocean. 1. Ross Sea, Austral Summer 1977–78. *Polar Biol.* **1**: 55–67.

Blackburn, M. (1977) Studies on pelagic animal biomasses, pp. 283–299. In: N.R. Andersen and B.J. Zuharec (Eds) *Oceanic Sound Scattering Prediction.* Plenum Press, New York.

Blackburn, M. (1981) Low latitude gyral regions, pp. 3–30. In: A.R. Longhurst (Ed.) *Analysis of Marine Ecosystems.* Academic Press, London.

Blanton, J.O., Atkinson, L.P., de Castillejo, F.F. and Montero, A.L. (1984) Coastal upwelling off the Rias Bajas, Galicia, Northwest Spain. 1. Hydrographic studies. *Rapp. P.-v. Réun. Cons. Int. Explor. Mer.* **183**: 79–90.

Blanton, J.O., Tenore, K.R., Castillejo, F., Atkinson, L.P., Schwing, F.B. and Lavin, A. (1987) The relationship of upwelling to mussel production in the rias on the western coast of Spain. *J. Mar. Res.* **45**: 497–511.

Boëtius, J. and Harding, E.F. (1985) A re-examination of Johannes Schmidt's Atlantic eel investigations. *Dana* **4**: 129–162.

Boje, R. and Tomczak, M. (Eds) (1978) *Upwelling Ecosystems.* Springer-Verlag, Berlin. 303 pp.

Bousfield, E.L. (1955) Ecological control of the occurrence of barnacles in the Miramichi Estuary. *Bull. Nat. Mus. Can.* **137**: 1–69.

Bowden, K.F. (1983) *Physical Oceanography of Coastal Waters.* Ellis Horwood, New York, 302 pp.

Bowden, K.F. (1977) Heat budget considerations in the study of upwelling, pp. 277–290. In: M.Angel (Ed.) *A Voyage of Discovery.* Pergamon Press, New York.

Bowers, D.G. and Simpson, J.H. (1987) Mean position of tidal fronts in European-shelf seas. *Continental Shelf Res.* **7**: 35–44.

Bowman, M.J. and Iverson, R. L. (1978) Estuarine and plume fronts, pp. 87–104. In: M.J. Bowman and W.E. Esaias (Eds) *Oceanic Fronts in Coastal Processes.* Springer-Verlag, New York.

Bowman, M.J., Yentsch, C.M. and Peterson, W.J. (Eds) (1986) *Tidal Mixing and Plankton Dynamics.* Springer-Verlag, Berlin.

Brandt, A., Sarabun, C.C., Seliger, H.H. and Tyler, M.A. (1986) The effects of a broad spectrum of physical activity on the biological processes in the Chesapeake Bay, pp. 361–384. In: J.C.L. Nihoul (Ed.) *Marine Interfaces Ecohydrodynamics*. Elsevier, Amsterdam.

Brink, K.H. (1983) The near-surface dynamics of coastal upwelling. *Prog. Oceanogr.* 12: 223–257.

Brink, K.H., Halpern, D. and Smith, R.L. (1980) Circulation in the Peru upwelling system near 15 deg. S. *J. Geophys. Res.* 85: 4036–4048.

Broecker, W.S., Peteet, D.M. and Rind, D. (1985) Does the ocean–atmosphere system have more than one stable mode of operation? *Nature* 315: 21–26.

Broenkow, W.W. (1965) The distribution of nutrients in the Costa Rica dome in the eastern tropical Pacific Ocean. *Limnol. Oceanogr.* 10: 40–52.

Bucklin, A., Brandt, A. and Orr, M. (1989) Small-scale biophysical coupling. Preliminary report of workshop sponsored by the Office of Naval Research. Washington, DC.

Budyko, M.I. (1974) *Climate and Life*. Academic Press, New York, 508pp.

Bugden, G.L., Hargrave, B.T., Sinclair, M.M., Tang, C.L., Therriault, J.-C. and Yeats, P.A. (1982) Fresh water run-off effects in the marine environment: the Gulf of St Lawrence example. Can. Tech. Rep. Fish. Aquat. Sci. 1078, 89 pp.

Bumpus, D.F. (1976) Review of the physical oceanography of George's Bank. ICNAF Res. Bull. No. 12: 119–134.

Butman, C.A. (1986) Larval settlement of soft-sediment invertebrates: Some predictions based on an analysis of near-bottom profiles, pp. 487–513. In: J.C.L. Nihoul (Ed.) *Marine Interfaces Ecohydrodynamics*. Elsevier, Amsterdam.

Canby, T.Y. (1984) El Niño's ill wind. *National Geographic* 165: 144–183.

Chapman, P. and Shannon, L.V. (1985) The Benguela ecosystem, Part 2. Chemistry and related processes. *Oceanogr. Mar. Biol. Ann. Rev.* 23: 183–251.

Charlson, R.J., Lovelock, J.E., Andreae, M.O. and Warren, S.G. (1987) Oceanic phytoplankton, atmospheric sulphur, cloud albedo and climate. *Nature* 326: 655–661.

Charney, J.G. (1955) The Gulf Stream as an inertial boundary layer. *Proc. Natl Acad. Sci.* 41: 731–740.

Chavez, F.P. and Barber, R.T. (1987) An estimate of new production in the equatorial Pacific. *Deep-Sea Res.* 34: 1229–1243.

Chelton, D.B., Bernal, P.A. and McGowan, J.A. (1982) Large-scale physical and biological interaction in the California Current. *J. Mar. Res.* 40: 1095–1125.

Chenoweth, S.B., Libby, D.A., Stephenson, R.L. and Power, M.J. (1989) Origin and dispersal of larval herring (*Clupea harengus* L.) in coastal waters of eastern Maine and southwestern New Brunswick. *Can. J. Fish. Aquat. Sci.* 46: 624–632.

Chereskin, T.K. (1983) Generation of internal waves in Massachusetts Bay. *J. Geophys. Res.* 88: 2649–2661.

Cho, B.C. and Azam, F. (1988) Major role of bacteria in biogeochemical fluxes in the ocean's interior. *Nature* 332: 441–443.

Chriss, T.M. and Caldwell, D.R. (1984) Universal similarity and thickness of the viscous sublayer at the ocean floor. *J. Geophys. Res.* 89: 6403–6414.

Clarke, R.A., Hill, H.W., Reiniger, R.F. and Warren, B.A. (1980) Current system south and east of the Grand Banks of Newfoundland. *J. Phys. Oceanogr.* 10: 25–65.

Clarke, R.A., Swift, J.H., Reid, J.L. and Koltermann, K.P. 1991. The formation of the Greenland Sea deep water: Double diffusion or deep convection? *Deep-Sea Res.* 37: 1385–1424.

Cloern, J.E., Alpine, A.E., Cole, B.E., Wong, R.L.J., Arthur, J.F. and Ball, M.D. (1983) River discharge controls phytoplankton dynamics in the north San Francisco Bay estuary. *Estuar. Coast. Shelf Sci.* 16: 415–429.

Codispoti, L.A., Dugdale, R.C. and Minas, H.J. (1982) A comparison of the nutrient regimes off northwest Africa, Peru and Baja California. *Rapp. P.-v. Réun. Cons. Int. Explor. Mer.* 180: 184–201.

Codispoti, L.A. and Friederich, G.E. (1978) Local and mesoscale influences on nutrient variability in the northwest African upwelling region near Cabo Corbeiro. *Deep-Sea Res.* **25**: 751–770.

Coelho, M.L. (1985) Review of the influence of oceanographic factors on cephalopod distribution and life history. *NAFO Sci. Coun. Studies* **9**: 47–57.

Colebrook, J.M. (1979) Continuous plankton records: seasonal cycles of phytoplankton and copepods in the North Atlantic Ocean and the North Sea. *Mar. Biol.* **51**: 23–32.

Colebrook, J.M. (1985) Continuous plankton records: overwintering and annual fluctuations in the abundance of zooplankton. *Mar. Biol.* **84**: 261–265.

Colebrook, J.M. (1986) Environmental influences on long-term variability in marine plankton. *Hydrobiologia* **142**: 309–325.

Colebrook, J.M. and Robinson, G.A. (1965) Continuous plankton records: seasonal cycles of phytoplankton and copepods in the north east Atlantic and the North Sea. *Bull. Mar. Ecol.* **6**: 123–129.

Conte, M.H., Bishop, J.K.B. and Backus, R.H. (1986) Nonmigratory, 12kHz, deep scattering layers of Sargasso Sea origin in warm-core rings. *Deep-Sea Res.* **33**: 1869–1884.

Corner, E.D.S., Head, R.N. and Kilvington, C.C. (1972) On the nutrition and metabolism of zooplankton. VIII. The grazing of *Biddulphia* cells by *Calanus helgolandicus. J. Mar. Biol. Ass. U.K.* **52**: 847–861.

Cowles, T.J., Roman, M.R., Gauzens, A.L. and Copley, N.J. (1987) Short-term changes in the biology of a warm-core ring: zooplankton biomass and grazing. *Limnol. Oceanogr.* **32**: 653–664.

Creutzberg, F. (1985) A persistent chlorophyll-a maximum coinciding with an enriched benthic zone, pp. 97–108. In: P.E. Gibbs (Ed.) *Proceedings of the Nineteenth European Marine Biology Symposium.* Cambridge University Press, Cambridge.

Crisp, D.J. (1960) Factors influencing growth rate in *Balanus balanoides. J. Anim. Ecol.* **29**: 95–116.

Csanady, G.T. (1979) The life and death of a warm-core ring. *J. Geophys. Res.* **84**: (C2): 777–780.

Csanady, G.T. (1981) Circulation in the coastal ocean. *Adv. Geophys.* **23**: 101–183.

Currie, J.T. (1984) Microscale nutrient patches: Do they matter to the plankton? *Limnol. Oceanogr.* **29**: 211–214.

Cushing, D.H. (1968) Grazing by herbivorous copepods in the sea. *J. Cons. Int. Explor. Mer.* **32**: 70–82.

Cushing, D.H. (1969) Upwelling and fish production. F.A.O. Fish. Tech. Paper 84: 38 pp.

Cushing, D.H. (1971) Upwelling and the production of fish. *Adv. Mar. Biol.* **9**: 255–334.

Cushing, D.H. (1975) *Marine Ecology and Fisheries.* Cambridge University Press, Cambridge, UK. 278 pp.

Cushing, D.H. (1978) Upper trophic levels in upwelling areas, pp. 101–110. In: R. Boje and M.Tomczak (Eds) *Upwelling Ecosystems.* Springer-Verlag, Berlin.

Cushing, D.H. (1979) Climatic variation and marine fisheries, pp. 608–627. In: *Proceedings of the World Climate Conference, Geneva 1979.* World Meteorological Organization, Geneva.

Cushing, D.H. (1982) *Climate and Fisheries.* Academic Press, London. 373 pp.

Cushing, D.H. (1986) The migration of larval and juvenile fish from spawning ground to nursery ground. *J. Cons. Int. Explor. Mer.* **43**: 43–49.

Cushing, D.H. (1988) The northerly wind, pp. 235–244. In: B.J. Rothschild (Ed.) *Towards a Theory on Biological–Physical Interactions in the World Ocean.* Kluwer, Amsterdam.

Cushing, D.H. and Dickson, R.R. (1976) The biological response in the sea to climate change. *Adv. Mar. Biol.* **14**: 1–122.

da Silva, A.J. (1986) River run-off and shrimp abundance in a tropical coastal ecosystem: The example of the Sofala Bank (Central Mozambique), pp. 329–344. In: S. Skreslet (Ed.) *The Role of Freshwater Outflow in Coastal Marine Ecosystems.* Springer-Verlag, Berlin.

Dandonneau, Y. (1988) Seasonal or aperiodic cessation of oligotrophy in the tropical Pacific Ocean, pp. 137–156. In: B.J. Rothschild (Ed.) *Towards a Theory on Biological–Physical Interactions in the World Ocean.* Kluwer, Dordrecht.

Davies, J.M. and Payne, R. (1984) Supply of organic matter to the sediment in the North Sea during a spring phytoplankton bloom. *Mar. Biol.* **78**: 313–324.

de Young, B. and Pond, S. (1988) The deepwater exchange cycle in Indian Arm, British Columbia. *Estuar. Coast. Shelf Sci.* **26**: 285–308.

Deardorff, J.W., Willis, G.E. and Lilly, D.K. (1969) Laboratory investigation of non-steady penetrative convection. *J. Fluid Mech.* **35**: 7–31.

Defant, A. (1958) *Ebb and Flow.* University of Michigan Press, Ann Arbor, Mich., 121 pp.

Denman, K.L. (1973) A time-dependent model of the upper ocean. *J. Phys. Oceanogr.* **3**: 173–184.

Denman, K.L. and Miyake, M. (1973) Upper layer modification at ocean station Papa: observations and simulation. *J.Phys. Oceanogr.* **3**: 185–196.

Dickson, R.R., Kelly, P.M., Colebrook, J.M., Wooster, W.S. and Cushing, D.H. (1988a) North winds and production in the eastern North Atlantic. *J. Plankt. Res.* **10**: 151–169.

Dickson, R.R., Lamb, H.H., Malmberg, S.-A. and Colebrook, J.M. (1975) Climatic reversal in the northern North Atlantic. *Nature* **256**: 479–482.

Dickson, R.R., Meincke, J., Malmberg, S.-A. and Lee, A.J. (1988b) The 'great salinity anomaly' in the northern North Atlantic 1968–1982. *Prog. Oceanogr.* **20**: 103–151.

Dietrich, G. (1950) Die anomale jahresschwankung des wärmeinhalts im Englischen Kanal, ihre Ursachen und auswirkungen. *Deutsche Hydrograph. Zeit.* **3**: 184–201.

Dietrich, G., Kalle, K., Krauss, W. and Siedler, G. (1980) *General Oceanography: An Introduction,* 2nd Ed. Wiley, New York, 626 pp.

Dole, R.M. (1989) Life cycles of persistent anomalies. Part I: Evolution of 500 mb height fields. *Monthly Weather Rev.* **117**: 177–211.

Doty, M.S. (1971) Antecedent event influence on benthic marine algal standing crops in Hawaii. *J. Exp. Mar. Biol. Ecol.* **6**: 161–166.

Doty, M.S. and Oguri, M. (1956) The island mass effect. *J. Cons. Int. Explor. Mer.* **22**: 33–37.

Drinkwater, K.F. (1986) On the role of freshwater outflow on coastal ecosystems — A workshop summary, pp. 429–438. In: S. Skreslet (Ed.) *The Role of Freshwater Outflow in Coastal Marine Ecosystems.* Springer-Verlag, Berlin.

Drinkwater, K.F. (1987) 'Sutcliffe revisited': Previously published correlations between fish stocks and environmental indices and their recent performance. pp. 41–61. In: R.I. Perry and K.T. Frank (Eds) *Environmental Effects on Recruitment to Canadian Atlantic Fish Stocks.* Can. Tech. Rep. Fish. Aquat. Sci. 1556.

Drinkwater, K.F. (1988) The effect of freshwater discharge on the marine environment, pp. 415–430. In: W. Nicholaichuk and F. Quinn (Eds) *Proceedings of a Symposium on the Interbasin Transfer of Water: Impact and Research Needs for Canada.* Department of the Environment, National Hydrology Research Centre, Saskatoon, Canada, 504 pp.

Dugdale, R.C. and Goering, J.J. (1967) Uptake of new and regenerated forms of nitrogen in primary productivity. *Limnol. Oceanogr.* **12**: 196–206.

Dyer, K. (1973) *Estuaries: A Physical Introduction.* Wiley, London.

El-Sayed, S.L. (1978) Primary productivity and estimates of potential yields of the Southern Ocean, pp. 141–160. In: M.A. McWhinnie (Ed.) *Polar Research to the Present, and the Future.* Westview Press, Boulder, Colorado.

Enfield, D.B. (1989) El Niño past and present. *Rev. Geophys.* **27**: 159–187.

Eppley, R.W. (1980) Estimating phytoplankton growth rates in the central oligotrophic oceans, pp. 231–242. In: P.G. Falkowski (Ed.) *Primary Productivity in the Sea.* Plenum, New York.

Eppley, R.W., Renger, E.H. and Harrison, W.G. (1979) Nitrate and phytoplankton production in southern California waters. *Limnol. Oceanogr.* **24**: 483–494.

Eppley, R.W. and Peterson, B.J. (1979) Particulate organic matter flux and planktonic new production in the deep ocean. *Nature* **282**: 677–680.

Ewing, G. (1950) Slicks, surface films and internal waves. *J. Mar. Res.* **9**: 161–187.

Falkowski, P.G. and Wirick, C.D. (1981) A simulation model of the effects of vertical mixing on primary productivity. *Mar. Biol.* **65**: 69–75.

Farmer, D.M. and Freeland, H.J. (1983) The physical oceanography of fjords. *Prog. Oceanogr.* **12**: 147–219.

Fasham, M.J.R. and Pugh, P.R. (1976) Observations on the horizontal coherence of chlorophyll-a and temperature. *Deep-Sea Res.* **23**: 527–538.

Flagg, C.N. and Beardsley, R.C. (1975) The 1974 M.I.T. New England shelf dynamics experiment (March 1974). Part 1. Hydrography. *M.I.T. Report* 75-1.

Flierl, G.R. and Wroblewski, J.S. (1985) The possible influence of warm core Gulf Stream rings upon shelf water larval fish distribution. *Fish. Bull.* **83**: 313–330.

Fofonoff, N.P. (1981) The Gulf Stream system, pp. 112–139 In: B.A. Warren and C. Wunsch (Eds) *Evolution of Physical Oceanography.* M.I.T. Press. Cambridge, Mass.

Fogg, G.E. (1985) Biological activities at a front in the western Irish Sea, pp. 87–96. In: P.E. Gibbs (Ed.) *Proceedings of the Nineteenth European Marine Biology Symposium.* Cambridge University Press, Cambridge.

Folland, C.K., Parker, D.E. and Kates, F.E. (1984) Worldwide marine temperature fluctuations 1856–1981. *Nature* **310**: 670–673.

Fortier, L. and Leggett, W.C. (1982) Fickian transport and the dispersal of fish larvae in estuaries. *Can. J. Fish. Aquat. Sci.* **39**: 1150–1163.

Fortier, L. and Leggett, W.C. (1983) Vertical migrations and transport of larval fish in a partially mixed estuary. *Can. J. Fish. Aquat. Sci.* **40**: 1543–1555.

Fortier, L. and Leggett, W.C. (1984) Small-scale covariability in the abundance of fish larvae and their prey. *Can. J. Fish. Aquat. Sci.* **41**: 502–512.

Fournier, R.O. (1978) Biological aspects of the Nova Scotian shelfbreak fronts, pp. 69–77. In: M.J. Bowman and W.E. Esaias. *Oceanic Fronts in Coastal Processes.* Springer-Verlag, New York.

Frank, K.T. (1986) Ecological significance of the ctenophore *Pleurobrachia pileus* off southwestern Nova Scotia. *Can. J. Fish. Aquat. Sci.* **43**: 211–222.

Frank, K.T., Perry, R.I., Drinkwater, K.F. and Lear, W.H. (1988) Changes in the fisheries of Atlantic Canada associated with global increases in atmospheric carbon dioxide: a preliminary report. *Can. Tech. Rep. Fish. Aquat. Sci.* 1652, 52 pp.

Franks, P.J.S., Wroblewski, J.S. and Flierl, G.R. (1986) Prediction of phytoplankton growth in response to the frictional decay of a warm-core ring. *J. Geophys. Res.* **91** (C6): 7603–7610.

Fransz, H.G. and Gieskes, W.W.C. (1984) The unbalance of phytoplankton and copepods in the North Sea. *Rapp. P. -v. Réun. Cons. Int. Explor. Mer.* **183**: 218–225.

Frechette, M. and Bourget, E. (1985a) Food-limited growth of *Mytilus edulis* L. in relation to the benthic boundary layer. *Can. J. Fish. Aquat. Sci.* **42**: 1166–1170.

Frechette, M. and Bourget, E. (1985b) Energy flow between the pelagic and benthic zones: Factors controlling particulate organic matter available to an intertidal mussel bed. *Can. J. Fish. Aquat. Sci.* **42**: 1158–1165.

Frechette, M., Butman, C.A. and Geyer, W.R. (1989) The importance of boundary-layer flows in supplying phytoplankton to the benthic suspension feeder *Mytilus edulis* L. *Limnol. Oceanogr.* **34**: 19–36.

Friedlander, A. and Smith, D. (1983) Sand Lance larvae found in entrainment feature associated with a warm core ring off Hudson Canyon. *Coast. Oceanogr. Climatol. News* **2**: 3–4. (Obtainable from Center for Ocean Management Studies, University of Rhode Island, Kingston RI 02881.)

Friedman, M.M. and Strickler, J.R. (1975) Chemoreceptors and feeding in calanoid copepods (Arthropoda: Crustacea). *Proc. Natl Acad. Sci.* **72**: 4185–4188.

Fuglister, F.C. (1960) *Atlantic Ocean Atlas,* Vol.1. Woods Hole Oceanographic Institution.

Fulton, J.D. and LeBrasseur, R.J. (1985) Interannual shifting of the subarctic boundary and some of the biotic effects on juvenile salmonids, pp. 237–252 In: W.S. Wooster and D.L. Fluharty (Eds) *El Niño North.* Washington Sea Grant Program. University of Washington, Seattle.

Gade, H.G. and Edwards, A. (1980) Deep water renewal in fjords, pp. 453–490 In: H.J. Freeland, D.M. Farmer and C.D. Levings (Eds) *Fjord Oceanography.* Plenum Press, New York.

Gallegos, C.L. and Platt, T. (1982) Phytoplankton production and water motion in surface mixed layers. *Deep-Sea Res.* **29**: 65–76.

Gargett, A.E. (1984) Vertical eddy diffusivity in the ocean interior. *J. Mar. Res.* **42**: 359–395.

Gargett, A.E., Osborn, T.R. and Nasmyth, P.W. (1984) Local isotropy and the decay of turbulence in a stratified fluid. *J. Fluid Mech.* **144**: 231–280.

Garrett, C.J.R. and J.W. Loder (1981) Dynamical aspects of shallow-sea fronts. *Phil. Trans. Roy. Soc. Lond. A* **302**: 562–581.

Garvine, R.W. (1986) The role of brackish plumes in open shelf waters, pp. 47–65. In: S. Skreslet (Ed.) *The Role of Freshwater Outflow in Coastal Marine Ecosystems.* Springer-Verlag, Berlin.

Garvine, R.W. and Monk, J.D. (1974) Frontal structure of a river plume. *J. Geophys. Res.* **79**: 2251–2259.

Gavis, J. (1976) Munk and Riley revisited: nutrient diffusion transport and rates of phytoplankton growth. *J. Mar. Res.* **34**: 161–179.

Geosecs (1981) *Atlantic Expedition,* Vol. 2. Sections and Profiles. *Pacific Expedition,* Vol. 4, Sections and Profiles. National Science Foundation, Washington.

Gerard, V.A. (1982) *In situ* water motion and nutrient uptake by the giant kelp *Macrocystis pyrifera. Mar. Biol.* **69**: 51–54.

Gerard, V.A. (1987) Hydrodynamic streamlining of *Laminaria saccharina* Lamour in response to mechanical stress. *J. Exp. Mar. Biol. Ecol.* **107**: 237–244.

Gerard, V.A. and Mann, K.H. (1979) Growth and production of *Laminaria longicruris* (Phaeophyta) populations exposed to different intensities of water movement. *J. Phycol.* **15**: 33–41.

Gerritsen, J. and Strickler, J.R. (1977) Encounter probabilities and community structure in zooplankton: a mathematical model. *J. Fish. Res. Board Can.* **34**: 73–82.

Gill, A.E. (1982) *Atmosphere–Ocean Dynamics.* International Geophysics Series 30. Academic Press, New York.

Glantz, M.H. (1984) Floods, fires and famine: is El Niño to blame? *Oceanus* **27** (2): 14–19.

Glantz, M.H. (1985) Climate and fisheries: a Peruvian case study. CPPS Boletin ERFEN **15**: 13–31.

Glémarec, M. (1973) The benthic community of the European North Atlantic continental shelf. *Oceanogr. Mar. Biol. Ann. Rev.* **11**: 263–289.

Glover, R.S. (1967) The continuous plankton recorder survey of the North Atlantic. *Symp. Zool. Soc. Lond.* **19**: 189–210.

Glynn, P.W. (1985) El Niño-associated disturbance to coral reefs and post-disturbance mortality by *Acanthaster planci. Mar. Ecol. Prog. Ser.* **26**: 295–300.

Goldman, J.C. (1984) Oceanic nutrient cycles, pp. 137–170. In: M.J.R. Fasham (Ed.) *Flows of Energy and Materials in Marine Ecosystems: Theory and Practice.* Plenum Press, New York.

Goldman, J.C. (1988) Spatial and temporal discontinuities of biological processes in surface waters, pp. 273–296. In: B.J. Rothschild (Ed.) *Towards a Theory on Biological-Physical Interactions in the World Ocean.* Kluwer Academic, Amsterdam.

Goodrich, D.M. (1988) On meteorologically-induced flushing in three U.S. east-coast estuaries. *Estuar. Coast. Shelf Sci.* **26**: 111–121.

Gordon, A.L. (1986) Interocean exchange of thermocline water. *J. Geophys. Res.* **91**: 5037–5046.

Gower, J.F.R., Denman, K.L. and Holyer, R.J. (1980) Phytoplankton patchiness indicates the fluctuations spectrum of mesoscale oceanic structure. *Nature* **288**: 157–159.

Graham, J.J. (1972) Retention of larval herring within the Sheepscot estuary of Maine. *Fish. Bull.* **70**: 299–305.

Graham, N.E. and White, W.B. (1988) The El Niño cycle: a natural oscillator of the Pacific ocean–atmosphere system. *Science* **240**: 1293–1302.

Gran, H.H. 1931. On the conditions for the production of plankton in the sea. *Rapp. P.-v. Réun. Cons. Int. Explor. Mer.* **75**: 37–46.

Gran, H.H. and Braarud, T. (1935) A quantitative study of the phytoplankton in the Bay of Fundy and the Gulf of Maine (including observations on hydrography, chemistry and turbidity). *J. Biol. Bd. Canada* **1**: 279–433.

Graybill, M.R. and Hodder, J. (1985) Effects of the 1982–83 El Niño on reproduction of six species of seabirds in Oregon. University of Washington, Seattle. pp. 205–210. In: W.S. Wooster and D.L. Fluharty (Eds) *El Niño North: Niño Effects in the Eastern Subarctic Pacific Ocean.* Washington Sea Grant Program, University of Washington, Seattle.

Greenberg, D.A. (1983) Modeling the mean barotropic circulation in the Bay of Fundy and Gulf of Maine. *J. Phys. Oceanogr.* **13**: 886–904.

Grindley, J.R. (1964) On the effect of low-salinity water on the vertical migration of estuarine zooplankton. *Nature (Lond.)* **203**: 781–782.

Hachey, H.B. (1935) The effect of a storm on an inshore area with markedly stratified waters. *J. Biol. Board Can.* **1**: 227–237.

Halpern, D. (1971) Observations on short-period internal waves in Massachusetts Bay. *J. Mar. Res.* **29**: 116–132.

Hamilton, K. and Mysak, L.A. (1986) Possible effect of the Sitka Eddy on sockeye (*Onchorhynchus nerka*) and pink salmon (*Onchorhynchus gorbusha*) migration off southeast Alaska. *Can. J. Fish. Aquat. Sci.* **43**: 498–504.

Haney, J.C. (1986) Seabird segregation at Gulf Stream frontal eddies. *Mar. Ecol. Prog. Ser.* **28**: 279–285.

Hansen, J. and Lebedeff, S. (1988) Global surface air temperature: update through 1987. *Geophys. Res. Lett.* **15**: 323–326,

Hansen, J., Fung, I., Lacis, A., Rind, D., Lebedeff, S., Ruedy, R., Russell, G. and Stone, P. (1988) Global climate changes as forecast by the Goddard Institute for Space Studies three dimensional model. *J. Geophys. Res.* **93**: 9341–9364.

Hare, F.K. (1988) The global greenhouse effect. *Conference Proceedings: The Changing Atmosphere: Implications for Global Security.* WMO No. 710. World Meteorological Organization. Geneva. 483p.

Hartline, B.K. (1980) Coastal upwelling: physical factors feed fish. *Science* **208**: 38–40.

Harvey, J. and Arhan, M. (1988) The water masses of the central North Atlantic in 1983–84. *J. Phys. Oceanogr.* **18**: 1855–1875.

Harvey, J.G. and Theodorou, A. (1986) The circulation of the Norwegian Sea overflow in the Eastern North Atlantic. *Oceanologica Acta* **9**: 393–402.

Hays, J.D., Imbrie, J. and Shackleton, N.J. (1976) Variations in the earth orbit: Pacemaker of the ice ages. *Science* **194**: 1121–1131.

Hayward, T.L. and Venrick, E.L. (1982) Relation between surface chlorophyll, integrated chlorophyll and integrated primary production. *Marine Biology* **69**: 247–252.

Heaps, N.S. (1980) A mechanism for local upwelling along the European continental slope. *Oceanologica Acta* **3**: 449–454.

Heath, R.A. (1973) Flushing of coastal embayments by changes in atmospheric conditions. *Limnol. Oceanogr.* **18**: 849–862.

Hellerman, S. and Rosenstein, M. (1983) Normal monthly wind stress over the world ocean with error estimates. *J. Phys. Oceanogr.* **13**: 1093–1104.

Hempel, G. (1982) The Canary current: studies of an upwelling system. *Rapp. P.-v. Réun. Cons. Int. Explor. Mer.* **180**: 1–455.

Hendershott, M.C. (1981) Long waves and ocean tides, pp. 292–341. In: B.A. Warren and C. Wunch. (Eds) *Evolution of Physical Oceanography*. M.I.T. Press, Cambridge, Mass.

Herbland, A. and Voituriez, B. (1979) Hydrological structure analysis for estimating the primary production in the tropical Atlantic Ocean. *J. Mar. Res.* **37**: 87–101.

Herman, A.W., Sameoto, D.D. and Longhurst, A.R. (1981) Vertical and horizontal distribution patterns of copepods near the shelf-break south of Nova Scotia. *Can. J. Fish. Aquat. Sci.* **38**: 1065–1076.

Hitchcock, G.L., Langdon, C. and Smayda, T.J. (1985) Seasonal variations in the phytoplankton biomass and productivity of a warm-core Gulf Stream ring. *Deep-Sea Res.* **32**: 1287–1300.

Hitchcock, G.L., Langdon, C. and Smayda, T.J. (1987) Short-term changes in the biology of a Gulf Stream warm-core ring: Phytoplankton biomass and productivity. *Limnol. Oceanogr.* **32**: 919–928.

Hogg, N.G., Pickart, R.S., Hendry, R.M. and Smethie, W.J. (1986) The northern recirculation gyre of the Gulf Stream. *Deep-Sea Res.* **33**: 1139–1165.

Holligan, P.M. (1981) Biological implications of fronts on the northwest European continental shelf. *Phil. Trans. Roy. Soc. A* **302**: 547–562.

Holligan, P.M., Harris, R.P., Newell, R.C., Harbour, D.S., Head, R.N., Linley, E.A.S., Lucas, M.I., Tranter, P.R.G. and Weekly, C.M. (1984a) Vertical distribution and partitioning of organic carbon in mixed, frontal and stratified waters of the English Channel. *Mar. Ecol. Prog. Ser.* **14**: 111–127.

Holligan, P.M., Williams, P.J. LeB., Purdie, D. and Harris, R.P. (1984b) Photosynthesis, respiration and nitrogen supply of plankton populations in stratified, frontal and tidally mixed shelf waters. *Mar. Ecol. Prog. Ser.* **17**: 201–213.

Holligan, P.M., Pingree, R.D. and Mardell, G.T. (1985) Oceanic solitons, nutrient pulses and phytoplankton growth. *Nature* **314**: 348–350.

Holm-Hansen, O. (1985) Nutrient cycles in Antarctic marine ecosystems. pp. 6–10. In: W.R. Siegfried, P.R. Condy and R.M. Laws (Eds) *Antarctic Nutrient Cycles and Food Webs*. Springer-Verlag, Berlin, 700 pp.

Holm-Hansen, O., El-Sayed, S.Z.; Franceschini, G.A. and Cuhel, R.L. (1977) Primary production and the factors controlling phytoplankton growth in the Southern Ocean. pp. 11–50. In: G.A. Llano (Ed.) *Adaptations within Antarctic Ecosystems*. Smithsonian Institution, Washington. 1252 pp.

Horel, J.D. and Wallace, J.M. (1981) Planetary-scale atmospheric phenomena associated with the southern oscillation. *Monthly Weather Rev.* **109**: 813–829.

Horne, E.P.W. (1978) Physical aspects of the Nova Scotian shelfbreak fronts, pp. 59–86. In: M.J. Bowman and W.E. Esaias (Eds) *Oceanic Fronts in Coastal Processes*. Springer-Verlag, New York.

Horne, E.P.W., Loder, J.W., Harrison, W.G., Mohn, R., Lewis, M.R., Irwin, B. and Platt, T. (1989) Nitrate supply and demand at the Georges Bank tidal front. *Scientia Marina* **53** (2–3): 145–158.

Hunter, J.R. (1972) Swimming and feeding behaviour of larval anchovy, *Engraulis mordax. Fish. Bull. US* **73**: 453–462.

Huntsman, S.A. and Barber, R.T. (1977) Primary production off northwest Africa: the relationship to wind and nutrient conditions. *Deep-Sea Res.* **24**: 25–33.

Hutchinson, G.E. (1967) *A Treatise on Limnology*, Vol. 2. Wiley, New York. 1116 pp.

Huthnance, J.M. (1973) Tidal current asymmetries over the Norfolk sandbanks. *Estuar.*

Coast. Mar. Sci. **1**: 89–99.

Huyer, A. (1976) A comparison of upwelling events in two locations: Oregon and northwest Africa. *J. Mar. Res.* 34: 531–546.

Huyer, A. (1983) Coastal upwelling in the California system. *Progr. Oceanogr.* **12**: 259–284.

Iles, T.D. and Sinclair, M. (1982) Atlantic herring: stock discreetness and abundance. *Science* **215**: 627–633.

Imbrie, J. and Kipp, N.G. (1971) A new micropalaeontological method for quantitative palaeoclimatology: application to a late Pleistocene Caribbean core, pp. 77–181. In: K.K.Turekian (Ed.) *The Late Cenozoic Glacial Ages.* Yale University Press, New Haven.

Isemer, H.J. and Hasse, L. (1987) *The Bunker Climate Atlas of the North Atlantic Ocean,* Vol.2: Air–sea interactions. Springer-Verlag, Berlin.

Ivanoff, A. (1977) Oceanic absorption of solar energy, pp. 47–71. In: E.B. Kraus (Ed.) Modelling and Prediction of the Upper Layers of the Ocean. Pergamon Press, New York.

Jackson, G.A. (1980) Phytoplankton growth and zooplankton grazing in oligotrophic oceans. *Nature* **284**: 439–441.

Jamart, B.M., Winter, D.F., Banse, K., Anderson, G.C. and Lam, R.K. (1977) A theoretical study of phytoplankton growth and nutrient distribution in the Pacific Ocean of the northwest U.S. coast. *Deep-Sea Res.* **24**: 753–773.

Jamart, B.M., Winter, D.F. and Banse, K. (1979) Sensitivity analysis of a mathematical model of phytoplankton growth and nutrient distribution in the Pacific Ocean off the northwest US coast. *J. Plankt. Res.* **1**: 267–290.

Jenkins, W.J. (1982) Oxygen utilization rates in North Atlantic subtropical gyre and primary production in oligotrophic systems. *Nature* **300**: 246–248.

Jenkins, W.J. and Goldman, J.C. (1985) Seasonal oxygen cycling and primary production in the Sargasso Sea. *J. Mar. Res.* **43**: 465–491.

Jerlov, N.J. (1976) *Marine Optics.* Elsevier Oceanography Series 14. Elsevier. Amsterdam.

Jones, B.H. and Halpern, D. (1981) Biological and physical aspects of a coastal upwelling event observed during March–April 1974 off northwest Africa. *Deep-Sea Res.* **28**: 71–81.

Jones, P.D., Wigley, T.M.L., Folland, C.K., Parker, D.E., Angell, J.K., Lebedeff, S. and Hansen, J.E. (1988) Evidence for global warming in the past decade. *Nature* **332**: 790.

Kamykowsky, D., McCollum, S.A. and Kirkpatrick, G.J. (1988) Observations and a model concerning the translational velocity of a photosynthetic marine dinoflagellate under variable environmental conditions. *Limnol. Oceanogr.* **33**: 57–65.

Karl, D.M., Knauer, G.A. and Martin, J.H. (1988) Downward flux of particulate organic matter in the ocean: a particle decomposition paradox. *Nature* **332**: 438–441.

Kerr, R.A. (1983) Orbital variation-ice age link strengthened. *Science* **219**: 272–274.

Kerr, R.A. (1985) Long-lived small eddies are criss-crossing the oceans carrying the effects of local mixing hundreds and even thousands of kilometres. *Science* **230**: 793.

Kerr, R.A. (1989) Greenhouse skeptic out in the cold. *Science* **246**: 1118–1119.

Kidd, R. and Sander, F. (1979) Influence of the Amazon River discharge on the marine production system off Barbados, West Indies. *J. Mar. Res.* **37**: 669–681.

Killworth, P.D. (1977) Mixing on the Weddell Sea continental slope. *Deep-Sea Res.* **24**: 427–448.

Kinder, T.H. and Bryden, H.L. (1988) Gibraltar experiment. Woods Hole Oceanogr. Inst. Tech. Rept. WHOI-88-30.

King, F.D. and Devol, A.H. (1979) Estimates of vertical eddy diffusion through the thermocline from phytoplankton nitrate uptake rates in the mixed layer of the eastern tropical Pacific. *Limnol. Oceanogr.* **24**: 645–651.

King, F.D., Cucci, T.L. and Townsend, D.W. (1987) Microzooplankton and macrozooplankton glutamate dehydrogenase as indices of the relative contribution of these fractions to ammonium regeneration in the Gulf of Maine. *J. Plankt. Res.* **9**: 277–289.

Kingsford, M.J. and Choat, J.H. (1986) Influence of surface slicks on the distribution and inshore movement of small fish. *Mar. Biol.* **91**: 161–171.

Kjelson, M.A., Raquel, P.F. and Fisher, F.W. (1982) Life history of fall-run juvenile Chinook salmon, *Oncorhynchus tshawytscha*, in the Sacramento–San Joaquin estuary, California, pp. 393–411. In: V.S. Kennedy (Ed.) *Estuarine Comparisons*. Academic Press, New York.

Kleckner, R.C. and McCleave, J.D. (1985) Spatial and temporal distribution of American eel larvae in relation to North Atlantic Ocean current systems. *Dana* **4**: 67–92.

Klein, B. and Siedler, G. (1989) On the origin of the Azores current. *J. Geophys. Res.* **94**: 6159–6168.

Koehl, M.A.R. (1984) Mechanisms of particle capture by copepods at low Reynolds numbers: possible modes of selective feeding, pp. 135–166. In: D.G. Meyers and J.R. Strickler (Eds) *Trophic Interaction within Aquatic Ecosystems*. American Association for the Advancement of Science. Washington, DC.

Koehl. M.A.R. (1986) Seaweeds on moving water: form and mechanical function, pp. 603–634. In: T.J. Givnish (Ed.) *On the Economy of Plant Form and Function*. Cambridge University Press, Cambridge, 717 pp.

Koehl, M.A.R. and Alberte, R.S. (1988) Flow, flapping and photosynthesis of *Nereocystis leutkeana*: a functional comparison of undulate and flat blade morphologies. *Mar. Biol.* **99**: 435–444.

Koehl, M.A.R. and Strickler, J.R. (1981) Copepod feeding currents: Food capture at low Reynolds number. *Limnol. Oceanogr.* **26**: 1062–1073.

Krauss, W. (1986) The North Atlantic current. *J. Geophys. Res.* **91**: 5061–5074.

Kremer. J.N. and Nixon, S.W. (1978) *A Coastal Marine Ecosystem: Simulation and Analysis*. Springer-Verlag, New York, 217 pp.

Lamb, H.H. (1969) The new look of climatology. *Nature* **223**: 1209–1215.

Lasker, R. (1975) Field criteria for the survival of anchovy larvae: the relation between inshore chlorophyll maximum layers and successful first feeding. *Fish. Bull. U.S.* **73**: 847–855.

Lasker, R. (1978) The relation between oceanographic conditions and larval anchovy food in the California current: identification of factors leading to recruitment failure. *Rapp. P.-v. Réun. Cons. Int. Explor. Mer.* **173**: 212–230.

Lasker, R. (1988) Food chains and fisheries: an assessment after 20 years, pp. 173–182. In: B.J. Rothschild (Ed.) *Towards a Theory on Biological–Physical Interactions in the World Ocean*. Kluwer, Dordrecht.

Laurs, R.M., Yuen, H.S.H. and Johnson, J.H. (1977) Small-scale movements of albacore, *Thunnus alalunga*, in relation to ocean features as indicated by ultrasonic tracking and oceanographic sampling. *Fish. Bull. US* **75**: 347–355.

Laws, E.A., DiTullio, G.R. and Redalje, D.G. (1987) High phytoplankton growth and production rates in the North Pacific subtropical gyre. *Limnol. Oceanogr.* **32**: 905–918.

Lazier, J.R.N. (1980) Oceanographic conditions at ocean weather ship Bravo, 1964–1974. *Atmosph.–Ocean* **18**: 227–238.

Lazier, J.R.N. (1988) Temperature and salinity changes in the deep Labrador Sea, 1962–1986. *Deep-Sea Res.* **35**: 1247–1253.

Lazier, J.R.N. and Mann, K.H. (1989) Turbulence and diffusive layers around small organisms. *Deep-Sea Res.* **36**: 1721–1733.

Le Fèvre, J. (1986) Aspects of the biology of frontal systems. *Adv. Mar. Biol.* **23**: 164–299.

Le Fèvre, J. and Grall J.R. (1970) On the relationships of *Noctiluca* swarming off the western coast of Brittany with hydrological features and plankton characteristics of the environment. *J. Exp. Mar. Biol. Ecol.* **4**: 287–306.

Le Fèvre, J. and Frontier, S. (1988) Influence of temporal characteristics of physical phenomena on plankton dynamics, as shown by northwest European marine ecosystems pp. 245–272. In: B.J. Rothchild (Ed.) *Towards a Theory on Biological–Physical Interactions in the World Ocean.* Kluwer, Dordrecht.

LeBlond, P.H., Hickey, B.M. and Thompson, R.E. (1986) Runoff-driven coastal flow off British Columbia, pp. 309–318. In: S. Skreslet (Ed) *The Role of Freshwater Outflow in Coastal Marine Systems.* Springer-Verlag, New York.

Lee, T.N., Atkinson L.P. and Legeckis, R. (1981) Observations on a Gulf Stream frontal eddy on the Georgia continental shelf, April 1977. *Deep-Sea Res.* 28: 347–378.

Leetma, A., McCreary, J.P. and Moore, D.W. (1981) Equatorial currents: observations and theory, pp. 186–196. In: B.A. Warren and C. Wunch (Eds) *Evolution of Physical Oceanography.* M.I.T. Press, Cambridge, Mass.

Legendre, L. (1981) Hydrodynamic control of marine phytoplankton production: the paradox of stability, pp. 191–207. In: J.C.J. Nihoul (Ed.) *Ecohydrodynamics. Proceedings of the 12th International Liège Colloquium on Ocean Hydrodynamics.* Elsevier, Amsterdam.

Legendre, L. and Demers, S. (1984) Towards dynamic biological oceanography and limnology. *Can. J. Fish. Aquat. Sci.* 41: 2–19.

Legendre, L. and Demers, S. (1985) Auxiliary energy, ergoclines and aquatic biological production. *Naturaliste Can. (Rev. écol. syst.)* 112: 5–14.

Legendre, L., Demers S. and LeFaivre, D. (1986) Biological production at marine ergoclines, pp. 1–54. In: J.C.J. Nihoul (Ed.) *Marine Interfaces Ecohydrodynamics.* Elsevier, Amsterdam.

Lehman, J.T. and Scavia, D. (1982a) Microscale patchiness of nutrients in plankton communities. *Science* 216: 729–730.

Lehman, J.T. and Scavia, D. (1982b) Microscale nutrient patches produced by zooplankton. *Proc. Natl Acad. Sci.* 79: 5001–5005.

Levasseur, M., Therriault, J.–C. and Legendre, L. (1984) Hierarchical control of phytoplankton succession by physical factors. *Mar. Ecol. Prog. Ser.* 19: 211–222.

Lewis, M.R. and Platt, T. (1982) Scales of variability in estuarine ecosystems, pp. 3–20. In: V.S. Kennedy (Ed.) *Estuarine Comparisons.* Academic Press, New York.

Lewis, M.R., Cullen, J.J. and Platt, T. (1983) Phytoplankton and thermal structure in the upper ocean: Consequences of nonuniformity in chlorophyll profile. *J. Geophys. Res.* 88 (C4): 2565–2570.

Lewis, M.R., Cullen, J.J. and Platt, T. (1984a) Relationships between vertical mixing and photoadaptation of phytoplankton: similarity criteria. *Mar. Ecol. Prog. Ser.* 15: 141–149.

Lewis, M.R., Horne, E.P.W., Cullen, J.J., Oakey, N.S. and Platt, T. (1984b) Turbulent motions may control phytoplankton photosynthesis in the upper ocean. *Nature* 311: 49–50.

Lewis, M.R., Harrison, W.G., Oakey, N.S., Hebert, D. and Platt, T. (1986) Vertical nitrate fluxes in the oligotrophic ocean. *Science* 234: 870–873.

Loder, J.W. (1980) Topographic rectification of tidal currents on the sides of George's Bank. *J. Phys. Oceanogr.* 10: 1399–1416.

Loder, J.W. and Greenberg, D.A. (1986) Predicted positions of tidal fronts in the Gulf of Maine region. *Continental Shelf Res.* 6: 397–414.

Loder, J.W. and Platt, T. (1985) Physical controls on phytoplankton production at tidal fronts, pp. 3–21. In: P.E. Gibbs (Ed.) *Proceedings of the 19th European Marine Biology Symposium.* Cambridge University Press, Cambridge.

Loder, J.W. and Wright, D.G. (1985) Tidal rectification and frontal circulation on the sides of George's Bank. *J. Mar. Res.* 43: 581–604.

Long, R.R. (1954) Some aspects of the flow of stratified fluids II. Experiments with a two-fluid system. *Tellus* 6: 97–115.

Longhurst, A.R. (1971) The clupeid resources of tropical seas. *Mar. Biol. Ann. Rev.* **9**: 349–385.

Longhurst, A.R. (1976) Interactions between zooplankton and phytoplankton profiles in the eastern tropical Pacific Ocean. *Deep-Sea Res.* **23**: 729–754.

Longhurst, A.R. (1981) Significance of spatial variability, pp. 415–441. In: A.R. Longhurst (Ed.) *Analysis of Marine Ecosystems.* Academic Press, London, 741 pp.

Longhurst, A.R. and Herman, A.W. (1981) Do oceanic zooplankton aggregate at, or near, the deep chlorophyll maximum? *J. Mar. Res.* **39**: 353–356.

Longhurst, A.R. and Williams, R. (1979) Materials for plankton modelling: vertical distribution of Atlantic zooplankton in summer. *J. Plankt. Res.* **1**: 1–28.

Lorius, C., Barkov, N.I., Jouzel, J., Korotkevich, Y.S., Kolyakov, V.M. and Raynaud, D. (1988) Antarctic ice core: CO_2 and climatic change over the last climatic cycle. EOS. *Trans. Am. Geophys. Union* **69**: 681–684.

Lovelock, J.E. (1979) *Gaia.* Oxford University Press, Oxford.

MacIsaac, J.J., Dugdale, R.C., Barber, R.T., Blasco, D. and Packard, T.T. (1985) Primary production cycle in an upwelling center. *Deep Sea Res.* **32**: 503–529.

Malone, T.C. (1982) Factors influencing the fate of sewage-derived nutrients in the lower Hudson estuary and New York Bight, pp. 301–320. In: G.F. Mayer (Ed.) *Ecological Stress in the New York Bight: Science and Management.* Estuarine Research Foundation, Columbia, South Carolina.

Malone, T.C. (1984) Anthropogenic nitrogen loading and assimilation capacity of the Hudson River estuarine system, USA, pp. 291–311. In: V.S. Kennedy (Ed.) *The Estuary as a Filter.* Academic Press, Orlando, Florida.

Mann, C.R. (1967) The termination of the Gulf Stream and the beginning of the North Atlantic Current. *J. Geophys. Res.* **14**: 337–359.

Mann, K.H. (1984) Fish production in open ocean ecosystems, pp. 435–458. In: M.J.R. Fasham (Ed.) *Flows of Energy and Materials in Marine Ecosystems: Theory and Practice.* Plenum Press, New York.

Mann. K.H. (1988) Towards predictive models for coastal marine ecosystems, pp. 291–316. In: L.R. Pomeroy and J.J. Alberts. (Eds) *Concepts of Ecosystem Ecology.* Springer-Verlag, New York.

Margalef, R. (1974) Asociacion o exclusion en le distribucion des especies del mimo género en algas unicelulares. *Mem. R. Acad. Cienc. Artes Barcelona* **42**: 353–372.

Margalef, R. (1978a) Life-forms of phytoplankton as survival alternatives in an unstable environment. *Oceanologica Acta* **1**: 493–509.

Margalef, R. (1978b) What is an upwelling system? pp. 12–14. In: R. Boje and M. Tomczak (Eds) *Upwelling Ecosystems.* Springer-Verlag, Berlin.

Markle D.F., Scott, W.B. and Kohler, A.C. (1980) New and rare records of Canadian fishes and the influence of hydrography on resident and nonresident Scotian shelf ichthyofauna. *Can. J. Fish. Aquat. Sci.* **37**: 49–65.

Marra, J. (1978) Effect of short-term variations in light intensity on photosynthesis of a marine phytoplankter: a laboratory simulation study. *Mar. Biol.* **46**: 191–202.

Marshall, S.M. and Orr, A.P. (1928) The photosynthesis of diatom cultures in the sea. *J. Mar. Biol. Assoc.* **15**: 321–364.

Martin, J.H., Knauer, G.A., Karl, D.M. and Broenkow, W.W. (1987) VERTEX: Carbon cycling in the northeast Pacific. *Deep-Sea Res.* **34**: 267–285.

Mathisen, O.A., Thorne, R.E., Trumble, R.J. and Blackburn, M. (1978) Food consumption of pelagic fish in an upwelling area, pp. 111–123. In: R.Boje and M.Tomczak. (Eds) *Upwelling Ecosystems.* Springer-Verlag, Berlin, 303 pp.

May, R.M. and Oster, G.F. (1976) Bifurcations and dynamic complexity in simple ecological models. *Am. Nat.* **110**: 573–599.

Mazé, R. (1983) Formation d'ondes internes induits dans un golfe par le passage d'un dépression et par la marée. Application au Golfe de Gascogne. *Ann. Hydrograph.* **8**: 45–58.

Mazé, R., Camus, Y. and Le Tareau, J.Y. (1986) Formation de gradients thermiques à la surface de l'ocean, au-dessus d'un talus, par interaction entre les ondes internes et le mélange dû au vent. *J. Cons. Perm. Int. Explor. Mer.* **42**: 221–240.

McCarthy, J.J. and Goldman, J.C. (1979) Nitrogenous nutrition of marine phytoplankton in nutrient-depleted waters. *Science* **203**: 670–672.

McCarthy, J.J. and Nevins, J.L. (1986) Sources of nitrogen for primary production in warm-core rings 79-E and 81-D. *Limnol. Oceanogr.* **31**: 690–700.

McClimans, T.A. (1986) Laboratory modelling of dynamic processes in fjords and shelf waters, pp. 67–84. In: S. Skreslet (Ed.) *The Role of Freshwater Outflow in Coastal Marine Ecosystems.* Springer-Verlag, New York.

McGowan, J.A. (1985) El Niño 1893 in the southern California bight, pp. 166–184. In: W.S. Wooster and D.L. Fluharty (Eds) *El Niño north: Niño Effects in the Eastern Subarctic Pacific Ocean.* Washington Sea Grant Program, University of Washington, Seattle.

McGowan, J.A. and Hayward, T.L. (1978) Mixing and oceanic productivity. *Deep-Sea Res.* **25**: 771–793.

Miller, G.R. (1966) The flux of tidal energy out of the deep oceans. *J. Geophys. Res.* **71**: 2485–2489.

Miller, C.B., Batchelder, H.P., Brodeur, R.D. and Pearcy, W.G. (1985) Response of zooplankton and ichthyoplankton off Oregon to the El Niño event of 1983. pp. 185–187 In: W.S. Wooster and D.L. Fluharty (Eds) *El Niño North: Niño Effects in the Eastern Subarctic Pacific Ocean.* Univ. of Washington, Seattle.

Milliman, J.D. and Boyle, E. (1975) Biological uptake of dissolved silica in the Amazon River estuary. *Science* **189**: 995–997.

Mills, E.L. (1989) *Biological Oceanography: An Early History, 1870–1960.* Cornell University Press, Ithaca, NY, 378 pp.

Minas, H.J., Codispoti, L.A. and Dugdale, R.C. (1982) Nutrients and primary production in the upwelling region off northwest Africa. *Rapp. P.-v. Réun. Cons. Int. Explor. Mer.* **180**: 148–183.

Minas, H.J., Minas, M. and Packard, T.T. (1986) Productivity in upwelling areas deduced from hydrographic and chemical fields. *Limnol. Oceanogr.* **31**: 1182–1206.

Mitchell, J.F.B. (1989) The 'greenhouse' effect and climate change. *Rev. Geophys.* **27**: 115–139.

Mix, A.C. (1989) Influence of productivity variations on long-term atmospheric CO_2. *Nature* **337**: 541–544.

Mommaerts, J.P., Pichot, G., Ozer, J., Adam. Y. and Baeyens, W. (1984) Nitrogen cycling and budget in Belgian coastal waters: North Sea areas with and without river inputs. *Rapp. P.-v. Réun. Cons. Int. Explor. Mer.* **183**: 57–69.

Mooers, C.N.K., Flagg, C.N. and Boicourt, W.C. (1978) Prograde and retrograde fronts, pp 43–58. In: M.J. Bowman and W.E. Esaias (Eds) *Oceanic Fronts in Coastal Processes.* Springer-Verlag, New York.

Muck, P. and Sanchez, G. (1987) The importance of mackerel and horse mackerel predation for the Peruvian anchoveta stock (a population and feeding model). pp. 279–293. In: D.Pauly and I. Tsukayama (Eds) *The Peruvian Anchoveta and its Upwelling System: Three Decades of Change.* International Center for Living Resources Management, Manila, the Phillipines.

Muench, R.D. and Heggie, D.T. (1978) Deep water exchange in Alaskan subarctic fjords, pp. 239–267. In: B. Kjerve (Ed.) *Estuarine Transport Processes.* Univ. South Carolina Press.

Müller-Navara, S. and Mittelstaedt, E. (1985) *Schadstoffbereitung und Schadstoffbelastung in der Nordsee. Eine Medellstudie.* Deutsches Hydrographisches Institut, Hamburg, 1–50.

Munk, W. and Riley, G.A. (1952) Absorption of nutrients by aquatic plants. *J. Mar. Res.* **11**: 215–240.

Muschenheim, D.K. (1987a) The dynamics of near-bed seston flux and suspension-feeding benthos. *J. Mar. Res.* **45**: 473–496.

Muschenheim, D.K. (1987b) The role of hydrodynamic sorting of seston in the nutrition of a benthic suspension feeder, *Spio setosa* (Polychaeta: Spionidae). *Biol. Oceanogr.* **4**: 265–288.

Myers, R.A. and Drinkwater, K. (1989) The influence of Gulf Stream warm core rings on recruitment of fish in the northwest Atlantic. *J. Mar. Res.* **47**: 635–656.

Mysak, L.A. (1986) El Niño, interannual variability and fisheries in the northeast Pacific Ocean. *Can. J. Fish. Aquat. Sci.* **43**: 464–497.

Namias, J. and Cayan, D.R. (1984) El Niño: Implications for forecasting. *Oceanus* **27** (2) 41–47.

Narimousa, S. and Maxworthy, T. (1985) Two-layer model of shear-driven coastal upwelling in the presence of bottom topography. *J. Fluid. Mech.* **159**: 503–531.

Narimousa, S. and Maxworthy, T. (1987) Coastal upwelling on a sloping bottom: the formation of plumes, jets and pinched-off cyclones. *J. Fluid. Mech.* **176**: 169–190.

Nehring, D. and Holzlöhner, S. (1982) Investigations on the relationship between environmental conditions and distribution of *Sardinella pilchardus* in the shelf area off northwest Africa. *Rapp. P.-v. Réun. Cons. Int. Explor. Mer.* **180**: 342–344.

New, A.L. (1988) Internal tidal mixing in the Bay of Biscay. *Deep-Sea Res.* **35**: 691–697.

Nichols, F.H. (1985) Increased benthic grazing: an alternative explanation for low phytoplankton biomass in northern San Francisco Bay during the 1976–1977 drought. *Estuar. Coast. Shelf Sci.* **21**: 379–388.

Nichols, F.H., Cloern, J.E., Luoma, S.N. and Peterson, D.H. (1986) The modification of an estuary. *Science* **231**: 567–573.

Nieland, H. (1982) The food of *Sardinella aurita* and *Sardinella eba* off the coast of Senegal. *Rapp. P.-v. Reun. Cons. Int. Explor. Mer.* **180**: 369–373.

Norcross, J.J. and Stanley, E.M. (1967) Inferred surface and bottom drift, June 1963 through October 1964. Circulation of shelf waters off the Chesapeake bight. *Prof. Pap. Environ. Sci. Serv. Admin.* **3**: 11–42.

O'Boyle, R.M., Sinclair, M., Conover, R.J., Mann, K.H. and Kohler, A.C. (1984) Temporal and spatial distribution of ichthyoplankton communities of the Scotian shelf in relation to biological, hydrological and physiographic features. *Rapp. P.-v. Reun. Cons. Int. Explor. Mer.* **183**: 27–40.

O'Reilly, J.E., Evans-Zetlin, C. and Thomas, J.P. (1981) The relationship between surface and average water column concentrations of chlorophyll-a in northwestern Atlantic shelf water. ICES CM 1981/L: 17, 19 pp. (mimeo).

O'Reilly, J.E. and Busch, D.A. (1984) Phytoplankton primary production on the northwest Atlantic shelf. *Rap. P.-v. Reun. Cons. Int. Explor. Mer.* **183**: 255–268.

Oakey, N.S. and Elliott, J.A. (1980) Dissipation in the mixed layer near Emerald Basin. pp. 123–133. In: J.C.L. Nihoul (Ed.) *Marine Turbulence.* Elsevier, Amsterdam, 270 pp.

Odum, H.T. (1967a) Biological circuits and marine systems of Texas, pp. 99–157. In: T.A. Olson and F.J. Burgess (Eds) *Pollution and Marine Ecology.* Wiley, New York.

Odum, H.T. (1967b) Energetics of world food production, pp. 55–94. In: I.L. Bennett (chairman). *The World Food Problem,* a report of the President's Science Advisory Committee, Vol. 3. The White House, Washington, DC.

Okubo, A. (1971) Horizontal and vertical mixing in the sea, pp. 89–168. In: D.W. Hood (Ed.) *Impingement of Man on the Oceans.* Wiley-Interscience, New York.

Okubo, A. (1987) Fantastic voyage into the deep: Marine biofluid mechanics, pp. 32–47. In: E.Teromoto and M. Yamaguti (Eds) *Mathematical Topics in Population Biology, Morphogenesis and Neurosciences.* Springer-Verlag, New York.

Oliver, J.K. and Willis, B.L. (1987) Coral-spawn slicks in the Great Barrier Reef: Preliminary observations. *Mar. Biol.* **94**: 521–529.

Olson, D.B. (1986) Lateral exchange with Gulf Stream warm-core ring surface layers.

Deep-Sea Res. **33**: 1691–1704.

Olson, D.B., Schott, F.A., Zantopp, R.J. and Leaman, K.D. (1984) The mean circulation east of the Bahamas as determined from a recent measurement program and historical XBT data. *J. Phys. Oceanogr.* **14**: 1470–1487.

Ortner, P.B., Wiebe, P.H. and Cox, J.L. (1980) Relationships between oceanic epizooplankton distributions and the seasonal deep chlorophyll maximum in the northwest Atlantic Ocean. *J. Mar. Res.* **38**: 507–531.

Ortner, P.B., Wiebe, P.H. and Cox, J.L. (1981) Reply to 'Do oceanic zooplankton aggregate at, or near, the deep chlorophyll maximum?' *J. Mar.Res.* **39**: 357–359.

Osborn, T.R. (1978) Measurements of energy dissipation adjacent to an island. *J. Geophys. Res.* **83**: 2939–2957.

Osborn, T.R. (1980) Estimates of the local rate of vertical diffusion from dissipation measurements. *J. Phys. Oceanogr.* **10**: 83–89.

Owen, R.W. (1968) Oceanographic conditions in the northeast Pacific and their relation to the albacore fishery. *Fish. Bull. U.S.* **66**: 503–526.

Packard, T.T., Blasco, D. and Barber, R.T. (1978) *Mesodinium rubrum* in the Baja California upwelling system, pp. 73–89. In: R. Boje and M. Tomczak. *Upwelling Ecosystems.* Springer-Verlag, Berlin.

Palomera, I. and Rubies, P. (1982) Kinds and distribution of fish eggs and larvae off northwest Africa in April/May 1973. *Rapp. P.-v. Réun. Cons. Int. Explor. Mer.* **180**: 356–358.

Pape, E.H. and Garvine, R.W. (1982) The subtidal circulation in Delaware Bay and adjacent shelf waters. *J. Geophys. Res.* **87**: 7955–7970.

Parker, C.E. (1971) Gulf Stream rings in the Sargasso Sea. *Deep-Sea Res.* **18**: 981–993.

Parrish, R.H., Bakun, A., Husby, D.M. and Nelson, C.S. (1983) Comparative climatology of selected environmental processes in relation to eastern boundary current pelagic fish production, pp. 731–777. In: G.D. Sharp and J. Csirke. (Eds) *Proceedings of the Expert Consultation to Examine Changes in Abundance and Species Composition of Neritic Fish Resources.* F.A.O. Fisheries Report 291: (3). Food and Agriculture Organization of the United Nations, Rome.

Parsons, T.R. and Lalli, C.M. (1988) Comparative oceanic ecology of the plankton communities of the subarctic Atlantic and Pacific Oceans. *Oceanogr. Mar. Biol. Ann. Rev.* **26**: 317–359.

Pasciak, W.J. and Gavis, J. (1974) Transport limitation of nutrient uptake in phytoplankton. *Limnol. Oceanogr.* **19**: 881–888.

Pauly, D. and Tsukayama, I. (1987) On the implementation of management-oriented fishery research: the case of the Peruvian anchoveta, pp. 1–13. In: D. Pauly and I. Tsukayama (Eds) *The Peruvian Anchoveta and its Upwelling Ecosystem: Three Decades of Change.* ICLARM Stud. Rev. 1987.

Payne, A.I.L., Gulland, J.A. and Brink, K.H. (Eds) (1987) The Benguela and comparable ecosystems. *S. Afr. J. Mar. Sci.* **5**: 1–956.

Pearcy, W. (1983) Abiotic variations in regional environments, pp. 30–34. In: W.S. Wooster (Ed.) *From Year to Year.* Washington Sea Grant Publication, University of Washington, Seattle.

Pearcy, W., Fisher, J., Brodeur, R. and Johnson, S. (1985) Effects of the 1983 El Niño on coastal nekton off Oregon and Washington, pp. 188–204. In: W.S. Wooster and D.L. Fluharty (Eds) *El Niño North: Niño Effects in the Eastern Subarctic Pacific Ocean.* Washington Sea Grant Program, University of Washington, Seattle.

Pearcy, W.G. and Keene, D.F. (1974) Remote sensing of water colour and sea surface temperatures off the Oregon coast. *Limnol. Oceanogr.* **19**: 573–583.

Pedlosky, J. (1990) The dynamics of the oceanic subtropical gyres. *Science* **248**: 316–322.

Peffley, M.B. and O'Brien, J.J. (1976) A three-dimensional simulation of coastal upwelling off Oregon. *J. Phys. Oceanogr.* **6**: 164–180.

Peinert, R. (1986) Production, grazing and sedimentation in the Norwegian coastal current, pp. 361–374. In: S. Skreslet (Ed.) *The Role of Freshwater Outflow in Coastal Marine Ecosystems.* Springer-Verlag, Berlin.

Peterson, W.T., Miller, C.B. and Hutchinson, A. (1979) Zonation and maintenance of copepod populations in the Oregon upwelling zone. *Deep-Sea Res.* **26**: 467–494.

Petrie, B., Topliss, B.J. and Wright, D.W. (1987) Coastal upwelling and eddy development off Nova Scotia. *J. Geophys. Res.* **29** (C12): 12979–12991.

Petterssen, S. (1969) *Introduction to Meteorology.* McGraw-Hill, New York.

Phillips, O.M. (1977) *The Dynamics of the Upper Ocean*, 2nd Edn. Cambridge University Press, Cambridge.

Pingree, R.D. (1978) Cyclonic eddies and cross-frontal mixing. *J. Mar. Biol. Ass. U.K.* **58**: 955–963.

Pingree, R.D. (1979) Baroclinic eddies bordering the Celtic Sea in late summer. *J. Mar. Biol. Ass. U.K.* **59**: 689–698.

Pingree, R.D., Bowman, M.J. and Esaias, W.E. (1978a) Headland fronts, pp. 78–86. In: M.J. Bowman and W.E. Esaias (Eds) *Oceanic Fronts in Coastal Processes.* Springer-Verlag, Berlin.

Pingree, R.D., Forster, G.R. and Morrison, G.K. (1974) Turbulent convergent tidal fronts. *J. Mar. Biol. Ass. U.K.* **54**: 469–479.

Pingree, R.D. and Griffiths, D.K. (1980) A numerical model of the M2 tide in the Gulf of St Lawrence. *Oceanologica Acta* **3**: 221–225.

Pingree, R.D., Holligan, P.M., Mardell, G.T. and Head, R.N. (1976) The influence of physical stability on spring, summer and autumn phytoplankton blooms in the Celtic Sea. *J. Mar. Biol. Ass. U.K.* **56**: 845–873.

Pingree, R.D., Holligan, P.M. and Mardell, G.T. (1978b). The effects of vertical stability on phytoplankton distributions in the summer on the northwest European shelf. *Deep-Sea Res.* **25**: 1011–1028.

Pingree, R.D. and Mardell, G.T. (1981) Slope turbulence, internal waves and phytoplankton growth at the Celtic Sea shelf-break. *Phil. Trans. Roy. Soc. Lond. A* **302**: 663–682.

Pingree, R.D. and Mardell, G.T. (1985) Solitary internal waves in the Celtic Sea. *Prog. Oceanogr.* **14**: 431–444.

Pingree, R.D., Mardell, G.T. and New, A.L. (1986) Propagation of the internal tides from the upper slopes of the Bay of Biscay. *Nature* **321**: 154–158.

Pingree, R.D. and Pennycuik, L. (1975) Transfer of heat, fresh water and nutrients through the seasonal thermocline. *J. Mar. Biol. Ass. U.K.* **55**: 261–274.

Pingree, R.D., Pugh, P.R., Holligan, P.M. and Forster, G.R. (1975) Summer phytoplankton blooms and red tides along tidal fronts in the approaches to the English Channel. *Nature* **258** 672–677.

Platt, T. (1984) Primary productivity in the central North Pacific: comparison of oxygen and carbon fluxes. *Deep-Sea Res.* **31**: 1311–1319.

Platt, T. and Harrison, W.G. (1986) Reconciliation of carbon and oxygen fluxes in the upper ocean. *Deep-Sea Res.* **33**: 55–58.

Platt, T., Harrison, W.G., Lewis, M.R., Li, W.K.W., Sathyendranath, S., Smith, R.E., Vezina, A.F. (1989) Biological production of the oceans: the case for consensus. *Mar. Ecol. Prog. Ser.* **52**: 77–88.

Platt, T. and Irwin, B. (1968) Primary production measurements in St Margaret's Bay, 1967. Fish. Res. Bd. Can. Tech. Rpt. 77. Dartmouth, N.S. 123 pp.

Platt, T., Prakash, A. and Irwin, B. (1972) Phytoplankton, nutrients and flushing of inlets on the coast of Nova Scotia. *Nat. Can.* **99**: 253–261.

Platt, T. and Subba Rao, D.V. (1975) Primary production of marine microphytes, pp. 249–80. In: J.P. Cooper (Ed.) *Photosynthesis and Productivity in Different Environments.* Cambridge University Press, Cambridge.

Platt, T. and Sathyendranath, S. (1988) Oceanic primary production: estimation by remote sensing at local and regional scales. *Science* **241**: 1613–1620.

Pond, S. and Pickard, G.L. (1983) *Introductory Dynamical Oceanography,* (2nd Edn). Pergamon Press, Oxford.

Prandtl, L. (1969) *The Essentials of Fluid Dynamics.* Blackie, London, 451 pp.

Price, H.J. and Paffenhofer, G.-A. (1980) Kinematographic analyses of the feeding of calanoid copepods over a range of algal sizes and concentrations. Abstr. Papers A.S.L.O. 3rd Winter Meeting.

Price, J.F., Weller, R.A. and Schudlich, R.R. (1987) Wind-driven ocean currents and Ekman transport. *Science* **238**:1534–1538.

Pritchard, D.W. (1967) What is an estuary: physical viewpoint, pp. 52–63. In: G.H. Lauff (Ed.) *Estuaries.* American Association for the Advancement of Science, Washington, DC.

Prospero, J.M. and Savoie, D.L. (1989) Comparison of oceanic and continental sources of non-sea-salt sulphate over the Pacific Ocean. *Nature* **339**: 685–687.

Provenzano, A.J., McConaugh, J.R., Philips, K.B., Johnson, D.F. and Clark, J. (1983) Vertical distribution of first stage larvae of the blue crab *Callinectes sapidus* at the mouth of Chesapeake Bay. *Estuar. Coast. Shelf Sci.* **16**: 489–499.

Purcell, E.M. (1977) Life at low Reynolds number. *Am. J. Physics* **45**: 3–11.

Purcell, E.M. (1978) The effect of fluid motions on the absorption of molecules by suspended particles. *J. Fluid Mech.* **84**: 551–559.

Rasmusson, E.M. and Carpenter, T.H. (1982) Variations in tropical sea surface temperature and surface wind fields associated with the southern oscillation/El Niño. *Monthly Weather Rev.* **110**: 354–384.

Rasmusson, E.M. and Wallace, J.M. (1983) Meteorological aspects of the El Niño/ southern oscillation. *Science* **222**: 1195–1202.

Rattray, M. and Hansen, D.V. (1962) A similarity solution for circulation in an estuary. *J. Mar. Res.* **20**: 121–133.

Reid, J.L. and Shulenberger, E. (1986) Oxygen saturation and carbon uptake near 28 N, 155 W. *Deep-Sea Res.* **33**: 267–271.

Revelante, N. and Gilmartin, M. (1976) The effects of Po River discharge on phytoplankton dynamics in the northern Adriatic Sea. *Mar. Biol.* **34**: 259–271.

Revelle, R. (1982) Carbon dioxide and world climate. *Sci. Am.* **247**: 35–43.

Revelle, R. (1983) The oceans and the carbon dioxide problem. Oceanus **26**: 3–9.

Rhines, P. (1986) Vorticity dynamics of the oceanic general circulation. *Ann. Rev. Fluid Mech.* **18**: 433–497.

Richards, F.A. (Ed.) (1981) *Coastal Upwelling.* American Geophysical Union. Washington, DC. 529 pp.

Richardson, C.A., Crisp, D.J. and Runham, N.W. (1980) Factors influencing shell growth in *Cerastoderma edule. Proc. R. Soc. Lond. B* **210**: 513–521.

Richardson, P.L. (1983) Gulf Stream Rings, pp. 19–45. In: A.R. Robinson (Ed.) *Eddies in Marine Science.* Springer-Verlag, New York.

Richardson, P.L., Cheney, R.E. and Worthington, L.V. (1978) A census of Gulf Stream rings, spring 1985. *J. Geophys. Res.* **83**: 6136–6145.

Richman, J.G., de Szoeke R.A. and Davis, R.E. (1987) Measurements of near-surface shear in the ocean. *J. Geophys. Res.* **92**: (C3): 2851–2858.

Riley, G.A. (1937) The significance of the Mississippi River drainage for biological conditions in the northern Gulf of Mexico. *J. Mar. Res.* **1**: 60–74.

Riley, G.A. (1941) Plankton studies IV: George's Bank. *Bull. Bingham Oceanogr. Coll.* **7**: 1–73.

Riley, G.A. (1942) The relationship of vertical turbulence and spring diatom flowering. *J. Mar. Res.* **5**: 67–87.

Riley, G.A. (1970) Particulate organic matter in sea water. *Adv. Mar. Biol.* **8**: 1–118.

Riley, G.A., Stommel, H. and Bumpus, D.F. (1949) Quantitative ecology of the plankton of the western North Atlantic. *Bull. Bingham Oceanogr. Coll.* **12**: 1–169.

Roberts, J. (1975) Internal gravity waves in the ocean. Marcell Dekker, New York.

Roberts, L. (1989) Global warming: blaming the sun. *Science* **246**: 992–993.

Robinson, A.R. (1983) Overview and summary of eddy science, pp. 3–15. In: A.R. Robinson (Ed.) *Eddies in Marine Science*. Springer-Verlag, New York.

Roden, G.I. (1961) On the wind-driven circulation in the Gulf of Tehuantepec and its effects on sea-surface temperature. *Geofis. Int.* **1**: 55–76.

Rogers, J.C. (1984) The association between the North Atlantic oscillation and the southern oscillation in the northern hemisphere. *Monthly Weather Rev.* **112**: 1999–2015.

Roman, M.R., Gauzens, A.L. and Cowles, T.J. (1985) Temporal and spatial changes in epipelagic microplankton and mesoplankton biomass in warm-core Gulf Stream Ring 82-B. *Deep-Sea Res.* **32**: 1007–1022.

Rothschild, B.J. (1988) Brodynamics of the sea: the ecology of high dimensionality systems, pp. 527–548. In: B.J. Rothschild (Ed.) *Toward a Theory on Biological–Physical Interactions in the World Ocean*. Kluwer, Dordrecht, 650 pp.

Rowe, G.T., Clifford, C.H. and Smith, K.L., Jr. (1977) Nitrogen regeneration in sediments off Cap Blanc, Spanish Sahara. *Deep-Sea Res.* **24**: 57–63.

Royce, W.F., Smith, L.S. and Hartt, A.C. (1968) Models of oceanic migrations of Pacific salmon and comments on guidance mechanisms. *Fish. Bull.* **66**: 441–462.

Rubenstein, D.I. and Koehl, M.A.R. (1977) The mechanism of filter feeding: Some theoretical considerations. *Am. Nat.* **111**: 981–994.

Russell, F.S. (1973) A summary of the observations of the occurrence of planktonic stages of fish off Plymouth 1924–72. *J. Mar. Biol. Ass. U.K.* **53**: 347–355.

Ryther, J.H. (1967) Occurrence of red water off Peru. *Nature*, **214**: 1318–1319.

Ryther, J.H. (1969) Photosynthesis and fish production in the sea. *Science* **166**: 72–80.

Sandstrom, H. and Elliott, J.A. (1984) Internal tides and solitons on the Scotian shelf: a nutrient pump at work. *J. Geophys. Res.* **89**: 6415–6426.

Sarnthein, M., Winn, K., Duplessy, J.-C. and Fontugne, M.R. (1988) Global variations of surface ocean productivity in low and mid-latitudes: influence on CO_2 reservoirs of the deep ocean and the atmosphere during the last 21,000 years. *Paleoceanography* **3**: 361–399.

Saunders, P.M. (1982) Circulation in the eastern North Atlantic. *J. Mar. Res.* **40**: (Suppl.) 641–657.

Savidge, G. (1976) A preliminary study of the distribution of chlorophyll-a in the vicinity of fronts in the Celtic and western Irish Seas. *Estuar. Coast. Mar. Sci.* **4**: 617–625.

Savidge, G. and Foster, P. (1978) Phytoplankton biology of a thermal front in the Celtic Sea. *Nature* **271**: 155–157.

Scheltema, R.S. (1966) Evidence for transatlantic transport of gastropod larvae belonging to the genus *Cymatium*. *Deep-Sea Res.* **13**: 83–95.

Schmidt, J. (1922) The breeding places of the eel. *Phil. Trans. Roy, Soc. B.* **211**: 178–208.

Schneider, S.H. (1989) The greenhouse effect: science and policy. *Science* **243**: 771–781.

Schumacher, G.J. and Whitford, L.A. (1965) Respiration and ^{32}P uptake in various species of freshwater algae as affected by current. *J. Phycol.* **1**: 78–80.

Schwartz, S.E. (1988) Are global cloud albedo and climate controlled by marine phytoplankton? *Nature* **336**: 441–445.

Shanks, A.L. (1983) Surface slicks associated with tidally forced internal waves may transport pelagic larvae of benthic invertebrates and fishes shoreward. *Mar. Ecol. Prog. Ser.* **13**: 311–315.

Shanks, A.L. (1985) Behavioural basis of internal-wave induced shoreward transport of megalopae of the crab *Pachygrapsus crassipes*. *Mar. Ecol. Prog. Ser.* **24**: 289–295.

Shanks, A.L. and Trent, J.D. (1979) Marine snow: Microscale nutrient patches. *Limnol. Oceanogr.* **24**: 850–854.

Shanks, A.L. and Wright, W.G. (1987) Internal-wave-mediated shoreward transport of cyprids, megalopae, and gammarids and correlated longshore differences in the settling rate of intertidal barnacles. *J. Exp. Mar. Biol. Ecol.* **114**: 1–13.

Shannon, L.V. (1985) The Benguela ecosystem Part 1. Evolution of the Benguela, physical features and processes. *Oceanogr. Mar. Biol. Ann. Rev.* **23**: 105–182.

Sheldon, R.A., Prakash, A. and Sutcliffe, W. (1972) The size distribution of particles in the ocean. *Limnol. Oceanogr.* **17**: 327–340.

Shelton, P.A. and Hutchings, L. (1982) Transport of anchovy, *Engraulis capensis* Gilchrist, eggs and early larvae by a frontal jet current. *J. Cons. Int. Explor. Mer.* **40**: 185–198.

Shulenberger, E. and Reid, J.L. (1981) The Pacific shallow oxygen maximum, deep chlorophyll maximum and primary productivity reconsidered. *Deep-Sea Res.* **28**: 901–919.

Shushkina, E.A. (1985) Production of principal ecological groups of plankton in the epipelagic zone of the ocean. *Oceanology* **25**: 653–658.

Silver, M.W. and Alldredge, A.L. (1981) Bathypelagic marine snow: deep-sea algal and detrital community. *J. Mar. Res.* **39**: 501–530.

Silver, M.W., Shanks, A.L. and Trent, J.D.F. (1978) Marine snow: microplankton habitat and source of small-scale patchiness in pelagic populations. *Science* **201**: 371–373.

Simpson, J.H. (1971) Density stratification and microstructure in the western Irish Sea. *Deep-Sea Res.* **18**: 309–319.

Simpson, J.H. (1981) The shelf-sea fronts: implications of their existence and behaviour. *Phil. Trans. R. Soc. Lond. A* **302**; 531–546.

Simpson, J.H. and Bowers, D. (1981) Models of stratification and frontal movement in shelf seas. *Deep-Sea Res.* **28**: 727–738.

Simpson, J.H. and Hunter, J.R. (1974) Fronts in the Irish Sea. *Nature* **250**: 404–406.

Simpson, J.H. and Pingree, R.D. (1978) Shallow sea fronts produced by tidal stirring. pp. 29–42. In: M.J. Bowman and W.E. Esaias (Eds) *Oceanic Fronts in Coastal Processes.* Springer-Verlag, New York.

Simpson, J.H. and Tett, P.B. (1986) Island stirring effects on phytoplankton growth. pp. 41–76. In: M.J. Bowman, C.M. Yentsch and W.T. Peterson (Eds) *Tidal Mixing and Plankton Dynamics.* Springer-Verlag, New York.

Sinclair, M. (1988) *Marine Populations: An Essay on Population Regulation and Speciation.* University of Washington Press, Seattle and London.

Sinclair, M., Bugden, G.L., Tang, C.L., Therriault, J.-C. and Yeats, P.A. (1986) Assessment of effects of freshwater runoff variability on fisheries production in coastal waters, pp. 139–160. In: S. Skreslet (Ed.) *The Role of Freshwater Outflow in Coastal Marine Ecosystems.* Springer-Verlag, Heidelberg.

Sinclair, M., Tremblay, M.J. and Bernal, P. (1985) El Niño events and variability in a Pacific Mackerel (*Scomber japonicus*) survival index: support for Hjort's second hypothesis. *Can. J. Fish. Aquat. Sci.* **42**: 602–608.

Sissenwine, M.P., Cohen, E.B. and Grosslein, M.D. (1984) Structure of the George's Bank ecosystem. *Rapp. P.-v. Réun. Cons. Int. Explor. Mer.* **183**: 243–254.

Skreslet, S. (1986) *The Role of Freshwater Outflow in Coastal Marine Ecosystems.* Springer-Verlag, Berlin, 453 pp.

Smayda, T.J. (1970) The suspension and sinking of phytoplankton in the sea. *Oceanogr. Mar. Biol. Ann. Rev.* **8**: 353–414.

Smith, P.C. (1989) Circulation and dispersal on Brown's Bank. *Can. J. Fish. Aquat. Sci.* **46**: 539–559.

Smith, P.E. and Eppley, R.W. (1982) Primary production and the anchovy population in the southern California bight: comparison of time series. *Limnol. Oceanogr.* **27**: 1–17.

Smith, P.E. and Lasker, R. (1978) Position of larval fish in an ecosystem. *Rapp. P.-v. Réun. Cons. Int. Explor. Mer.* **173**: 77–84.

Smith, S.D (1980) Wind stress and heat flux over the ocean in gale force winds. *J. Phys. Oceanogr.* **10**: 709–726.

Smith, S.D. and Dobson, F.W. (1984) The heat budget at ocean weather ship Bravo. *Atmosphere–Ocean* **22**: 1–22.

Smith, S.L. and Whitledge, T.E. (1977) The role of zooplankton in the regeneration of nitrogen in a coastal upwelling system off NW Africa. *Deep-Sea Res.* **24**: 49–56.

Sommer, U. (1988) Some size relationships in phytoplankton motility. *Hydrobiologia* **161**: 125–131.

Sorokin, Yu. I., Sukhanova, I.N., Konolova, G.V. and Pavelyeva, E.B. (1975) Primary production and phytoplankton in the area of equatorial divergence in the equatorial Pacific. *Trans. Inst. Oceanol.* **102**: 108–122.

Soutar, A. and Isaacs, J.D. (1969) History of fish populations inferred from fish scales in anaerobic sediments off California. *Calif. Coop. Ocean. Fish. Invest. Rep.* **13**: 63–70.

Soutar, A. and Isaacs, J.D. (1974) Abundance of pelagic fish during the 19th and 20th centuries as recorded in anaerobic sediment off California. *Fish. Bull.* **72**: 257–273.

Stacey, M.W., Pond, S. and LeBlond, P. H. (1986) A wind-forced Ekman spiral as a good statistical fit to low-frequency currents in a coastal strait. *Science* **233**: 470–472.

Steele, J.H. (1981) Some varieties of biological oceanography. In: Warren, B.A. and Wunsch, C. (Eds) *Evolution of Physical Oceanography*. M.I.T. Press, Cambridge, Mass.

Steele, J.H. (1988) Scale selection for biodynamic theories, pp. 513–526. In: B.J. Rothschild (Ed.) *Towards a Theory on Biological-Physical Interactions in the World Ocean.* Kluwer, Amsterdam.

Steele, J.H. (1989) Discussion: Scale and coupling in ecological systems, pp. 177–180. In: J. Roughgarden, R.M. May and S.A. Levin (Eds) *Perspectives in Ecological Theory.* Princeton University Press, Princeton, New Jersey.

Steele, J.H. and Henderson, E.W. (1984) Modeling long-term fluctuations in fish stocks. *Science* **224**: 985–987.

Steele, J.H. and Yentsch, C.S. (1960) The vertical distribution of chlorophyll. *J. Mar. Biol. Ass. U.K.* **39**: 217–226.

Stevens, D.E., Kohlhorst, D.W., Miller, L.W. and Kelly, D.W. (1985) The decline of striped bass in the Sacramento–San Joaquin Estuary, California. *Trans. Am. Fish. Soc.* **114**: 12–30.

Stommel, H. (1948) The westward intensification of wind-driven ocean currents. *Trans. Am. Geophys. Union.* **29**: 202–206.

Stommel, H. (1958) The abyssal circulation. *Deep-Sea Res.* **5**: 80–82.

Stommel, H. (1965) *The Gulf Stream − A Physical Description,* 2nd Edn. Cambridge University Press, Cambridge.

Stommel, H., Niiler, P. and Anati, D. (1978) Dynamic topography and recirculation of the North Atlantic. *J. Mar. Res.* **36**: 449–468.

Strass, V. and Woods, J.D. (1988) Horizontal and seasonal variation of density and chlorophyll profiles between the Azores and Greenland, pp. 113–136. In: B.J. Rothschild (Ed.) *Towards a Theory on Biological–Physical Interactions in the World Ocean.* Kluwer, Dordrecht.

Strickler, J.R. (1984) Sticky water: a selective force in copepod evolution, pp. 187–242. In: D.G. Meyers and J.R. Strickler (Eds) *Trophic Interactions within Aquatic Ecosystems.* American Association for the Advancement of Science. Washington, DC.

Sutcliffe, W.H. (1972) Some relations of land drainage, nutrients, particulate material and fish catches in two eastern Canadian bays. *J. Fish. Res. Bd. Can.* **29**: 357–362.

Sutcliffe, W.H. (1973) Correlations between seasonal river discharge and local landings of the American lobster (*Homarus americanus*) and Atlantic halibut (*Hippoglossus hippoglossus*) in the Gulf of St Lawrence. *J. Fish. Res. Bd. Can.* **30**: 856–859.

Suthers, I.M. and Frank, K.T. (1989) Inter-annual distributions of larval and pelagic juvenile cod (*Gadus morhua*) in southwestern Nova Scotia determined with two different gear types. *Can. J. Fish. Aquat. Sci.* **46**: 591–602.

Sverdrup, H.U. (1953) On conditions for the vernal blooming of phytoplankton. *J. Cons. Perm. Int. Exp. Mer.* **18**: 287–295.

Sverdrup, H.U., Johnson, M.W. and Fleming, R.H. (1942) *The Oceans: Their Physics, Chemistry and General Biology.* Prentice-Hall, Englewood Cliffs, New Jersey. 1060 pp.

Swallow, J.C. (1976) Variable currents in mid-ocean. *Oceanus* **19**: 18–25.

Swallow, J.C. and Hamon, B.V. (1960) Some measurements of deep-sea currents in the eastern North Atlantic. *Deep-Sea Res.* **6**: 155–168.

Swift, J.H., Aagaard, K. and Malmberg, S.-A. (1980) The contribution of the Denmark Strait overflow to the deep North Atlantic. *Deep-Sea Res.* **27**: 29–42.

Sy, A. (1988) Investigation of large-scale circulation patterns in the central North Atlantic: the North Atlantic current, the Azores current and the Mediterranean water plume in the area of the mid-Atlantic ridge. *Deep-Sea Res.* **35**: 383–413.

Takahashi, T. (1989) The carbon dioxide puzzle. *Oceanus* **32**(2): 22–29.

Taylor, A.H., Harris, J.R.W. and Aiken, J. (1986) The interaction of physical and biological processes in a model of the vertical distribution of phytoplankton under stratification, pp 313–330. In: J.C. Nihoul (Ed.) *Marine Interfaces Ecohydrodynamics.* Elsevier, New York.

Templeman, W. (1972) Year class success in some North Atlantic stocks of cod and haddock. *Spec. Publ. Int. Comm. NW Atl. Fish* **8**: 223–239.

Tenore, K.R. (and 18 others) (1982) Coastal upwelling in the Rias Bajas, N.W. Spain: Contrasting the benthic regimes of the Rias de Arosa and de Duras. *J. Mar. Res.* **40**: 701–772.

Tett, P. (1981) Modelling phytoplankton production at shelf-sea fronts. *Phils. Trans. Roy. Soc. Lond. A* **302**: 605–615.

The Ring Group (1981) Gulf Stream cold-core rings: their physics, chemistry and biology. *Science* **212**: 1091–1100.

Therriault, J.-C. and Levasseur, M. (1985) Control of phytoplankton production in the lower St Lawrence estuary: Light and freshwater runoff. *Naturaliste Can.* **112**: 77–96.

Thompson, J.D. (1977) Ocean deserts and ocean oases. In: M.N. Glantz (Ed.) *Desertification.* Westview Press, Boulder, Colorado.

Thompson, J.D. (1978) Role of mixing in the dynamics of upwelling systems, pp. 203–222. In: R. Boje and M. Tomczak (Eds) *Upwelling Ecosystems.* Springer-Verlag, Berlin.

Thordardottir, T. (1986) Timing and duration of spring blooming south and southwest of Iceland, pp. 345–360. In: S. Skreslet (Ed.) *The Role of Freshwater Outflow in Coastal Marine Ecosystems.* Springer-Verlag, Berlin.

Tolmazin, D. (1985) Changing coastal oceanography of the Black Sea. 1: Northwest shelf. *Prog. Oceanogr.* **15**: 217–276.

Townsend, D.W., Graham, J.J. and Stevenson, D.K. (1986) Dynamics of larval herring (*Clupea harengus*) production in tidally mixed waters of the eastern coastal Gulf of Maine, pp. 253–277. In: M.J. Bowman, C.M. Yentsch and W.D. Peterson (Eds) *Tidal Mixing and Plankton Dynamics.* Springer-Verlag, New York.

Traganza, E.D., Conrad, J.C. and Breaker, L.C. (1981) Satellite observations of a cyclonic upwelling system and giant plume in the California current, pp. 228–241. In: F.A. Richards (Ed.) *Coastal Upwelling.* Am. Geophys. Union, Washington, DC.

Traganza, E.D., Redalje, D.G. and Garwood, R.W. (1987) Chemical flux, mixed layer entrainment and phytoplankton blooms at upwelling fronts in the California coastal zone. *Continental Shelf Res.* **7**: 89–105.

Traganza, E.D., Silva, V.M., Austin, D.M., Hanson, W.L. and Bronsink, S.H. (1983) Nutrient mapping and recurrence of coastal upwelling centers by satellite remote sensing: its implication to primary production and the sediment record, pp. 61–83. In: E. Suess and J. Thiede (Eds) *Coastal Upwelling: its Sediment Record.* Part A. Plenum Press, New York.

Tranter, D.J., Carpenter, D.J. and Leech, G.S. (1986) The coastal enrichment effect of the east Australian current eddy field. *Deep-Sea Res.* **33**: 1705–1728.

Tranter, D.J., Parker, R.R. and Creswell, G.R. (1980) Are warm-core eddies unproductive? *Nature* **284**: 540–542.

Trites, R.W. (1983) Physical oceanographic features and processes relevant to *Illex illecebrosus* spawning in the western North Atlantic and subsequent larval distribution.

NAFO Sci. Coun. Studies **6**: 39–55.

Trumble, R.J., Mathisen, O.A. and Stuart, D.W. (1981) Seasonal food production and consumption by nekton in the northwest African upwelling system, pp. 458–463. In: F.A. Richards (Ed.) *Coastal Upwelling.* Am. Geophys. Union, Washington, DC.

Tsujita, T. (1957) The fisheries oceanography of the east China Sea and Tsuchima Strait. 1. The structure and ecological character of the fishing grounds. *Bull. Seikai Reg. Fish. Lab.* **13**: 1–47.

Turner, J.S. (1973) *Buoyancy Effects in Fluids.* University Press, Cambridge.

Tyler, A.V. and Westerheim, S.J. (1986) Effect of transport, temperature and stock size on recruitment of Pacific cod. *Int. N. Pac. Fish. Comm. Bull.* **47**: 175–189.

Tyler, M.A. and Seliger, H.H. (1978) Annual subsurface transport of a red tide dinoflagellate to its bloom area: water circulation patterns and organism distributions in the Chesapeake Bay. *Limnol. Oceanogr.* **23**: 227–246.

Tyler, M.A. and Seliger, H.H. (1981) Selection for a red tide organism: Physiological responses to the physical environment. *Limnol. Oceanogr.* **26**: 310–324.

Valdivia, J.E. (1978) The anchoveta and El Niño. *Rapp. P.-v. Reun. Cons. Int. Explor. Mer.* **173**: 196–202.

van Loon, H. and Rogers, J.C. (1978) The seesaw in winter temperatures between Greenland and northern Europe. Part I. General description. *Monthly Weather Rev.* **106**: 296–310.

Vincent, A. and Kurc, G. (1969) Hydrologie: Variations saisonnieres de la situation hermique du Golfe de Gascogne en 1967. *Rév. Trav. Inst. Peches Maritimes.* **33**: 203–212.

Vinogradov, M.E. (1981) Ecosystems of equatorial upwellings, pp. 69–93. In: A.R. Longhurst (Ed.) *Analysis of Marine Ecosystems.* Academic Press, London, 741 pp.

Vogel, S. (1981) Life in moving fluids: the physical biology of flow. Willard Grant Press, Boston, 352 pp.

Von Schwind, J.J. (1980) *Geophysical Fluid Dynamics for Oceanographers.* Prentice-Hall, New York, 307 pp.

Walsh, J.J. (1981) A carbon budget for overfishing off Peru. *Nature* **290**: 300–304.

Walsh, J.J. (1983) Death in the sea: enigmatic phytoplankton losses. *Prog. Oceanogr.* **12**: 1–86.

Walsh, J.J. and Dieterle, D.A. (1988) Use of satellite ocean colour observations to refine understanding of global geochemical cycles, pp. 287–317. In: T.Rosswall, R.G. Woodmansee and P.G. Risser. *Scales and Global Change.* Wiley, New York.

Walsh, J.J., Rowe, G.T., Iverson R.L. and McRoy, C.P. (1981) Biological export of shelf carbon is a sink of the global CO_2 cycle. *Nature* **291**: 196–201.

Walsh, J.J., Whitledge, T.E., Esaias, W., Smith, R.L., Huntsman, S.A., Santander, H. and de Mendiola, B.R. (1980) The spawning habitat of the Peruvian anchovy, *Engraulis ringens. Deep-Sea Res.* **27**: 1–27.

Warren, B. (1983) Why is no deep water formed in the North Pacific? *J. Mar. Res.* **41**: 327–347.

Warren, B.A. (1981) Deep circulation of the world ocean. In: B.A. Warren and C. Wunsch (Eds) *Evolution of Physical Oceanography.* Scientific Surveys in Honour of Henry Stommel. M.I.T. Press, Cambridge, Mass., 620 pp.

Webb, K.L. and D'Elia, C.F. (1980) Nutrient and oxygen redistribution during a spring neap tidal cycle in a temperate estuary. *Science* **207**:983–985.

Webster, P.J. (1983) Large scale structure of the tropical atmosphere, pp. 235–276. In: B.Hoskins and R.Pearce (Eds) *Large Scale Dynamical Processes in the Atmosphere.* Academic Press, New York, 397 pp.

Weinstein, M.P., Weiss, S.L., Hodson, R.G. and Gerry, L.R. (1980) Retention of three taxa of postlarval fishes in an intensely flushed tidal estuary, Cape Fear River, North Carolina. *Fish. Bull.* **78**: 419–435.

Westlake, D.F. (1967) Some effects of low-velocity currents on the metabolism of aquatic macrophytes. *J. Exp. Bot.* **18**: 187–205.

Wheeler, W.N. (1978) Ecophysiological studies on the giant kelp *Macrocystis*, Ph.D. Thesis. University of California, Santa Barbara.

Wheeler, W.N. (1980) Effect of boundary layer transport on the fixation of carbon by the giant kelp *Macrocystis pyrifera. Mar. Biol* **56**: 103–110.

Whitford, L.A. and Schumacher, G.J. (1961) Effect of current on mineral uptake and respiration by a freshwater alga. *Limnol. Oceanogr.* **6**: 423–425.

Wildish, D.J. and Peer, D.L. (1983) Tidal current speed and production of benthic macrofauna in the lower Bay of Fundy. *Can. J. Fish. Aquat. Sci.* **40** (Suppl. 1.): 309–321.

Wildish, D.J. and Kristmanson, D.D. (1979) Tidal energy and sublittoral macrobenthic animals in estuaries. *J. Fish. Res. Board. Can.* **36**: 1197–1206.

Wildish, D.J. and Kristmanson, D.D. (1984) Importance to mussels of the benthic boundary layer. *Can. J. Fish. Aquat. Sci.* **41**: 1618–1625.

Williams, P.J. Le B. and Muir, L.R. (1981) Diffusion as a constraint on the biological importance of microzones in the sea, pp. 209–218. In: J.C. Nihoul (Ed.) *Ecohydrodynamics.* Elsevier, Amsterdam.

Wolanski, E. and Hamner, W.M. (1988) Topographically controlled fronts in the ocean and their biological influence. *Science* **241**: 177–181.

Wolanski, E., Drew, E., Abel, K.M. and O'Brien, J. (1988) Tidal jets, nutrient upwelling and their influence on the productivity of the alga *Halimeda* in the Ribbon Reefs, Great Barrier Reef. *Estuar. Coast. Shelf Sci.* **26**: 169–201.

Wolf, K.U. and Woods, J.D. (1988) Lagrangian simulation of primary production in the physical environment — the deep chlorophyll maximum and nutricline, pp. 51–70. In: B.J. Rothschild (Ed.) *Towards a Theory on Biological–Physical Interactions in the World Ocean.* Kluwer, Dordrecht.

Wolff, W.J. (1977) A benthic food budget for the Gravelingen Estuary, the Netherlands, and a consideration of the mechanisms causing high benthic secondary production in estuaries, pp. 267–280. In: B.C. Coull (Ed.) *Ecology of Marine Benthos.* Belle W. Baruch Library in Marine Science No. 6. University of South Carolina Press.

Wong, C.K. (1980) Copepod predation and prey defense. M.Sc. Thesis, University of Ottawa, Ontario.

Wood, L. and Hargis, W.J. (1971) Transport of bivalve larvae in a tidal estuary, pp. 21–44. In: D.J. Crisp (Ed.) *4th European Marine Biology Symposium.* Cambridge University Press, Cambridge.

Woods, J.D. (1968) Wave-induced shear instability in summer thermocline. *J. Fluid Mech.* **32**: 791–800.

Woods, J.D. (1977) Parameterization of unresolved motions, pp. 118–140. In: E.B. Kraus (Ed.) *Modelling and Prediction of the Upper Layers of the Ocean.* Pergamon Press, New York.

Woods, J.D. (1988) Scale upwelling and primary production, pp. 7–38 In: B.J. Rothschild (Ed.) *Towards a Theory on Biological–Physical Interactions in the World Ocean.* Kluwer, Dordrecht.

Woods, J.D. and Barkmann, W. (1986) The response of the upper ocean to solar heating. I. The mixed layer. *Quart. J. R. Met. Soc.* **112**: 1–27.

Woods, J.D. and Onken, R. (1982) Diurnal variation and primary production in the ocean — preliminary results of a Lagrangian ensemble model. *J. Plankton Res.* **4**: 735–756.

Wooster, W.S., Bakun, A. and McLain, D.R. (1976) The seasonal upwelling cycle along the eastern boundary of the North Atlantic. *J. Mar. Res.* **34**: 131–141.

Worthington, L.V. (1976) On the North Atlantic circulation. *Johns Hopkins Studies in Oceanography,* No. 6, 110 pp. Johns Hopkins Marine Laboratory, Monterey, Calif.

Wright, D.G., Hendry, R.M., Loder, J.W. and Dobson, F.W. (1986) Oceanic changes associated with global increase in atmospheric carbon dioxide: a preliminary report for the Atlantic coast of Canada. *Can. Tech. Rep. Fish. Aquat. Sci.* 1426, 78 pp.

Wroblewski, J.S. (1982) Interaction of currents and vertical migration in maintaining

Calanus marshallae in the Oregon upwelling zone — a simulation. *Deep-Sea Res.* **29**: 665–686.

Wroblewski, J.S. and Cheney, J. (1984) Ichthyoplankton associated with a warm-core ring off the Scotian shelf. *Can. J. Fish. Aquat. Sci.* **41**: 294–303.

Wyrtki, K. (1962) The upwelling in the region between Java and Australia during the south-east monsoon. *Aust. J. Mar. Freshwater Res.* **13**: 217–225.

Wyrtki, K. (1964) Upwelling in the Costa Rica dome. *Fish. Bull. U.S.* **63**: 335–372.

Xie, L. and Hsieh, W.W. (1989) Predicting the return migration routes of the Fraser River sockeye salmon (*Onchorhynchus nerka*). *Can. J. Fish. Aquat. Sci.* **46**: 1287–1292.

Yamanaka, H. (1985) Effect of El Niño on fish migration and yield in the western Pacific Ocean. pp. VI-16 to VI-22 In: International Conference on the TOGA Scientific Programme. WMO/TD No. 65. World Climate Research Programme Publication Series No. 4.

Yentsch, C.S. and Phinney, D.A. (1985) Rotary motions and convection as a means of regulating primary production in warm core rings. *J. Geophys. Res.* **90**: (C2) 3237–3248.

Yoder, J.A., Atkinson, L.P., Bishop, S.S., Hofmann, E.E. and Lee, T.N. (1983) Effect of upwelling on phytoplankton productivity of the outer southeastern United States continental shelf. *Continental Shelf Res.* **4**: 385–404.

Yoder, J.A., Atkinson, L.P., Lee, T.N., Kim, H.H. and McLain, C.R. (1981) Role of Gulf Stream frontal eddies in forming phytoplankton patches on the outer southeast shelf. *Limnol. Oceanogr.* **26**: 1103–1110.

Zeldis, J.R. and Jillett, J.B. (1982) Aggregation of pelagic *Munida gregaria* (Fabricius) (Decapoda, Anomura) by coastal fronts and internal waves. *J. Plankt. Res.* **4**: 839–857.

Index

461